Birds of Western Ghats,
Kokan & Malabar
(Including Birds of Goa)

इला–पृथ्वी

A book by **Ela** Foundation

Birds of Western Ghats,
Kokan & Malabar
(Including Birds of Goa)

Satish Pande
Saleel Tambe / Clement Francis M. / Niranjan Sant

BOMBAY NATURAL HISTORY SOCIETY

OXFORD
UNIVERSITY PRESS

OXFORD

UNIVERSITY PRESS

Oxford University Press is a department of the University of Oxford.
It furthers the University's objective of excellence in research, scholarship,
and education by publishing worldwide in

Oxford   New York   Auckland   Bangkok   Buenos Aires   Cape Town   Chennai
Dar es Salaam   Delhi   Hong Kong   Istanbul   Karachi   Kolkata
Kuala Lumpur   Madrid   Melbourne   Mexico City   Mumbai   Nairobi
São Paulo   Shanghai   Taipei   Tokyo   Toronto

Oxford is a registered trademark of Oxford University Press
in the UK and in certain other countries

ISBN 019 566878 2

Printed By : Ravindra Joshi, UNITED PRINTERS, 264/4, River View, Shaniwar Peth, Pune 411 030
PUBLISHED BY THE BOMBAY NATURAL HISTORY   SOCIETY, HORNBILL HOUSE, SHAHEED
BHAGAT SINGH ROAD, MUMBAI   400   023, AND CO-PUBLISHED BY MANZAR KHAN, OXFORD
UNIVERSITY PRESS, YMCA LIBRARY BUILDING, JAI SINGH ROAD,NEW DELHI 110 001

Warli art.

Acknowledgement :
Photographers -
Abrar Ahmad. Adrian Webb. Ajit Kulkarni. Amit Pawashe. Amol Mithari. Amol Warange. Anant Zanzale. Anil Damle. Anil Kashyap. Anil Mahabal. Ansar Khan. Arnaon T-Sairi. Asad Rahmani. Ashfaq A. Zarri. Ashley Fisher. Atul Dhamankar. Avinash Nangare. Banda Pednekar. Bas van De Meulengraf. Bharat Cheda. Bill Johnson. Bjorn Johansson. Bruce Craig. Chaitra M. R; Chaiyan Kosorndorkbue. Chan Kai Soon. Chandrashekhar Bapat. Chandrashas Kolhatkar. Clement Francis M. Datta Ugaokar. Dave Brokman. Dave Curtis. Deepak Joshi. Devendra Singh. Dick Newell. Dilip Yardi. Erach Bharucha. Frank O' Connor. Gehan D' Silva Wijeyeratne. George & Lindsey Swann. George McCarthy. H. S. Ananth. Harish Ingawale. Hema Gupte. Hemant Shinde. Herman Van Oosten. Hira Punjabi. Ido Tsurim. J. Judge. James Flynn. Jayant Tadphale. Jean Sebastien Rosseau-Piot. Jijo Mathews. Joanna Van Gruisen. Job K. Joseph. John Edwards. Jugal Tiwari. K. S. Rajshekhara. Kees Bakker. Khaleed Rafeek. Kim Hyun Tae. Kiran Purandare. Kishor Joshi. Krys Kazmierczak. Lawrence Poh. Lean Yen Loong. Lok Wan Tho. M. G. Kanitkar; M. P. Nagendra. M. P. S. Prasad. Mahesh Mahajan. Manoj Kulshreshtha. Marek Kosinski. Michael Gelinas. Mike Danzenbaker. Mike Shipman. Milind Bendale. Mohan Panse. Mukund Deshpande. Mukund Jere. Naoto Kitagawa. Navendu Lad. Nazim Siddiqui. Nico-De-Regge. Niranjan Sant. Ooi Beng Yean. Otto Pfizer. P. M. Lad. P. N. Papanna. Parag Dandage. Paul Gale. Peter Draper. Peter Jones. Prabhakar Kukdolkar. Prakash Joglekar. Pramod Bansode. Pramod Nargolkar. Pramod Pawashe. Prashant Kanvinde. Premsagar Mestri. R. S. Dayanand. R. S. Suresh. Raghunandan Kulkarni. Rahul Warange. Raja Purohit. Rajat Bhargava. Rajesh B. P; Rajesh Pardeshi. Rajkumar Vijaykumar Thondaman. Raju Kasambe. Ram Mone. Ramchandran. Rev. Eric J. Lott. Rishad Naoroji. Rudd Kampf. S. Balchandran. S. G. Neginhal. Sachin Palkar. Saleel Tambe. Sandeep Labade. Sanjay Karkare. Sanjay Shegaokar. Sanjeev Nalavade. Satish Pande. Satish Ranade. Satoshi Maenishi. Sattyasheel Naik. Shamita Harishchandra. Shivaji Jaware. Shreesh Kshirsagar. Shrikant Ingalhalikar. Steven Falk. Subhash Puranik. Sudhakar Kurhade. Suppalak Kladbee. Suresh Elamon. Suresh Pardeshi. T. N. A. Perumal. T. S. U. de Zylva. Tetsu. Tim Loseby. Tony Palliser. Ulhas Rane. Valeri Moseikin. Venkatswamappa M. Vijay Cavale. Vijay Tuljapurkar. Vinay Thakar. Vishwas Katdare. Vivek Sinha. William Hague. Yashodhan Bhatia. Yves Adams.
Illustrators -
Carl D' Silva. Jayant Tadphale. John Henry Dick. Laxmi Joshi.
Photo credits of initial pages :
Flying Page Front-M. V. Deshpande-Peacock, Back-R. V. Thondaman-Poacher
Title / Authors / Publishers / Page-Satish Pande-Long Legged Buzzard
Acknowledgement Page-Front-Abrar Ahmad-Bulbuls Back-Clement Francis-Brown Fish Owl
Preface Page-Yashodhan Bhatia-Lesser Flamingo Foreword-Clement Francis-Dabchicks,
Dedication Page-Satish Pande-Flamingos
Topography Page-R. V. Thondaman-Blackbird, Satish Pande-Striolated Bunting
Introduction Page-Niranjan Sant-Shikra
Scope of the book Page-Satish Pande-Marine Terns
Index Page-Clement Francis-Stone Chat

FOREWORD

"The study of Indian Ornithology was founded on a bedrock of amateur interest. T. C. Jerdon, the father of Indian Ornithology was a doctor. Allan O. Hume, whose collections and journal of ornithology *Stray Feathers* laid the foundation of scientific ornithology, was a bureaucrat. Stuart Baker who wrote the *Fauna of India* volumes on birds was a police officer; and Salim Ali, who strode the field of Indian Ornithology as a colossus had no degree to his name. The unique tradition, fortunately, continues. This delightfully innovative book *Birds of Western Ghats, Kokan and Malabar (including birds of Goa)* is written by amateur ornithologists. The author Satish Pande is a doctor.
The co-authors Saleel Tambe, Niranjan Sant and Clement Francis are also from various other professions.
Focused on describing one of the best bird fauna areas of the Indian Subcontinent and also of a Hotspot of global importance, the book takes a new look at the life and habits of the birds of the Western Ghats. There is particular emphasis on the rescue and rehabilitation of injured birds.
Photographs of all endemic and threatened species are provided. The book is the result of the passionate interest in birds of the authors who have woven an illustrated natural history of the birds of the Western Ghats, particularly of the lush countryside of the land between the hills and the sea - Kokan and Malabar, and also on the sea. Birds do inspire and the excellent results in this case are truly delightful."

J. C. Daniel
Honorary Secretary
Bombay Natural History Society

PREFACE

Birds......

Melody of their song, majesty of their flight and magic of their colour, have symbolized the infinite spirit of happiness and freedom since ancient times. The whistling groves, the whispering grasslands, the rustling forests, the chanting streams, the silent sea are ever ready to give. Our noble tradition teaches us to accept with humility. The sublime imprint of birds in Indian culture since Vedic times, and the tradition of reverence for living beings, is probably for the first time amalgamated in a scientific text, as in this book. I hope that this shall bring birds and nature closer to our hearts.

This book on birds intends to document, with unique photographs, the current status of rich bird life of an Indian Hotspot of global importance-the Western Ghats, including Kokan and Malabar. Several committed and knowledgeable bird watchers are studying local bird life. I am thankful to them for sharing with me their meticulously documented data and rare photographs. Having seriously indulged in bird photography, I appreciate that these represent a part of their very life, the time spent in search of facts about the enigmatic feathered bipeds. The larger canvas could be painted due to the strokes of several of these smaller brushes, each important and essential. The splendour of pelagic birds; the resplendent breeding plumage of winter migrating avian species; the elusive endemic and threatened birds and several aspects of more than 580 species could be photographically presented in this book, due to unconditional contributions from ornithologists and bird watchers, not only from India but also from Europe, North Asian, East Asian and Austral-Asian countries. This book is a truly international collaboration. Another unique aspect of this endeavour is that our organisation, ELA Foundation, has met the cost of this no-profit publication through donations from several nature lovers.

I feel honoured that Bombay Natural History Society, the premier organization devoted to the study and conservation of nature and wild life in Asia, and Oxford University Press are publishing this book. It is a small drop in the vast ocean of avian knowledge, from which we have drawn time and again. I am grateful to Mr. J. C. Daniel, Honorary Secretary BNHS, for kindly writing the foreword. I express my gratitude to Dr. Asad Rahmani, Director, BNHS and Dr. Gayatri Ugra for continued inspiration and support. I am most grateful to my friends Saleel Tambe, Niranjan Sant and Clement Francis M. for all their help, without which the book would not have been possible. I thank R. Vijaykumar Thondaman, Amit Pawashe, Gehan D'Silva Wijeyeratne, Rev. E. J. Lott, P.M.Lad, Kishor Joshi and Yashodhan Bhatia. I thank Mr. Jayant Tadphale for excellent designing of the book. I thank my colleagues from the medical profession, who have encouraged me in every manner. I thank my wife Suruchi and my family, for due to their tolerant encouragement, a dream is coming true.

It gives me great satisfaction to give back to people what was given to me generously by Nature and nature lovers. I feel honored to be worthy of their trust in receiving from them this avian treasure and humbly presenting to the readers the glory of bird life, in this book. If this book inspires the common man to protect wildlife and the students of science, whose knowledge can truly conserve nature, it shall serve its purpose.

If only we realize that man too is a part of this natural heritage and not its master ! Several avifaunal species have silently vanished, before science had even seen them well. I wish that we should succeed in preserving the once bountiful and rapidly depleting natural treasure of Mother Earth, not only for our future generations but also for its own sake.

- Satish A. Pande

To my parents
For initiating me
To the singing sky….

To Nivedita, Nikhil, and
All the children,
Who hear the
Voice of the vibrant wings…

- Satish

BIRDS OF PREY

Falcons : Slim. Wings Angular, pointed. Tail Long. Soar. Stoop.

Buteos : (Buzzards) Plump. Wings-Broad, Tail - Short, broad. Soar.

Accipiters : (Hawks) Sleek. Wings - Round, short. Tail Long. Wing beats.

Vultures : Large. Wings Long, narrow, fingered. Head bare. Soar.

Osprey : Stout, pied. Wings Long, narrow, bent at the wrist. Soar & hover.

Eagles : Large, stout. Wings Broad, long, fingered. Tail Broad, fanlike. Soar.

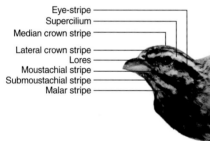

Eye-stripe
Supercilium
Median crown stripe
Lateral crown stripe
Lores
Moustachial stripe
Submoustachial stripe
Malar stripe

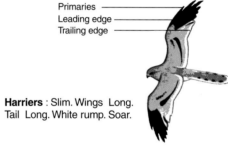

Primaries
Leading edge
Trailing edge

Harriers : Slim. Wings Long. Tail Long. White rump. Soar.

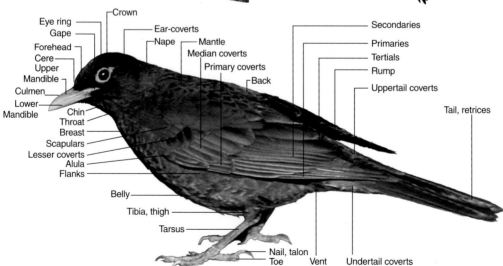

Eye ring
Gape
Forehead
Cere
Upper Mandible
Culmen
Lower Mandible
Chin
Throat
Breast
Scapulars
Lesser coverts
Alula
Flanks
Belly
Tibia, thigh
Tarsus
Crown
Ear-coverts
Nape
Mantle
Median coverts
Primary coverts
Back
Nail, talon
Toe
Vent
Undertail coverts
Secondaries
Primaries
Tertials
Rump
Uppertail coverts
Tail, retrices

Habitat :

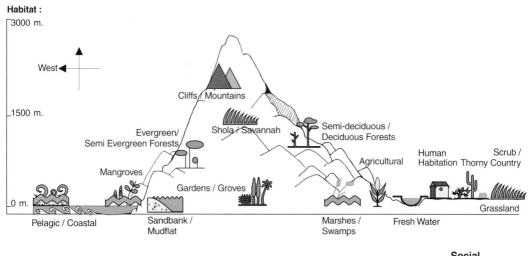

3000 m.

West ◄

1500 m.

Cliffs / Mountains

Evergreen/ Semi Evergreen Forests

Shola / Savannah

Semi-deciduous / Deciduous Forests

Mangroves

Agricultural

Human Habitation

Scrub / Thorny Country

Gardens / Groves

Grassland

0 m.

Pelagic / Coastal

Sandbank / Mudflat

Marshes / Swamps

Fresh Water

Habit :

Aerial

Arboreal

Terrestrial

Aquatic / Semi-aquatic

Food :

Small Mammals

Fish

Seeds / Nuts

Eggs / Birds

Aquatic Invertebrates

Fruits / Berries

Reptiles

Terrestrial Invertebrates

Grass / Shoots / Algae

Amphibians

Insects

Flower / Nectar

Time :

Crepuscular

Nocturnal

Diurnal - when not specified

Social Organization :

Single

Pair

Flock

Flight :

Straight

Undulating

Sudden Takeoff

Soaring

Hovering

Zig-Zag

Acrobatic

Up & Down

Nest Type :

Open Cup

Communal

Floating

Platform Pad

Half-Cup on a vertical surface

Tree hole

Ground Scrape

Hanging

Hole / Crevice in a wall

Enclosed

Tunnel in an earth-bank

Cavity in cliff / ledge

Ground Cup

Mound

Hammock

KEY TO SYMBOLS & ABBREVIATIONS :
(Commonest habits, habitats, food preferences, flight patterns & nest types are given.
Other preferences can be observed.
Symbols appear on each page in the following sequence viz.
Habit, Habitat, Food, Flight, Nest, Social Organisation.)

About The Book :

The avian systemic order, taxonomy and standardized common names (in bold type) & scientific names (in italics) are as per *Buceros Vol. 6, No. 1 (2001)*, an ENVIS publication by Bombay Natural History Society. Birds seen in Goa are in blue type. Alternate English names are given. 79 avian Families from 20 Orders are mentioned. Ripley's Synopsis numbers from *'A Synopsis of the Birds of India and Pakistan'* are stated after the scientific names to help distinguish races (except in cases of new additions to the bird list). This shall also preserve the umbilical cord with the *'Handbook of Birds of India and Pakistan'* by Salim Ali & S. D. Ripley. Bird sizes in mm. are an average length from tip of the beak to the tip of the tail. Egg sizes are that given by E.C.Stuart Baker. Sanskrit names (in bold & italic font) are critically taken from several sources but the main reference is *'Birds In Sanskrit Literature'* by K.N. Dave. The chronology from the Vedic period onwards, given in the book, is as per *'An Encyclopedic Dictionary of Sanskrit On Historical Principles Vol. 1'*, general editor A. M. Ghatage, Deccan College Postgraduate & Research Institute, Pune. Vernacuar Marathi names (in grey & bold font) are mentioned below the main photograph. Most popular names are selected from several local checklists & books, directly collected by the authors in the field and from *Buceros Vol. 3, No. 1, April, 1998*. Distribution of birds within the scope of this book and their migratory habits, an essentially dynamic entity, are based on direct observations by the authors, from confirmed sightings by reliable local birdwatchers and extensive survey of ornithological literature. New observations are welcome & readers are requested to convey the same to the authors (Satish Pande, C-9, Bhosale Park, Sahakarnagar-2, Pune 411009, India). The records have been updated to 2003 and cross checked with records and lists in various authentic publications like the Journal of Bombay Natural History Society, Oriental Bird Club Bulletins, Newsletter for Birdwatchers, Forktail, Zoological Survey of India publications and several others. Field Guides, books and papers especially by Salim Ali, S. Ripley; J. Ranjit Daniels; Krys Kazmierczak; Richard Grimmet, Carol & Tim Inskipp; Ash & Shafeeg, Roberts, Lamfuss, Gehan De Silva Wijeyeratne, Deepal Warakagoda & T. S. U. De Zylva; John Harrison & Tim Worfolk and G. M. Henry (last three for Sri Lankan birds) have been referred. Several new geographical distribution records for our region can be seen in this book. Threatened birds in the region (in red type) have been identified from *'Threatened Birds of Asia'* by BirdLife International and from *'Threatened Birds of India'*, *Buceros Vol.7, No. 1 & 2, 2002*. Designation of Endemic or Restricted Range Species (in green type) is based on *Buceros, Vol. 6, No. 2, 2001*. Western Ghats is a home to endemic species & is identified as a HotSpot by BirdLife International.

On each page, a main species is described with a mention of specific field identification marks, and related species, usually with lesser occurrence than the main species, is briefly described. Any look alike species (from neighboring Sri Lanka or other regions) are photographically illustrated on the same page. 580 avian species are illustrated. Extreme care has been taken to reflect true avian colours & avian peculiarities in the photographs. Photographs are presented in an appealing manner and are not necessarily to scale since the avian sizes are mentioned. Peculiar bird calls, mnemonics & repertoires of local birds are transcribed into English, based on recordings by several others and us. These are to be used with caution due to their inherent limitations (geographical, individual variations & interspecies similarities). Altitudinal range of several birds is given. Ecological notes are intended to highlight protection measures and risks faced by avian species. Successful rehabilitations are photographically depicted. Cultural Notes are intended to bring the birds closer to our hearts by linking the avian life to our culture and heritage. Wherever possible, sexual dimorphism, plumage variations, nesting habits, nidification, food choices, flight patterns, sexual behaviour, habitat preferences, ringing recoveries and other related avian information is photographically illustrated. Additional chapters are annexed at the end of the book to highlight related subjects. Index of English, Scientific Marathi, and Sanskrit names is given. Several photo-plates are inserted liberally to present to the reader a glimpse of the enchanting & fragile avian kingdom. It is hoped that the book shall inspire readers to conserve our rich natural heritage for the future generations.

Geographical Scope of the Book

The area covered is as follows:

The entire Western Ghats from Navapur (20.12 degrees N) in the north till Kanyakumari (8.06 degrees N, 77.35 degrees E) in the south. The eastern offshoots of the main range are also included and these are: Satmala-Ajanta, Balaghat and Mahadeva ranges (Maharashtra), Baba Budan ranges (Karnataka), Nilgiri and associated hills like BR Hills (Karnataka,Tamilnadu), Palani and Cardamom Hills (Tamilnadu) and Anaimalai Hills (Kerala,Tamilnadu). For convinience this region is divided into: North Western Ghats (Navapur till Goa), Central Western Ghats (Goa to Palghat Gap) and South Western Ghats (Palghat Gap southwards).

Tableland between Narmada and Krishna rivers is considered as the Deccan. Western edge of the Deccan from Nandurbar through Nashik, Ahmadnagar, Pune, Satara districts is included. In the same arbitrary line Kolhapur, Belgaum, through Dharwad, Shimoga till Mysore districts are included. Malnad, the hilly country along the eastern slope of the Western Ghats in Karnataka is included.

Coastal areas included are from Surat (Gujarat) in the north, through Kokan (or Konkan- coastal Maharashtra, Goa and north Karnataka), Malabar (SW coast of India comprising south Karnataka and Kerala states between the Western Ghats and Arabian Sea), till Kanyakumari in the south.

Adjacent territorial sea and marine islands like Elephanta Is., Vengurla Rocks, Oyster Rocks, etc. are included.

The western most limit is 73.50 degrees E (around Mumbai) and the eastern most extent is 77.35 degrees E (Kanyakumari).

The states covered are southern Gujarat, Western Maharashtra, Goa, Western Karnataka, Kerala and part of Tamilnadu. The union territories of Daman and Dadra and Nagar Haveli are also included.

The approximate area under the scope of the book is 200,000 sq.Km. This is around seven percent of the area of our country and harbors almost 50 percent of its avifauna.

QUICK INDEX TO CONTENTS :

Laggar Falcons

Pheasant - tailed Jacanas.

These rare and realistic lithographs in vibrant colours are taken from
' The Birds of Asia ' (1850-1877), painted by John Gould, Henry C. Richter and and William Hart,
and *' The Game Birds of India, Burmah and Ceylon '* (1879), by Hume and Marshall
(Painted by E. Neale). They are from the archives of Bombay Natural History Society.

Little Grebe

Dabchick

Podicipedidae

Tachybaptus ruficollis (5)

Vanjulak

Size: 330 mm.   M / F : Alike   **R, LM.**

Distribution : The entire region, up to 1400 m. in the Western Ghats.

Nest-site: Half submerged nest in reeds or on floating aquatic plants. Nest may tilt and spill eggs if water level recedes abruptly. *Apr.-Sept.*
Material: Lined with wet weeds and rushes.
Parental care: Both.
Eggs: 3-7, white, becoming muddy through contact with decaying vegetation. 36 x 25 mm.
Call: A sharp tittering, during the breeding season or when disturbed. Vocal when playing. Wing flaps are also noisy.
Ecological notes: Their presence on a water body indicates a rich aquatic fauna like aquatic insects, crustaceans and amphibia. Chicks face the risk from water-snakes and Marsh Harriers.
Cultural notes: Being the smallest Indian grebe, it is called Tibukli, the little one !
Status: Common.

Related species:
*****G**reat Crested Grebe (Podiceps cristatus-3) Vagrant WM to our region. White wing patch and wing-edges are seen in flight. 500 mm.
*****B**lack-necked Grebe (P.nigricollis-4). Rare WM. 330 mm. Silvery flanks and black head, neck. Upcurved bill.
Look alike: Coot chicks have red crown and may appear like adult Dabchicks, when seen from a distance.

Breeding.   **Tibukli, Lande-Badak**

Description: Short billed tailless duck-like bird. Head and neck become dark brown to red when breeding. An excellent swimmer and diver. Gets food underwater. These playful grebes are often seen chasing one another, half-running, half-swimming in the water. Cover eggs in the nest with aquatic vegetation, when leaving the nest unattended. Chicks are carried on the back or under the wings, when alarmed.

Great Crested Grebe - Breeding

Black-necked Grebe - Breeding.

Black-necked Grebe (Non-Br.)

Little Grebes (Non. Br.)

Great Crested Grebe incubating & with chick.

Little Grebe feeding fish to the chick.

Little Grebe (Non. Br.)

Little Grebe with chicks.

Flesh-footed Shearwater

Procellariidae

Puffinus carneiceps (8)

No Sanskrit name recorded

Size: 410-450 mm   M / F : Alike   **WM, MM**

Distribution : Rarely on our marine waters in the Arabian Sea and Indian Ocean.

Nest-site: Islands off Australia and New Zealand. Communal. *Oct.-Jan.*
Parental care: Both. Birds visit the nest burroughs at night.
Eggs: White.
Call: Silent in winter.
Ecological notes: Eat surface marine fauna like flying fish, squid. Also scavenge on refuse from ships thrown overboard.
Cultural notes: The name Vadali- Pakhru implies its sporadic occurence when the bird is blown to the shore during rain storms.
Status: Vagrant.

Other Shearwaters: All rare, pelagic.
*Streaked (Calonectris leucomelas-7) 480 mm. Pale streaked head, white belly, dark back, dark tipped pale bill.
*Audubon's (Puffinus lherminieri-11) 300 mm. Breed in Maldives. Sooty above with grey breast bands on sides, white below, pied undertail coverts.
Persian (Puffinus persicus) 330 mm. Brown axillaries, flanks. Dark brown above, long bill, muddy underwings. **Near Threatened.*
Petrels: (All pelagic vagrants to our region)
*Barau's (Pterodroma barauii-N) 380 mm. Grey back with dark rump, white forehead, grey cap, white belly, underwing-black edged.
*Bulwer's (Bulweria bulwerii-136) 260 mm. Longtailed dark body. Light band on greater coverts.
*Jouanin's (B.fallax-13a) 320 mm. Larger replica of Bulwer's Petrel. Bill stouter.

Vadali-Pakhru

Description: Dark brown, sooty. Head and neck pale grey. Flesh coloured feet, bill. Glide over sea and also run on water.

Streaked Shearwater
(Calonectris leucomelas-7)

Wedge-Tailed Shearwater
(Puffinus pacificus-9)

Bulwer's Petrel

Jouanin's Petrel

White-Bellied Storm-Petrel

Barau's Petrel

Audubon's Shearwater

Bulwer's Petrel

Cape Petrel

Flesh-Footed Shearwater

Persian Shearwater

Sooty Shearwater

Wilson's Storm-Petrel

Mother Carey's Chicken

Hydrobatidae

Oceanites oceanicus (14)

Vichikak

Size: 190 mm   M / F : Alike   **MM, PM**

Distribution : On our marine waters during the monsoon probably on migration to Socotra.

Vadal Tiwale

Nest-site: In Antarctica. Crevices, cliffs.
Nov.-Jan.

Call: None recorded during migration.

Ecological notes: Surface feeder. Equally at ease in calm and stormy weather. Good swimmer. Pelagic.

Cultural notes: Their habit of apparently walking on water has given them the name Petrel after Saint Peter, who as per a biblical story tried to walk on water!

Status: Common monsoon passage migrant on off shore marine waters.

Description: Black with white rump. Yellow webbed feet extend beyond square tail. These are conspicuous when it flutters with feet typically dangling. Perch on rocks and take off with difficulty. Fly between waves picking animal matter from troughs as if walking on water with head touching the surface.

Grey-backed Tropicbird 480 mm.

Related species: All pelagic vagrants.
*White-faced Storm-Petrel (Pelagodroma marina-N) White eyebrow, dark above, white below. Dark mask, grey rump.
*Swinhoe's Storm-Petrel (Oceanodroma monorchis-16) Dark overall, gently forked tail. Feet short of tail.
Family-Phaethontidae:
*Grey-backed or Red-billed Tropicbird (Phaethon aethereus-17) Grey back, black primaries, white tail, red bill. Juvenile-yellow bill. Rare pelagic MM.
Cape Petrel (Daption capense-6) 400 mm.
Pied plumage separates from others.
Vagrant near SriLanka.

Swinhoe's Storm-Petrel 200 mm.

Wilson's Storm-Petrel 190 mm.

White-faced Storm-Petrel 200 mm.

A flock of Rosy Pelicans. They are a vagrant to our region. These birds can be seen in urban areas.

Rosy Pelican has prosopis twig pierced in the beak pouch.
An assembly of Rosy Pelicans.

Dalmatian Pelican is not recorded in our region.

Spot-billed Pelican

Grey Pelican

Pelicanidae

Pelacanus philippensis (21)

Kesari, Jalsinha

Size: 1520 mm   M / F : Alike   **R, LM**

Distribution : Rarely on water bodies along the eastern side of the Ghats.

Nest-site: Around Kokkare-Bellur near Bangalore, on tall trees, palms. *Sept.-Mar.*
Material: Sticks, twigs, leaves, wet reed, weed.
Parental care: Both. Incubation 30 days.
Eggs: 3-4, white, round. 78.8x53.4 mm.
Call: Loud yelps, croaks, grunts. Bill clapping.
Ecological notes: Colonial nests on traditional trees. Fresh water bodies with abundant fish essential. Also eat small ducks. Chicks often fall from nest and they can be rescued.
Cultural notes: The beak pouch arouses much interest. Charak, ancient Indian physician, called this bird **Kesari**-Water Lion!
Status: Uncommon. **Vulnerable.**

Zoliwala

Description: Black spotted upper mandible, grey head, neck, wings, brown nuchal crest and pinkish pouch. Juvenile is brown with unspotted bill. Chicks feed by inserting heads in bill pouches of parents. Strong flier and good swimmer. Fishes in parties by driving fish to shallow water and then scooping them in bill pouches. Pelecanries are noisy and traditional sites need protection.

Spot-billed Pelican in Flight

Spot-billed Pelican with Fish

Nesting Spot-billed Pelicans

Burkhya Samudra-kawala

Description: White with black primaries, tail and mask. Unstriped underwing coverts. Webbed digits. Juvenile has black head and scaly black wings. Plunge head first in water from 25-30 m. up in air to take food.

Brown Booby

Related species: All pelagic vagrants.
Brown Booby (Sula leucogaster-25) Dark brown with white belly and underwing coverts. Thick conical beak. Recorded during the NE monsoon.
Red-footed Booby (Sula sula-24) Vagrant to our marine waters off West coast. 660-770 mm. Red feet, pink facial skin, blue bill. White morphs are seen in our region. Juveniles have grey legs.

Masked Boobies

Masked Booby

Bluefaced Booby

Sulidae

Sula dactylatra (23)

No Sanskrit name recorded

Size: 800 mm   M / F : Alike   **WM, V**

Distribution : On our marine waters and off shore islands during cyclones and gales.

Nest-site: Colonial. On off shore islands, cliff faces on ledges. Maldives. *April-May-June.*
Call: None recorded during migration.
Ecological notes: Marine bird. Hunts by deep diving. A booby entangled in fish net set 27 m. under water is recorded(Handbook,Salim Ali). Fish, squids are taken by diving hence they require clean transparent water.
Cultural notes: An un-noticed bird, known only to a few fishermen!
Status: Vagrant to off shore marine waters.

Red-footed Booby

Little Cormorant

Phalacrocoracidae

Phalacrocorax niger(28)

Paniyakakika

Size: 510 mm. M / F : Alike **R, LM**

Distribution : In the entire region.

Nest-site: Usually in a heronry, near or away from water on a medium sized tree. *Jun.- Oct.*
Material: Sticks, twigs, leafy branches.
Parental care: Both.
Eggs: 3-5, bluish-white, turning yellow-brown by the time of hatching. 44.8 x 29 mm.
Call: Noisy during nesting. Loud croaks are uttered. Otherwise silent.
Ecological notes: Solitary nesting is rare. Require safe sites and trees for heronries. The birds are seen on polluted water. Chicks fall from their nests, but fair well if reinstated and if not seriously injured. Dead or unattended chicks are eaten by cats, crows and other predators.
Cultural notes: Manu, in the ancient text 'Manusmruti' forbids eating the cormorants flesh. A water thief is said to be born as a cormorant ! Large Cormorants are used by the fishermen in the Far-East, to catch fish with a ring placed on their neck to prevent them from swallowing the catch.
Status: Common.

Pankawla

Description: The bill is hooked at the tip. Toes are webbed. An expert diver, underwater swimmer. Seen sitting upright on a favourite riverside perch, with outstretched wings. Oily wings get water-logged after prolonged immersion in water, and the wings are thus dried before the next dive. The heronries are often marauded by large birds of prey, when parents resist communally.

Nesting Cormorants.

Large Cormorants on the nest.

Communal roosting.

Little Cormorants - Sun drying.

Indian Shag

Indian Cormorant

Phalacrocoracidae

Phalacrocorax fuscicollis (27)

Jalkak, Plav, Kakmadgu

Size: 630 mm.   M / F : Alike   **R, LM**

Distribution : Widespread in the entire region.

Pankawla

Description: Glossy black-brown, long tailed and necked, slender billed cormorant with whitish throat and brown oval head. White tuft behind eyes, filoplumes on neck when breeding. Fish is taken by deep diving and flocks operate in unison. They are seen perched on low branches, stakes or rocks with wings spread out for sun-drying. The shag holds the long neck in a serpentine manner and can be mistaken for the Darter. The latter has a longer pointed unhooked beak and white neck.

Large Cormorant (Br.)

Nest-site: On trees in colonies. *Jun. - Sep.*
Material: Leaves, twigs, sticks.
Parental care: Both.
Eggs: 3-6, blue-green. 60.6 x 39.3 mm.
Call: A silent bird. Croaks and guttural hoarse cries uttered when alarmed and during breeding.
Ecological notes: Aquatic bird. Subsists entirely on fish. Weed-free water is required for feeding.
Cultural notes: An ancient Vedic rite Abhiplav-Shadaha was named after the large cormorant(**Abhiplav**). It was performed before and after a function. This is in keeping with the white patch on the either end of this bird (both flanks and throat)!
Status: Uncommon.

Cormorant fishing in the Far-East.

Related Species: *Great or Large Cormorant* *(Phalacrocorax carbo-26) Breeding birds have white cheeks, throat, thighs, white head plumes and an orange-yellow gular patch. Non-breeding birds are black, lack the head patch. Immatures have white underparts and adult plumage is attained after about 4 years. Rare. Size - 800 mm.*

Large Cormorant (Br.)   *Large Cormorant (Non-Br.)*
With nesting material.

Darter

Oriental Darter

Anhingidae

Anhinga melanogaster (29)

Madgu

Size: 900 mm.   M / F : Alike **R, LM**

Distribution : Patchily throughout the region.
Rare in the Kokan. Commoner in the
Deccan, very rare in the Western Ghats.

Nest-site: Branches of tall trees near water, in
mixed colonies with cormorants, egrets,
herons. *Jun. - Oct.*
Material: Leaves, grass, sticks, twigs.
Parental care: Both.
Eggs: 3-4, pale blue-green, elongated.
52.9 x 33.5 mm.
Call: A hoarse disyllabic croak.
Ecological notes: Often with little cormorants.
Large number of Darters indicates an ample
fish population.
Cultural notes: It is thought that Darters
pierce fish by the dagger-like beak. On the
contrary, they always catch fish, toss it in the
air and then swallow! Darter is said to
represent the Sun-God, in ancient Sanskrit
texts, owing to the silver in its wings.
Status: Near Threatened.

*Look Alike: Cormorants also have the habit of
swimming in water with neck stretched out in a
Darter like manner. Darter-chicks are white in
colour initially, when they may be mistaken for
Egret chicks, but have a longer neck.*

Darter chicks.

Tirandaz, Sapmanya, Sarpa-pakshi

Description: The snake like neck is kept
outside water, and the long tail is fanned
beneath the surface, while swimming. Toes are
webbed. Also a good diver. A special neck
muscle helps the Darter to throw it's neck at a
fish, like a javelin, and catch it with lightning
speed under water! Perch on rocks, trees to
dry their wet wings. Drop in water when
alarmed and surface some distance away.

Darter fishing.

Probing the oil-gland.

Christmas Island Frigatebird (M)   **Chorkawala**

Lesser Frigatebird

Least Frigatebird

Fregatidae

Fregata ariel (32)

No Sanskrit name recorded

Size: 800 mm   M / F : Dimorphic   **V**

Distribution : Vagrant to the West coast during cyclonic storms.

Lesser Frigatebird & Kite.

Description: Streamlined pointed wings, deeply forked tail, long hooked beak. White belly with a conical spur extending in both underwings. Yellow gular skin in male. Female with red eye ring and white neck straps. Juveniles with dark breast band, rufous to white head. Toes webbed, tarsus feathered. Incapable of taking off from a flat surface hence do not alight on sea. In flight can be mistaken for a kite.

Nest-site: On ground. Atolls in the Maldives. *Nov.-Feb.*
Material: Sticks.
Parental care: Both. Incubation 6 weeks.
Eggs: 1-2, white chicks are born blind, naked.
Call: None recorded in our region.
Ecological notes: Pan-tropical marine bird. Surface feeder on flying fish, squids. Pirates food from boobies, tropicbirds.
Cultural notes: Nothing particular in the region. However the inflated brightly coloured gular pouch of the breeding birds make them popular subjects of wildlife films.
Status: Vagrant.

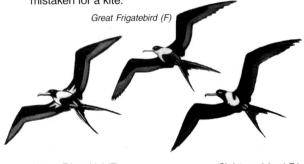

Great Frigatebird (F)

Lesser Frigatebird (F)
Great Frigatebird-Breeding (M)

Christmas Island Frigatebird (F)

Lesser Frigatebirds.

*Related Species: *Great Frigatebird (Fregata minor-31) 850-1050 mm. Adult male all black with red gular skin. Female pink eye ring, black cap, white belly and black neck straps. Lacks white wing spur. *Christmas Island Frigatebird (Fregata andrewsi-30) 900 mm. Black except white patch on lower belly, red gular pouch. Female pink eye ring, black throat, white belly with black wing spurs on belly. Juvenile buff head, dark breast band. Likely to be encountered on the West coast marine waters.*

Little Egret

Ardeidae

Egretta garzetta (49)

Balakika

Size: 630 mm.   M / F : Alike   **R, LM**

Distribution : Throughout the region and up to about 1400 m. in the Ghats.

Nest-site: 2 to 5 m. up in trees in North Kokan, near wet-land in mixed heronries. Large scale local migration towards such heronries is observed at breeding time. *Jun. - Sep.*
Material: Leaves, grass, sticks.
Parental care: Both.
Eggs: 3-5,pale, blue-green. 44.4 x 31.7 mm.
Call: Loud croaks. Vocal when breeding.
Ecological notes: Hunts in and around relatively clean or less polluted water.
Cultural notes: The ornamental aigrettes were much sought after by women in the western countries, for milinery purposes. The trade is no more allowed. Heronries cause much nuisance to people, if they are situated on private land, due to fallen rotten fish and noise. They are not allowed to flourish.
Status: Common.

Lahan Bagla, Chota Bagla, Morbagla

Description: White bird with black bill, black legs, yellow toes. During breeding, a drooping crest of two narrow plumes, ornamental fine feathers on the back, breast (Hence the name Morbagla-Peacock Egret)! Hunting at the edge of water in a typical heron like fashion.

Related Species: *Large Egret (Casmerodius albus-46)650-720 mm. Solitary, has black legs and toes and the dark line from the yellow (black when non-breeding) beak extends behind the eyes. Ornamental feathers only on the back.*
Median Egret (Mesophoyx intermedia-48)650-720 mm. Black legs, toes. Smaller than above. Plumes during breeding time on the back and breast.

Large Egret.

Little Egret (Br.)   Large Egret (Br.)
Median Egrets.

Median Egret preening

*The beautiful filo-plumes or aigrettes
were used for ornamental purposes in the past.*

Little Egret (Br.)

Median Egret

Large Egrets (Br.)

Western Reef-Egret

Indian Reef Egret or Heron

Ardeidae

Egretta gularis (50)

Wakruka

Size: 630 mm.   M / F : Alike   **R, LM**

Distribution : In Kokan, Malabar on coasts.

Rarely inland.

Nest-site: Nests in mixed colonies on Mangrove, other trees near coasts or swamps. *Mar. - Jul.*
Material: Twigs, sticks, branches.
Parental care: Both.
Eggs: 3-4, sea-green, pale, plain. 44.9 x 34.3 mm.
Call: Croaks uttered on the nest. Silent.
Ecological notes: A coastal egret. Aquatic invertebrates, mud-crawlers and fish eaten. Rocky seashores are preferred.
Cultural notes: When the Reef-Egret is seen on inland waters, it arouses interest, since a black egret is a distinct rarity! To the common man, an egret is a white bird. Its black colour has featured in folk songs.
Status: Common on the coasts, rare inland.

Dark morph.   *Intermediate morph.*

Dark morph fishing.

Kala Bagla

Description: Similar to Little Egret in the white phase. In the dark phase, slaty-blue with a white throat patch. Intermediate phases exist. Yellow toes, lores, bill typical. White filoplumes on head, neck, tail and black bill when breeding. Roost at high tide, at noon in a hunched position. Actively run about, jumping from rock to boulder, chasing the waves while feeding. Wings are flapped to flush out fish. Prey is battered and devoured.

Little Green Heron

Related species: *Little Green Heron (Butorides striatus-38) Green, streaked above, grey below. Iris is yellow, red in Night Heron. Legs are green and a dark streak extends below and behind the eyes. Solitary. Throughout our area. Keeps to cover. Call 'Kek, kek, ke, yow'. Common on coast than inland.*

White & Dark morphs.

Grey Heron

Rakhi Bagla, Kudal, Dhok

Ardeidae

Ardea cinerea (36)

Krushachanchu, Kank

Size: 1000 mm.   M / F : Alike   **R, LM**

Distribution : The entire region.

Nest-site: *Tamarindus indicus, Ficus* spp. etc. trees at a considerable height. With Ibises, Cormorants, Painted Storks. *Mar. - Sep.*
Material: Lined with feathers, green leaves. Made of freshly broken branches, twigs, sticks. The nests are continually repaired.
Parental care: Both perform domestic duties.
Eggs: 3-7, greenish-white, oval.
58.6 x 43.5 mm.
Call: A loud harsh *Frank* uttered in flight.
Ecological notes: Prefers large water bodies, marine and fresh. Nest platform intactness is necessary for feeding the chicks since they are fed by regurgitation in the nest.
Cultural notes: In the epic Harivansha, the dark colour of the Grey Heron is likened to the sable colour of Lord Krishna! The name **Krushachanchu** means one with a pointed bill.
Status: Occasional.

*Related species: *Purple Heron (Ardea purpurea-37) 870 mm. Crepuscular, solitary. A blue-grey bird with rufous head, neck. Belly is chestnut-black. Secretive. Prefers dense reed beds. Almost invisible to a casual observer, due to the habit of remaining stationary with the neck outstretched. Congregate at breeding time.*

Description: S-shaped streaked neck, yellow beak conspicuous even in flight. This master of infinite patience stands motionless in shallow water, with the head poised for instant action. Also wades with outstretched neck. Flies strongly with the head held back and legs stretched in the typical heron fashion.

Purple Heron.

Courtship display.

Purple Herons on nest.

Night Heron exhibiting fracture of the left wing
(Radius & Ulna) on this Roentgenograph.
Also note the larger size of the eyes as compared
to the size of the skull vault.

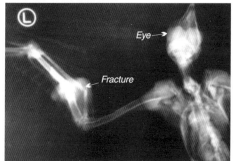

Grey Heron in breeding plumage.

Key-hole iris.
An unusual injury sustained
by the eye of a Grey Heron.
This injury was not surgically
corrected.

Ringed Grey Heron Juveniles on their way to freedom.

Purple Heron

Vanchak, Bhura-Bagla, Kok

Indian Pond-Heron

Pond Heron, Paddybird

Ardeidae

Ardeola grayii (42)

Swalpakank, Andhabak

Size: 450 mm.   M / F : Alike   **R, LM**

Distribution : The entire region and up to 2000 m. in the Ghats.

Nest-site: Trees at a fair height, often near human dwellings, in heronries. *May - Sep.*
Material: Sticks and twigs, lined with down feathers and leaves.
Parental care: Both.
Eggs: 3-6, blue-green, 38 x 28.5 mm.
Call: A harsh *Frank* is uttered when disturbed. *Waku, waku* in low tone and a variety of mumbling sounds during breeding.
Ecological notes: Indicator of water-logged areas. Controller of insects.
Cultural notes: Aptly called **Andhabak** due to its invisibility. Habit of the Pond Heron of standing immobile, as if in meditation, is mentioned as the **Bakdhyan**-a state of apparent meditation, when to the contrary, the mind thinks only of a possible kill! The **Bakdhyan** is the epitome of hypocrisy.
Status: Common.

Description: The back becomes deep maroon and a filamentous plume develop when breeding. Alights with a croak, with a flash of the white wings. It is seen hunched at the edge of every possible water body, ready to strike at a suitable prey when it comes in the range. Moves slowly in search of a quarry.

Pond Heron (Br.)

Pond Heron (Non-Br.)

Look Alike: The immature ***N**ight Heron (N.nycticorax) looks similar to the adult non-breeding Pond Heron. Eyes of the former are larger, redder, the bill heavier and greenish.*

Night Heron (Imm.)

Cattle Egret

Ardeidae

Bubulcus ibis (44)

Pingalika

Size: 510 mm.   M / F : Alike   **R, LM**

Distribution : The entire region and up to 2100 m. in the Ghats.

Nest-site: Trees, considerable height from ground, unless on a branch overhanging the water. Near human dwellings. Heronries are mainly in the North Kokan. *Jun. - Sep.*
Material: Leaves, grass, sticks.
Parental care: Both.
Eggs: 4-5, white, pale blue, oval.
44.1 x 36.5 mm.
Call: Loud croaks while foraging. Vocal during breeding when territorial fights are common.
Ecological notes: Follows farmers while ploughing and disinfests fields by eating insects. Rids grazing cattle of flies, ticks, lice, leeches. Egret guano under the heronries is a good fertilizer.
Cultural notes: The name ***Pingalika*** refers to the golden yellow plumage of the breeding bird. A story in the epic Mahabharat refers to a ***Balak***, who having angered a sage sleeping under a tree by soiling him with droppings, was reduced to ashes with fire rendered from his angry eyes.
Status: Common.

*Look Alike: *Little Egret and white phase of *Reef heron are similar to the Cattle egret. In the breeding phase can be differentiated from the latter by the buff appearance of Cattle Egret. Little Egret has yellow toes, and Reef heron is seen near the coasts where Cattle Egret is rare.*

Intermediate Egret.

Gaibagla, Dhorbagla, Gochidkhaoo

Description: Yellow bill and dark legs. In the breeding season head, neck and back turn orange-buff. Often seen attending grazing cattle, either riding on their backs, pecking ticks, or catching insects that are disturbed as the cattle graze in pastures.

Cattle Egret (Non-Br.)

Cattle Egret (Br.) with chick.

Threat display of
the Night Heron near it's nest.
Kites, Shikras, Langurs, etc.
predate on herons nests.

Night Heron

New born chick & two eggs.

Nursing the Juvenile Night Heron.

Night Herons on the nest.

Black-crowned Night-Heron

Night-Heron

Ardeidae

Nycticorax nycticorax (52)

Naktakraunch, Vakaschandravihangam

Size: 580 mm.   M / F : Alike   **R, LM**

Distribution : Throughout the region.

Nest-site: Large trees-*Holoptelea integrifolia, Syzygium cumini, Delonix regia, Mangifera indica, Ficus* etc spp. often away from water. *Nov. - Apr.*

Material: Leafy branches, sticks, lined with leaves. Nests are constantly strengthened.

Parental care: Both.

Eggs: 3-5, pale sea-green. 49 x 35.1 mm.

Call: Loud *Kwaak, kwaark*. Squaks and harsh croaks uttered. Very noisy when breeding.

Ecological notes: Nesting is in unmixed or mixed heronries, when Cormorants share the trees. Egg predation by birds, langurs is known. Fallen chicks can be rescued and released. Fractures heal well.(Satish Pande)

Cultural notes: Heronries in human settlements cause nuisance from littering from half- eaten fish, egg-shells, droppings and noise which bird lovers find difficult to tolerate if present near houses. Official recognition and compensation may go a long way in conserving the heronries on private grounds.

Status: Uncommon.

Malayan Night Heron.

Raatbagla, Raj-Kok

Description: Night Herons leave their roosts in thickly foliaged lofty trees at dusk and flocks fly towards water bodies, where they feed throughout the night, returning by dawn. They are active at daytime when breeding. A heron defending attacks on it's nest from a langur, by jabbing at it with beak is recorded. Birds with both pink and yellow legs have been seen in the same heronry. A chick has been recorded blind born; it died after a fall from the nest (Satish Pande,NLBW). It is likely to have been evicted by siblings or parents.

Night Heron (Br.)

*Malayan Night-Heron (Gorsachius melanolophus-53) Tiger Bittern. 510 mm. Black crown, buff tipped dark primaries, white throat, cinnamon with dark streaks. Immature-dark headed, spotted, greyish. Patchily in the region in marshes.

Lal-Bagla, Lal-Tapas

Chestnut Bittern

Cinnamon Bittern

Ardeidae

Ixobrychus cinnamomeus (56)

Jyotsnabaka

Size: 380 mm.   M / F : Alike   **R, MM**

Distribution : The entire region up to 900 m.

Nest-site: About a meter or less above water in a swamp, dense reed-bed, aquatic vegetation. *Jun. -Aug.*
Material: Aquatic plants, reed, grass, shoots.
Parental care: Both.
Eggs: 4-6, white, blue-green, 36.5 x 26.4 mm.
Call: Silent. Booming call at times.
Ecological notes: Crepuscular. Reed-beds are essential for survival. Active when the sky is cloudy or overcast.
Cultural notes: Jyotsnabaka (Jyotsna-moonlight, **Baka**-egret, bittern) sublimely indicates the nocturnal and secretive nature of this bird! Boom of the Bittern on the moors of England, is immortalized in English literature.
Status: Rare. Likely to be commoner than thought. Status difficult to evaluate due to the secretive nature of the bittern.

Description: Chestnut wings, upperparts, white centrally striped throat. Cryptically coloured defying spotting, seen mostly in flight. If noticed assumes 'On Guard' posture, freezing, almost disappearing in reeds with the neck stretched, beak pointing towards the sky! Display during the SW Monsoon.

Black Bittern (Female)

*Related species: *Yellow Bittern (Ixobrychus sinensis-57) Drab yellow-brown-fawn with black wings seen in flight. *Black Bittern (Dupetor flavicollis-58) Black head, back, wings and an orange patch on sides of the neck. Rare. Northern limits of our area. Recently recorded at Dighi, (Dist.Raigad) in Kokan and and Near Pune. (P. Mestri, J.Khasgiwale, A.Pawashe.) *Little Bittern (Ixobrychus minutus-55) Buff, streaked belly, dark primaries, black crown. Female brown. 330-380 mm. *Great Bittern (Botaurus stellaris-59) Buff, streaked. 710 mm. Last two WM, Vagrant.*

Yellow Bittern.

Black Bittern

Little Bitterns (M & F)

Great Bittern
Cinnamon or Chestnut Bittern (Juv.)

Black Bittern "On Guard"

Yellow Bittern

Chitrabalak, Rangeet-karkocha

Ciconiidae

Mycteria leucocephala(60)

Kachaksha, Pingalaksha, Bruhadbak

Size: 1000 mm.   M / F : Alike   **R, LM**

Distribution : The entire region, commonly on the eastern margins of the Ghats.

Description: The pink shoulder patch and wings are typical of this large stork. Greenish black glistening wing coverts, black tail and a black breast-band give it a unique identity. Yellow curved heavy bill and bare face. Slowly and cautiously wades in marshy water, with the open bill immersed and moving from side to side. During the day, it stands hunched up and motionless. Often seen with openbill storks, spoonbills, ibises and other waders.

Painted Stork.

Nest-site: Large lofty trees at 15-20 m. Always in water logged areas. In heronries. Mixed heronries-near Indapur, Dist. Pune. *Jun. - Sep.*

Material: Sticks, twigs, lined with grass, leaves.

Parental care: Both. Feeding is by regurgitation, but the chicks also actively take food from the parent's beaks.

Eggs: 2-5, dull-white, oval, eventually soiled to brown as incubation proceeds. 69.5 x 49 mm.

Call: A loud harsh croak. Chicks are vocal when begging for food.

Ecological notes: Require shallow water for feeding. Breeding is in very large mixed heronries, for which, large protected trees and healthy water bodies are essential.

Cultural notes: The Sanskrit name **Bruhadbak** indicates the large size of the stork and **Pingalaksha** is descriptive of the orange coloured iris.

Status: Near Threatened.

Look Alike: The immature Painted Storks can be confused with the immature **O*penbill Storks, (*A.oscitans*), when observed from a distance. The bill of the latter, shows the narrow gap.

Openbill Stork Imm.

Painted Storks (Juveniles)

Painted Storks on nest.

Asian Openbill-Stork

Openbill Stork

Ciconiidae

Anastomus oscitans (61)

Avabhanjan, Ghonk, Shambukbhanjan

Size: 800 mm .   M / F : Alike   **R, LM**

Distribution : Throughout the region, up to 600 m. in the Ghats.

Nest-site: In mixed heronries on lofty trees. *Jun. - Sep.*

Material: Sticks, twigs, branches are used to make the large platform. Lined with leaves.

Parental care: Both.

Eggs: 2-5, creamy-white. 57.9 x 41.2 mm.

Call: The stork is incapable of producing true laryngeal sound. Hoarse croaks and clapping of the bills are heard.

Ecological notes: Eats molluscs and *Ampullaria* snails with its specialized open-bill.

Cultural notes: This silent bird is called the Mugdhabalak (silent stork). The name *Shambukbhanjan* describes the feeding habit of prying open molluscs, snails prior to eating!

Status: A common and widely distributed stork.

Mugdhabalak, Ughdyachochicha karkocha

Description: A white stork with black wing coverts, tail, and beak with arching mandibles which leave an open gap, when the beak is closed. This strong flier can soar high and is often seen in flocks. In mixed congregations.

Oriental Stork - Endangered.

White Stork.

Openbill Stork.

Openbill Stork, Chicks.

Related Species: *White Stork (Ciconia ciconia-63)* and the *Oriental Stork (C.boyciana-64)* **Endangered,** *can be mistaken for the Openbill if seen from afar. Both are rare WM. C.ciconia is larger, has a red bill without a gap and has red legs. C.boyciana has a black bill. All storks are fond of marshes. C.ciconia has been recorded at Veer near Mahad in Kokan In 1997, 98.*(P. Mestri, S.Pande. Also one record in 1880-Gazetteer)

Kala Karkocha

Description: Pairs, small flocks are seen wading in marshes, on the edge of water body. Often with shanks, sandpipers, whitenecked and painted storks. Wade in knee deep water probing the mud with long beaks, at times submerging the head fully. Wary, taking to wings when alarmed. Flight is typical of storks with legs and neck extended. The beak and legs are red in adults, grey in Immatures.

Lesser Adjutant-Stork.

Related Species:
Black-necked Stork (Ephippiorhynchus asiaticus-66) WM. Rarely on secluded waters along the eastern fringes of the Ghats. Eat aquatic food. Pied wings, metalic multihued glistening dark neck, heavy bill and red legs. Near Threatened.
Lesser Adjutant-Stork (Leptoptilos javanicus-68) 1130 mm. Pied. Massive yellowish beak, naked orange neck. Rare. Up to 500 m. in the region. Near Threatened.

Black Stork

Ciconiidae
Ciconia nigra (65)
Kalbaka

Size: 1060 mm.   M / F : Alike   **WM**

Distribution : Rarely in the region along the eastern side of the Ghats. Not on the coasts.

Nest-site: Nests in Central Europe. Large trees are used for building the nests. *Apr. - May.*
Material: Sticks, branches. Old nests re-used.
Parental care: Both.
Eggs: 3-5, white. 65.3 x 48.7 mm.
Call: Typical harsh *Croak, croak* is uttered when taking off or if disturbed.
Ecological notes: Feeds in water. Also eats small sick birds, chicks, eggs, lizards, if easily available. Inhabits undisturbed waters.
Cultural notes: Less known bird, but the nuptial dance of the related Black-necked Stork, which breeds in India has been aptly glorified in old Sanskrit texts, where it is called the Actor who dances and also plays music by the clapping of the bill. The enormous plough-share like bill has given it the name **Karalphal Baka** (**Karal**-large, **Phal**-plough, **Baka**-stork).
Status: Rare.

Black-necked Stork.

Greater Adjutant Storks are not recorded in our region.

Black-necked Stork (M)

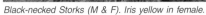

Black-necked Storks (M & F). Iris yellow in female.

Painted Stork protecting its chicks from sunlight.

Black-necked Stork (M) with a Little Grebe in the beak.

Kandesar, Kaurav, Kardhok

Description: Glossy black plumage is well appreciated in sunshine. White neck and abdomen, black crown, red-black bill, red legs conspicuous. Prefers waterlogged, marshy inundated terrain, paddy fields and tidal flood zones. Soars high in a vulture-like fashion. Commonly seen on village tanks, apparently oblivious of the company of women, who perform daily washing activities! Fish kept for drying near the fishing hamlets on the coasts are eaten, when fishermen are not around.

Risks:
The young storks sometimes fall from their nests. They can be successfully relocated. If the nests are very high, or if the chicks are injured, they can be reared in captivity. Readily accept fresh fish and feed by themselves after a few days. These storks, if not allowed to be domesticated, can be set free when their flight is well established (Amit Pawshe, S.Pande).

To avoid the risk from electrocution, nest platforms can be constructed on the top of electric poles.

Woolynecked Stork

Ciconiidae

Ciconia episcopus (62)

Shitikanth

Size: 900 mm.   M / F : Alike   **R, LM**

Distribution : The entire region up to 700 m.

Nest-site: Tall tress near village temples. Often away from water. Solitary. *May - Sep.*
Material: Twigs, sticks, freshly broken branches.
Parental care: Both.
Eggs: 3-4, bluish-white. 62.9 x 47.4 mm.
Call: Loud croaks and honks uttered when approaching the nest. Usually silent.
Ecological notes: Large *Tamarindus indicus* and *Ficus* spp. trees are commonly used for nesting purposes. Due to the Increasing non-availibility of such trees, nests are sometimes seen on electrical pylons,towers. Electrocution of a bird which nested on a pylon has been seen (Satish Pande).
Cultural notes: The nest tree's proximity to a temple is noteworthy. The traditional nest tree is usually of *ficus* spp. and is therefore protected. Parakeets, mynas, tits, owlets often use the same tree for nesting.
Status: Common.

Black Ibis

Red-naped Ibis

Threskiornithidae

Pseudibis papillosa (70)

Raktashirsha, Dhawalskandha, Aati

Size: 680 mm.   M / F : Alike   **R**

Distribution : Deccan and eastern margins of the Western Ghats.

Nest-site: Tall trees often away from water, singly or in mixed or exclusive colonies. Also amidst towns or villages. *Jun. - Oct.*
Material: Sticks, twigs, lined with straw, feathers.
Parental care: Both.
Eggs: 2-4, greenish, spotted, streaked. 63 x 44 mm.
Call: Loud screams are uttered in flight. The heronries are noisy. Also bill clapping noise.
Ecological notes: Discarded vulture's and eagle's nests are used for nesting (Salim Ali). Less dependent on water and hence is seen in drier areas. Also feeds on stagnant, polluted, sewage water and garbage dumps. Trees with heronries need protection. A heronry shared with bats is recorded near Niphad, Nasik.
Cultural notes: Black ibis was a favourite prey of the falconers (Shyainik-Shastra), due to the Ibis's strong and high rising flight. The long surgical tong, invented by Sushrut (Aatimukh Yantra, 400 AD), was after the beak of the black Ibis-Aati. The Black Ibis carried spiritual significance as per the Yajurveda. The white shoulder patch signified the Sun, the bald head indicated spiritual wisdom and red crown symbolized sacrificial fire! What the sacred Ibis or Father Sickle was to the ancient Egyptians, the Black Ibis was to the ancient Indians.
Status: Uncommon.

Black Ibis.

Kala Awaak, Kala Sharati

Description: Black with white shoulder patch (Sanskrit name **Dhawalskandha**, **Dhawal**-white, **Skandha**-shoulder), red crown and nape. The birds roost at noon. Crepuscular. Flocks fly in V and other formations.

Surgical instrument resembles the beak of an Ibis.

Related Species: *Glossy Ibis (Plegadis falcinellus-71)* Our smallest ibis. Breeding birds-chestnut with a green gloss on wings and white bars near lores. Head streaked brown in non-breediding birds. Juveniles-dull brown. Forage in belly deep water, taking aquatic invertebrates with the thin curved bill. Locally common. Nesting is in heronries. 520 mm.

Glossy Ibis.

Oriental White Ibis

Pandhara Sharati, Kudal

White Ibis, Blackheaded Ibis

Threskiornithidae

Threskiornis melanocephalus (69)

Sharatika, Shwetakak

Size: 750 mm.   **M / F :** Alike   **R, LM**

Distribution : The entire region. In the Ghats up to 1000 m.

Nest-site: On tall trees near water or even in the midst of villages. In mixed heronries. *Jun. - Oct.*

Material: Sticks, freshly broken branches.

Parental care: Both.

Eggs: 2-4, bluish, greenish-white. 63 x 43.8 mm.

Call: Lacks true voice-producing mechanism hence is silent. Grunting sounds when nesting.

Ecological notes: Devour insects in the standing paddy fields. Keep to fresh water.

Cultural notes: Vernacular name Kudal means a pickaxe, indicating the curved beak. In ancient Egypt the Ibis was considered sacred. Red underwing patch is often mistakenly thought to be blood and is anxiously reported.

Status: Near Threatened.

Description: Long down curved black beak is striking even in flight. Actively feed in shallow water, paddy fields, marshes, mud banks with other waders. Red wing patch is seen in flight. Immatures have grey neck. During breeding, the tail feathers become grey and upper breast feathers become fluffy.

Chick feeding from the mouth of parent.

Immatures.

Glossy Ibis (Adult).

Black Ibis.

Glossy Ibis (Juv.)

Thoth-Ancient Egyptian Ibis headed God.

Spoonbills on the nest with chicks.

Spoonbill in breeding plumage.

Eurasian Spoonbill

Spoonbill

Threskiornithidae

Platalea leucorodia (72)

Darvida, Khajaka

Size: 850 mm.   M / F : Alike   **R, LM**

Distribution : The entire region up to 600 m.

Chamcha, Dabel

Description: Snow-white bird with a spoon shaped bill, black legs and yellow neck-ring. A drooping crest in the breeding season. Active in the morning and evening. During the day, flocks rests at the water's edge. While feeding, the bird half immerses the bill, turns it from side to side in a semicircular fashion, ready to catch food perceived by the tactile beak-tip sensors. Take to air in a spiral flight.

Nest-site: Trees in or near water, in mixed heronries. These are scattered in the Deccan. *Jun. - Oct.*

Material: Dry sticks, branches, twigs, repaired from time to time. The same nest is used year after year if undisturbed or if the tree is not felled!

Parental care: Both.

Eggs: 4-5, dull-white, spotted. 65.6 x 44.2 mm.

Call: A low grunting note is uttered. Clapping of the bill is also heard.

Ecological notes: Extralimital migration also occurs. A bird of the nominate race ringed in Turkey, was captured at the mouth of the river Indus (SA Handbook.)

Cultural notes: Representative of the Wind-God, as per a belief. The names **Darvida** and **Darvitund** are derived from sanskrit **Darvi**- a spoon. **Khajaka** means a churning spoon indicating the swinging action of the birds neck.

Status: Rare. Patchily common only in the preferred habitat.

Juvenile.

Mixed flock.

Greater Flamingo

Flamingo

Phoenicopteridae

Phoenicopterus ruber (73)

Mrunalkanth, Raktak, Vakrachanchu

Size: 1400 mm.   M / F : Alike   **LM**

Distribution : Patchily in the entire region. Not in the Western Ghats.

Nest-site: Earth mound nests in the Great Rann of Kutch, communally (Flamingo Cities). *Sep. - Oct.*

Material: Mud lined with down feathers.

Parental care: Both.

Eggs: 1-2, faint blue. 88.8 x 54.5 mm.

Call: A loud goose-like honking note. Babbling sounds when feeding.

Ecological notes: The nesting time and their arrival to extra-nesting areas like ours, depends on the SWM. Exhausted juveniles are sometimes left behind, with an adult or two to attend, while returning from migration sites. Require water with salinity. Migrate extralimitally-a bird ringed in Iran, recovered in Thanjavur in S.India (Handbook, Salim Ali).

Cultural notes: *Agnipankhi* and *Raktak*, respectively mean, one with wings on fire, stained with blood. Feature in literature and movies.

Status: Uncommon. Some traditional migratory grounds are disturbed due to irrigation and construction activities.

Related Species: **Lesser Flamingo (Phoenicopterus minor-74) Two-third the size of the Greater Flamingo. Short neck, dark bill. Plumage rosy. Young birds of both the species are grey-brown. Lesser is seen in smaller numbers mixed with Greater. On coasts. A favourite site is Uran, Mumbai.*

Rohit, Pandav, Agnipankhi

Description: Curved heavy bill, long pink legs, white bird with pink and black wings. Feeds in shallow saline water with bill and head almost submerged. Aquatic food is seived through the comb edges of the bill. Toes webbed hence can swim, when it looks like a swan!

Comb edged beak.

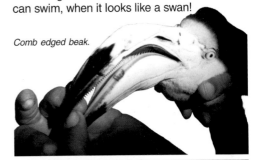

Immature Lesser & Greater Flamingos.

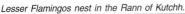

Lesser Flamingos nest in the Rann of Kutchh.

Greater Flamingo.

Lesser Flamingo.

Dancing Lesser Flamingos.

Mating Lesser Flamingos-cloacal kiss.

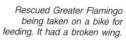

Rescued Greater Flamingo being taken on a bike for feeding. It had a broken wing.

Lesser Whistling-Duck

Lesser Whistling Teal

Anatidae

Dendrocygna javanica(88)

Tamra Hansa, Sharali, Ravi Hans

Size: 420 mm.   M / F : Alike   **R, LM**

Distribution : Water bodies in the low lying areas of Kokan, Malabar and foothills of the Western Ghats and Deccan.

Maral

Nest-site: In tree holes and also in reed beds. May also nest in trees away from water bodies. *Jun. - Sep.*

Material: The tree hole is lined with down feathers and grass. Nest platforms on water are firmed with reeds and leaves of water plants.

Parental care: Both.

Eggs: 7-13, ivory-white, turning brown as the incubation progresses. Oval. 46.9 x 36.8 mm.

Call: A three syllabled harsh, shrill whistle is uttered in flight.

Ecological notes: It is commonly encountered on water bodies amidst human habitation, even if the water is partially polluted. Being a good diver and grazer, it is seen on water bodies and marshes alike. Nesting coincides with the South West Monsoon.(SWM)

Cultural notes: Still hunted for flesh!

Status: Common.

Description: A pale brown teal with a dark brown crown and blackish wings; a chestnut patch on the forewing and on the upper tail coverts. To some extent a confiding bird. Seen in small flocks even on smaller water bodies. If approached they retreat to the safety of the center of the pond or lake or take flight.

Related species:
*Large Whistling-Duck (D.bicolor-89) is similar in appearances to the Lesser Whistling Duck, but is larger in size. It is distinct from *D.javanica* by its white upper tail-coverts and a dark line on the back of the neck. Size-510 mm. Resident. Rarely to our southern area.

Large & Lesser Whistling-Ducks.

Lesser Whistling-Duck.

Lesser Whistling-Ducks with chicks.

Lesser Whistling-Ducks.

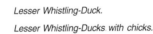

Geese are depicted on temple walls since Mauryan period.

Mute Swan.

Fresh droppings of Bar-headed Geese are green in colour since they eat tender shoots.

Noose traps are still used for capturing ducks & geese. These should be destroyed when found.

Bar-headed Geese.

Bar-headed Goose

Anatidae

Anser indicus(82)

Kadambahansa

Size: 750 mm.   M / F : Alike   **WM**

Distribution : Patchy. Constant sightings on a reservoir near Phonda Ghat in south Kokan (RM-per.com.). Also in the Deccan; not Malabar.

Nest-site: On floating vegetation in mountain lakes. Breeds in Ladakh, and not in our region. *May - Jul.*
Material: Leaves, grass, sticks, etc.
Parental care: Both.
Eggs: 3-6,white,round. 84.4 x 55.1 mm.
Call: Loud honking. Utters babbling noises while grazing. The vibrant call *Aang, aang* forever remains etched in memory !
Ecological notes: They cause considerable damage to winter crops especially in the areas along the eastern margin of the Western Ghats which they frequent in large numbers. Young birds are seldom seen in the flocks coming to our region, raising a doubt about their successful breeding. They are one of the highest flying birds.
Cultural notes: These are supposedly the revered and much admired swans or the **Hans** of the Manasarovar, described since ancient times by pilgrims undertaking the holy Kailas-Manas pilgrimage. Their sightings make newspaper headlines till this date. Depicted on temple walls since Mauryan period(400AD).
Status: Uncommon.

*Related Species: *Greylag Goose (Anser anser-81) 750-900mm. A few scattered flocks are seen along the coasts in our northern range. These nocturnal feeders rest during the day. Call is a loud 'Aang'.*
*Lesser White-fronted Goose (Anser erythropus-80). White forehead and yellow eye ring. Grey-brown, vermiculated. Vagrant near Mumbai. 530-660 mm. **Vulnerable**.*
Bar-headed Geese.

Patta-kadamb

Description: Overall a pale grey goose with white head face and throat. Two dark bands on the back of the head are conspicuous. Juveniles lack these bands. Almost always seen in flocks. An extremely elusive wary bird, seen on the sandbanks of rivers and fields..
Greylag Goose.

Lesser White-fronted Goose

Brahminy Shelduck

Anatidae

Tadorna ferruginea(90)

Chakrawak

Size: 640 mm.   M / F : Alike   **WM**

Distribution : Patchy. The Western Ghats, Kokan and north coastal Karnataka. Not in Malabar.

Bramhani-Badak

Description: A large chestnut brown duck with buff head and neck, white upper wing and under wing coverts, with green flashing in flight. Breeding male has a black neck band. One of the first winter visitors to arrive and the last one to leave. Prefer to graze along river banks. Large mixed congregations are seen on favoured, undisturbed waters. In summer, it shares the habitat with Blacknecked Cranes in some of its range.

Brahminy Shelduck.

Common Shelduck.

Nest-site: Holes in cliffs, often away from water; Nests in Ladakh, Nepal and Tibet. *Apr. - Jul.*
Material: Lined with down feathers.
Parental care: Both.
Eggs: 6-10,white. 67 x 47 mm.
Call: A loud metallic *Aang, aang.* The grazing grounds are very noisy at dawn and dusk.
Ecological notes: Rarely scavenges on flesh (Salim Ali-Handbook), a fact also recorded in the epic Mahabharata! Also seen on polluted water bodies. The duck is still secretly hunted with horse-hair snares for its flesh.
Cultural notes: Features in the Vedic, Epic and Sanskrit literature. Much is written in admiration of its beauty, fidelity and as a symbol of grief during separation. It is however strongly suspected that the birds form new pairs each year!
Status: Common to occasional. The numbers appear to have remained stable in the past three decades.

Related species:
*Common or Redbilled Shelduck (T.tadorna-91) Uncommon winter visitor on water bodies around Pune. White with a greenish- black head, neck and scapular stripe. Chestnut breast band and red bill conspicuous.

Anatidae

Sarkidiornis melanotus(115)

Nandimukh, Nasachhinna

Size: 760 mm.   M / F : Dimorphic   **R, LM**

Distribution : Patchy. Kokan, Goa and coastal Karnataka. Rarely in the Deccan.

Naktebadak

Nest-site: Natural hole in an old tree standing in or near water. Rarely in holes in walls.
Jun. - Sep.
Material: The nest is lined with leaves, grass, sticks, feathers etc.
Parental care: Incubation probably by female.
Eggs: 7-15, creamy white. 61.8 x 43.3 mm.
Call: Mostly a silent duck but utters loud honks in the breeding season. The drake utters a few croaks.
Ecological notes: A good wader and diver, it inhabits marshy as well as open water bodies. It also perches very well on trees.
Cultural notes: The duck generates much awe due to the peculiar shape of its bill. Sanskrit names **Nandimukh** and **Nasachhina** are descriptive of the shape of the bill.
Status: Occasional. Seen singly or in pairs, rarely in small flocks.

Description: Overall a whitish coloured large duck with speckled head and neck and bronze wings. The male has a fleshy knob on the bill. The duck is quite fond of playing in water with noisy vigorous flapping of the wings, thereby splashing a liberal quantity of water, apparently enjoying the effect!

Related Species: *Cotton Teal (Nettapus coromandelianus-114) Appears from a distance like a miniature comb-duck. This small duck is our widespread resident and is gregarious in habit. The male has a white head and neck with a black cap and breast-band. The female is duller. Locally, throughout the region.*
Look Alike: *White-Winged Wood Duck (Cairina scutulata) seen in NE India may be mistaken for the Comb-Duck.*

White-Winged Wood Duck.

Cotton Teal.

Comb Duck - preening.

Ducks and pochards are still sold in the fish market.

Baikal Teal (Anas formosa) is not seen in our region.

Red-necked Phalarope (Phalaropus lobatus) can be encountered in our region.

Duck hunting - a technique extensively used in the past. From : Tribes of My Frontier- E. H. Atkins

Spot-billed Duck

Grey Duck

Anatidae

Anas poecilorhyncha(97)

Varmukhi Hans, Nandimukhi

Size: 610 mm.   M / F : Alike   **R, LM**

Distribution : Widespread in the Deccan and the Western Ghats up to 1400 m. Rarely in Kokan and Malabar.

Nest-site: A pad concealed in short vegetation and bushes on the margins of water bodies. *Jun. - Oct.*

Material: The nest pad is made of grass and water weeds and is lined with feathers.

Parental care: ? Both.

Eggs: 7-9, greyish-white,56 x 42 mm.

Call: Generally silent, hence often overlooked, but utters loud quacks when suddenly alarmed and while taking flight. Occasional croaks.

Ecological notes: Occasionally cause damage to the standing paddy by eating and trampling the standing crop.

Cultural notes: One of the best known duck species. The red and yellow spots on the bill have given it a vernacular name 'Haldi-Kunku', the pigments traditionally applied on the forehead by married women.

Status: Rare in Kokan, common elsewhere.

Plava, Dhanwar, Sarja, Haldi-Kunku

Description: Dark crowned with spotted breast, yellow tipped bill with red lores. Green speculum and white tertials are best seen in flight. Scalloped wings are seen when swimming. Congregations are seen on favoured water bodies.

Eurasian Wigeon (F & M)

Mallard (F & M)

Related species: ***M**allard (*A.platyrhynchos-100*),**Gadwall* (*A.strepera-101*), **Wigeon* (*A.penelope-103*) *more or less share the same water bodies with the Spotbilled Duck. All are uncommon. Mallard has a yellow bill, green head and chestnut breast with a white neck ring. Male Wigeon has a yellow forehead and chestnut head. Male Gadwall is drab. Females of all species appear almost similar from a distance.*

Spot-billed Ducks mating.

Gadwall (M & F)

Pintail Ducks upending.
Garganey (M)

Wigeon (M) in partial moult.
Garganey (F) in partial moult.

Shovelers (M & F)

Common Merganser (F) with chicks. Vagrant to our region.

Northern Pintail

Pintail

Anatidae

Anas acuta(93)

Deerghapuchha, Shankuhansa

Size: 560-740 mm.   M / F : Dimorphic   **WM**

Distribution : On fresh and brackish waters in Kokan, Malabar and the Western Ghats.

Nest-site: The nest is placed in a marshy grassland. Breeding in the Palaearctic region of Europe and North and Central Asia. *May- Jul.*

Material: Compactly lined with aquatic weeds and down feathers.

Parental care: Both.

Eggs: 7-12, sea-green.

Call: Quacks and croaks. Silent in winter.

Ecological notes: A vegetarian dabbling duck. Also feeds by upending and grazing in marshes and wet fields. Ringed birds have lived for more than 11 years (S.A.Handbook).

Cultural notes: The duck is mentioned in the Shivpuran (1000 AD). It was extensively hunted in the days of the Raj. The practice sadly is still in vogue.

Status: Common. Large flocks are seen on open water bodies.

Sararuchi

Description: A long necked, wary bird. Male with upward pointing long pins to central tail feathers. Plain buffish head of the female lacks the eyebrow, which is distinct from similar female mallard. Flocks feed in shallow water. Around noon they congregate with other ducks in deep water, away from the shore.

Marbled Teal.

Northern Shoveller (M)

Common Teal (M)

Garganey (M)

Related Species: *Garganey or Bluewinged Teal (Anas querquedula-104) Brown drake with white eye stripe, silvery grey flanks. Duck has pale and dark eye stripes and dark cheek bar. Common, widespread WM to our region.*

Common Teal (A.crecca-94) WM to our fresh and brackish waters. Chestnut and green head of the drake is striking. Duck-orange based bill, white streaked undertail-coverts.

Northern Shoveller (A.clypeata-105) WM. Drake has green head, white breast, chestnut underparts. Buff duck has orange base to bill. Strains food with spatulate, toothed bill.

Female.   **Nayansari**

Description: This brown diving duck inhabits deep waters; flocks rest at noon in the center of a lake or beyond the surf zone in the sea. Swims with grace and ease but takes off the water with effort and much wing flapping. White wing bars and abdomen are seen in flight. Iris is white in adult males and brown in females and young males.

Common Pochard (M)

Related species:
Common Pochard (Aythya ferina-108)
Chestnut head, grey vermiculated back, grey breast. Upperwings grey and not white in flight. Banded grey-black bill. WM to our lakes, rivers except in Malabar.
*Red-crested Pochard (Rhodonessa rufina-107) Drake-Square,golden-orange head, red bill (Sanskrit name-**Raktachanchu**), black breast, white flanks. Duck with brown head, pale cheeks, grey bill. Diving, dabbling WM to our Northern region. Is often mistaken for the extinct *Pinkheaded Duck(R.caryophyllacea-106).*Tufted Pochard (A.fuligula-111) Drake-pied, with drooping occipital tuft, glossy black head, breast, back. Black replaced by brown and white wing bar paler in female. Diver. WM.*

Red-crested Pochard (M)

Ferruginous Pochard

White-Eyed Pochard
Anatidae
Aythya nyroca(109)
Mallikaksha
Size: 410 mm.   M / F : Dimorphic   **WM**
Distribution : Patchy. Marine and fresh waters in Kokan, North and Central Western Ghats. Unrecorded in Malabar and Southern Ghats.

Nest-site: Within India in Kashmir. Breeds in marshy areas in the reed-beds. *May - Jun.*
Material: Lined with grass and down feathers. Built with rushes and aquatic weeds.
Parental care: Both.
Eggs: 6-10, buff. 51.7 x 37.9 mm.
Call: A loud *Koor, ker, ker, kerr* is uttered on the wing or when feeding. Also croaks and honks.
Ecological notes: Feeds by diving and also grazes in marshes. Seen in marine and fresh water alike.
Cultural notes: The pochard is mentioned as a waterbird by Sushrut (400 AD).
Status: Occasional.

Tufted Pochard (M)

Greater Scaup (Aythya marila-112) WM, Vagrant (M)

Black Baza

Blackcrested Baza, Lizard Hawk

Accipitridae

Aviceda leuphotes (127)

Shyam

Sitana, one of the prey species.

Size: 330 mm   M / F : Alike   **R**

Distribution : Western Ghats Nilgiri south in foothills and in Wynaad.

Nest-site: Forks of leafy trees. *Mar. - Apr.*
Material: Leaves, grass, sticks, twigs. Eggs are laid on green leaves.
Parental care: Both.
Eggs: 2-3, oval, greyish, white spotted. 37.4x31.1 mm.
Call: Plaintive whistle.
Ecological notes: Predator. Lizards, frogs, insects eaten. Crepuscular.
Cultural notes: Features on logos of wildlife organisations. However, taxidermists found them repulsive due to the disaggreable rotten frog odour of a freshly killed bird that lingered for quite some time!
Status: Rare.

*Related Species: *Jerdon's or Legge's Baza (Aviceda jerdoni-126) 480 mm. Larger than Black Baza.In Ghats Coorg south,Wynaad up to 900 m. Brown,crested, three bands on tail-male, four to five bands-female. Rufous bands on white breast. Upper mandible serrated. Palm squirrel like call 'Kip, Kip'. Aerial displays prior to breeding. Rare. **Near Threatened.***

Kala Baaz

Description: A long-crested black bird with white breast band and rufous bars on the belly. Undersides of primaries grey and underwing coverts black. Perch upright in the canopy and is usually seen in flight when it can be mistaken for a crow. Soars in circles. Flocks are encountered in winter.

Jerdon's Baza.

Black Baza.

Male.

Oriental Honey-Buzzard

Crested Honey Buzzard

Accipitridae

Pernis ptilorhynchus (130)

Madhuha

Size: 680 mm   M / F : Dimorphic   **LM, R**

Distribution : The entire region and in the Western Ghats up to 1000 m.

Nest-site: Large leafy trees, 6-20 m. up. *Feb. - Jun.*
Material: Sticks, lined with leaves.
Parental care: Both. Incubation 32 days.
Eggs: 2,red brown, oval. 52.8x42.8 mm..
Call: High pitched *Whee-ew.*
Ecological notes: Eat honey, wax and bee-larvae even those of the ferocious Rock-Bees. Scale like feathers are protective. Also eat lizards, mice, birds, poultry. Nematode worm infestation of their toes is reported(JBNHS 44:23,1943).
Cultural notes: This buzzard is supposed to be one of the offspring of Garuda, the vehicle of Lord Vishnu.
Status: Rare.

Madhadya Garud

Description: Perches erect in leafy groves. Plumage variable. Long narrow neck. Male has two bands on tail, female has three. Prominent moustachial and gular stripes. Dark nuchal crest is erected when alarmed.

Male.

Female.

Dark Morph.

Juvenile.

Black-shouldered Kite

Blackwinged Kite

Accipitridae

Elanus caeruleus(124)

Kumud, Shabalika

Size: 330 mm.   M / F : Alike   **R, LM**

Distribution : Kokan, Malabar and the Western Ghats in deciduous terrain up to 1200 m.

Nest-site: Generally 5-10 mt. up in trees amidst farmland, in groves or forests.
Mar. - Aug.

Material: Sticks, twigs, branches; lined with grass, leaves, roots.

Parental care: Both.

Eggs: 3-4,pale-yellow, brown blotched. 39.3 x 30.9 mm.

Call: Usually silent. Emits a shrill whistle.

Ecological notes: Predator. Eats reptiles, rats, locusts and aquatic food during the SWM from the flooded paddy fields.

Cultural notes: There is a superstition that, should this kite alight on the head of a person (an improbable occurrence!) kingship is bestowed upon him. Its name **Kumud** in Mahabharata means a white lilly, for such is the frontal appearance of this kite.

Status: Common.

Kapshi

Description: Hovering flight, descending in steps, is typical of this kite. The scanning focus is shifted from time to time, till prey is seized. Often seen hanging in mid air at the edge of a cliff with wings held above the head staying at the same spot, without moving the wings, with the help of the upward draught. Partial to cultivated areas. Red eyes, black shoulders and ventral wing-tips.

Juvenile.   Adult.

Subadult.

Juvenile.

Look Alikes: In flight, the male *Pied, *Pallid and *Montagu's Harriers can be mistaken for the Blackwinged Kite. The harriers are long tailed and larger birds, the Pied has a black head, neck; Pallid lacks black shoulder patch and Montagu's has a grey breast.

Ghaar

Description: Turning and twisting amidst wires and plying vehicles, this kite dives at the overflowing urban garbage dumps and lifts it's food unerringly. Crows pursue and pirate their food. Deeply forked tail, brown colour, black primaries and an underwing crescent at the base of primaries in flight. They soar with the mid-morning thermals.

Related sub-species: *Black-eared Kite (M.m.lineatus-134) Differs from the common kite by shallow fork to tail, broader white underwing patch, larger size and black ears. Less common. Keeps to fringes of towns near farmland, hills, within our entire range. Sometimes pecked by crows. Such birds are often rescued by people. Also WM.

Black-eared Kite.

Adults (L) & Juvenile (R) Black Kites.

Black Kite

Common Pariah Kite

Accipitridae

Milvus migrans govinda (133)

Shakuni, Kshudragrudhri

Size: 610 mm.   M / F : Alike   **R**

Distribution : Seen in the entire range. Rarely in forests.

Nest-site: In trees amidst towns, on roof-tops. *Oct. - May.*

Material: Twigs, sticks, branches, wires, rubbish.

Parental care: Both. Kite carrying a stick in its beak to the nest is a common urban sight.

Eggs: 2-4, pink-white, spotted. 52.7 x 42.7 mm.

Call: High pitched whistle *Chili-li-li-wirrr.* Vocal when fighting over a disputed morsel.

Ecological notes: This urban kite is indicative of the presence of litter. Fish markets, nullahs are its favoured haunts. Scavenger in urban areas and predator.

Cultural notes: In the Shukla-Yajurveda (1500 BC), the kite, owing to its unerring excellence at picking eatables from the ground, is devoted to the deity of Marksmanship!

Status: Common in urban areas.

Black-shouldered Kite can be seen with Black-eared Kite.

Brahminy Kite

Accipitridae

Haliastur indus (135)

Khakamini, Kshemankari, Lohaprusthkank

Size: 480 mm   M / F : Alike   **R**

Distribution : The entire region near water bodies.

Brahmany Ghar

Description: A beautiful kite with rounded tail, rusty back and white head. Often surrender food to pirating black kites, crows, being timid by nature. Unerringly catch prey in mid air a feat best demonstrated at the 'Raptor Show' held in Jurong Bird Park, Singapore!

Nest-site: 6-10 m. up in a lofty tree, ruins. Near water. *Dec. - Apr.*

Material: Leaves, grass, sticks, twigs, rags.

Parental care: Both. Incubation 26-27 days.

Eggs: 3-4, white,round. 50.7x40.2 mm.

Call: Whistles, squeaks.

Ecological notes: Nesting period depends on rains. Scavenger of waterside refuse. Also predates lizards, crabs, birds, fish.

Cultural notes: Padmapuran (400 BC) mentions it as a bird of good omen- ***Kshemankari***, a belief shared by Muslims who call it Rumubarik-lucky faced. ***Khakamini*** means one who loves the sky, ***Lohaprushtha*** means rusty backed.

Status: Common.

Sheshari, Samudra Garud

Description: Territorial. Fish, snakes are lifted from the surface of the sea with strong feet, sharp talons and powerful wing-beats. Males engage in aerial fights in the breeding season, and may suffer a fall. Two such incidences have been noted. Treated eagles can be released after first-aid (C. Salunkhe). Fish-piracy from a gull has been recorded, when the eagle's nest was active(SatishPande). when the eagle lifted the fallen fish from the sea shore.

Related species: *Pallas's or Ringtail Fish-Eagle (H.leucoryphus-174) Rare. Seen in Goa. Brown with pale longish head, neck, back, white, banded wedge shaped tail. Fishes by diving. Prey is also pirated from other eagles. Vulnerable.*
Greater Grey-headed Fish-Eagle (Ichthyophaga ichthyaetus-175) Rarely on coasts and inland lakes. White tail with terminal black band is seen in flight. Rufous breast, grey head, white belly and vent; brown above. Fish lifted from the surface and not by diving. Near Threatened.
White-tailed Sea-Eagle (Haliaeetus albicilla-172a)Vagrant, WM. 700-900 mm. Yellow bill, white wedged tail. Adult brown, juvenile darker with pale axillaries, bill, tail. Near Threatened.

Ringtail Fish-Eagle.

White-bellied Sea-Eagle

White-bellied Fish-Eagle
Accipitride
Haliaeetus leucogaster(173)

Utkrosh, Saptaraav, Kohasa

Size: 660-710 mm.   M / F : Alike   **R**

Distribution : All along the coast and the off-shore islands. Also in tidal creeks, lagoons and estuaries. Sometimes over the Ghats.

Nest-site: Mango, *Casuarina, Ficus* spp., Palm, Tamarind, etc. trees along the coast or a few km. inland. 64 active nests have been identified in coastal Ratnagiri (Vishwas Katdare, Ram Mone). More than one nest known.
Oct. - Apr.

Material: Branches of mango, *ficus*, bamboo etc. Eggs are laid on a bed of green leaves.

Parental care: Both build and repair the nest, incubate and feed the chicks. The nest is not left unattended for long. Nest is abandoned if disturbed before the eggs are laid.

Eggs: 2(1-3),white. Low success rate. Incubation period is 41 days.(V.Katdare) 77.7 x 53.4 mm.

Call: A loud, nasal, resonant *Kenk, kenk*. One bird joins the other and the vicinity of the nest resounds with the commotion. Call is given when changing the guard on the nest during incubation and chick rearing.

Ecological notes: Crabs, gulls, chicks of marine terns, fish and sea-snakes are eaten. If undisturbed, the same nest is used over generations. Nest-trees on private land may be felled for wood or to avoid the nuisance of littering.

Cultural notes: The eagle-*Utkrosh* is mentioned in the Skandpuran (400 BC) as one who lifts sea-snakes!

Status: Common.

Greater Grey-headed Fish-Eagles.

White-bellied Sea-Eagle - A sculpture in Jurong Bird Park.

White-bellied Sea-Eagle chick.

White-tailed Sea-Eagle.

White-bellied Sea-Eagle on the nest with chicks.

*Lesser Fish Eagle (Icthyophaga humilis-177) is a vagrant.
Brownish tail with white base and rufous brown breast.*

Pandhare Gidhad

Description: Untidy looking kite-like yellow faced white vulture with black primaries and secondaries. Juveniles darker. Often at village side garbage dumps, slaughter houses, near carcasses, with others of it's kind. Less aggressive than other vultures. Frogs, rotting fish are taken from drying ponds. Also rummage in cow dung heaps and hence are known as Gobargiddha-Dung vultures in Hindi.

Cinereous Vulture.

Related species:
Red-headed, Black or King Vulture (Sarcogyps calvus-178) Solitary, rare vulture. Feeds on carcass by stealing a piece now and then, when the going is slow. Looks fierce due to the red head but is timid by nature. Nests in forest trees. Rare. Near threatened.
Cinereous Vulture (Aegypius monachus-179) Fiercest and rarest of the vultures in our area, respected by others, who promptly make way for it at the carcass. Perch on hill slopes. Is seen in drier, hilly areas, soaring in high heavens. Dark, except for the pale head. Near threatened.

Egyptian Vulture

White Scavenger Vulture

Accipitridae

Neophron percnopterus (187)

Bhas, Goshtha-Kukkut

Size: 640 mm.   M / F : Alike   **R, LM**

Distribution : The entire region and up to 2000 m. in the Ghats.

Nest-site: Cliff ledges, platforms in ruins or in tree forks. Mixed nesting colonies may be seen on a cliff-face. Such mountains have been called **Grudhrakut**-vulture mountains, in Sanskrit texts. *Feb. - May.*
Material: Twigs, sticks, rubbish.
Parental care: Both.
Eggs: 2, pink-white, blotched. 64.3 x 49.3 mm.
Call: Harsh screams and hissing sounds are given when a rare fight breaks over a carcass.
Ecological notes: Scavenger. Feed mainly at garbage dumps, refuse and offal.
Cultural notes: Goshtha-kukkut means one feeding like a cock on village garbage, a name reminiscent of its other well known name-Pharaoh's chicken! These are the well known daily visitors to Pakshi-Teertha, Thirukalikundram,near Madras. The fierce Cinereous Vulture, is a carcass scavenger and as per Hindu mythology, is one of the assistants of Yama, lord of death, who tortures the guilty condemned to hell!
Status: Rare.

Red-headed Vulture.

Red-headed vultures.

Egyptian Vultures - Adult and juvenile at nest in a cliff face.

Red-headed Vultures.

Egyptian Vulture - Immature.

Red-headed Vulture on a carcass.

Such flocks of White-backed or any other vultures are now a very rare sight.

Chick of the
Indian White-backed vulture.

A honey gatherer on a cliff face
with a bee-hive and a vultures nest
in Indian rock Art.

Vulture in Indian Rock Art (30,000 BP)

Falcon & Long-billed Vulture.

Indian White-backed Vulture.

Indian White-backed Vulture

White-rumped or Bengal Vulture

Accipitridae

Gyps bengalensis (185)

Grudhra, Shitikaksha, Hiranyakaksha

Size: 900 mm.   M / F : Alike   **R, LM**

Distribution : Throughout the region.

Pandharpathi Gidhad

Nest-site: Lofty trees in agricultural areas, cliffs. Colonial nesting on trees is known. *Nov. - Mar.*

Material: Sticks, branches, twigs.

Parental care: Both.

Eggs: 1, white, speckled. 85.8 x 64.2 mm. Juvenile birds lack the white rump and have streaked underparts.

Call: Loud screaming, screeching, hissing when feeding on a carcass. Otherwise silent.

Ecological notes: A useful scavenger. Also feeds on putrified flesh. Carcasses are quickly devoured by the flock of vultures and cleared to the bones, which are eaten by other scavengers.

Cultural notes: A person afflicted with the ailment of sciatica, develops a waddling gait, similar to that of the vulture's, hence sciatica was known as **Grudhrasi** in ancient Indian medicine.(**Grudhra**-vulture). Jatayu and Sampati, in the epic Ramayana, who helped Rama when his queen wife Sita was abducted by demon king Ravana, were vultures.

Status: Critical. Commonest of all vultures. Presently threatened due to suspected disease and changing methods of carcass disposal.

Description: Bare black head, neck, white rump, back and underwing-coverts. Masters of patience. When caracass is sighted the soaring vultures descend one by one, congregate for scavenging. Chunks are torn and gulped. After a heavy meal many birds find it difficult to take off immediately. Slaughter houses are their favourite haunts. Several species of vultures also have a keen sense of smell.

Eurasian Griffon.

Long-billed Vulture.

Related Species: *Long-billed Vulture (Gyps indicus-182) Pale brown body, cere, upper wing-coverts, black head, neck, tail and white collar. Pale below with a dark broad band to trailing wing edges. Nests on cliff edges on several hill forts. **Critical.***

Eurasian Griffon (Gyps fulvus-180) Pale head, neck, cinnamon-brown overall. Dark cere, streaked below. Pale brown underwing coverts reveal a white central band on medians in flight. Rarely to our northern area in winter.*

This Crested Serpent Eagle was found with a possible snake bite on the leg. It succumbed to this after profuse salivation.

Threat display of the Crested Serpent Eagle.

Crested Serpent Eagle's chick.

Black Eagles.

Short-toed Eagles - in flight and at the nest.

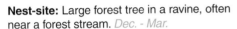

Crested Serpent-Eagle

Acciptridae

Spilornis cheela (196)

Sarpari, Nagashi, Sarpaant, Maalaaya

Size: 740 mm.   M / F : Alike   **R, LM**

Distribution : In Kokan, Malabar and the Western Ghats up to 1500 m.

Pannagad

Nest-site: Large forest tree in a ravine, often near a forest stream. *Dec. - Mar.*
Material: Branches, twigs. Lined with leaves.
Parental care: Both. The chick has a small nuchal crest.
Eggs: 1,cream, red blotched. 71.8 x 56.2 mm.
Call: A far reaching, shrill whistle *Kek, kek, ki, ki, kii-ii.* Call is also uttered from the nest.
Ecological notes: Prefers vicinity of water. Predator of reptiles and birds. Poisonous snakes are eaten after decapitation. Can die from snake bite. Requires large forest trees for nesting which are now felled in increasing numbers.
Cultural notes: The Sanskrit names of this eagle are descriptive of the snake-eating habit. The name **Maalaya** refers to it's crest.
Status: Uncommon.

Description: A hefty, powerful, elegant eagle. White wing-bars, tail-bars, drooping crest with white tipped ocellations on the nape. Yellow cere, lores and white-spotted brownish underparts. Perches near a water-hole or a forest clearing waiting for the prey. Known to attack peafowl.

Short-toed Snake-Eagle

Black Eagle.

Related Species: *Short-toed Snake-Eagle (Circaetus gallicus-195) A hovering eagle. Dark head, breast, barred belly. Dark trailing edge to pale underwings, bars on longish tail. Pale phase is known. Indulges in tumbling and other aerobatics prior to breeding. R.*
Black Eagle (Ictinaetus malayensis-172) Long tail barred on both surfaces. Yellow cere, legs. Glides with wings held in a V. Underwings show pale patches. R. In Ghats up to 2000 m.
Look Alike: *Other similar dark, brown-black eagle is the dark morph of Crested Hawk-Eagle. Has un-barred tail.*

Crested Serpent-Eagle.

Western Marsh-Harrier

Eurasian Marsh Harrier

Accipitridae

Circus aeruginosus (193)

Saras (One frequenting marshes).

Size: 540-590 mm.   M / F : Dimorphic   **WM**

Distribution : In the entire region.

Nest-site: The nest pad is placed on ground in marshes amongst reed. Extralimital. *Apr. - Jul.*

Material: Reed, grass, aquatic vegetation.
Parental care: Both.
Eggs: 4-6,white.
Call: Silent on the wintering grounds.
Ecological notes: Predator, scavenger. Inhabits marshy areas and shores of water bodies. Aquatic, semiaquatic food, water birds are eaten.
Cultural notes: The white head of the female Marsh Harrier is conspicuous in the green grass. This has bestowed the vernacular name Hareen,upon this harrier!
Status: Common.

Marsh Harrier (F) **Daldal Sasana, Daldal Hareen**

Marsh Harrier (M)

Description: Flies low over water's edge, marshes, mangroves, lazily beating the wings and gliding with wings held in a V. Excites commotion amongst water birds. Prey is taken in the talons of the anteriorly thrust legs. Roosts in marshes on ground for prolonged periods. Night roosts are communal, probably to avoid predators. Brown male has a pale head and black primaries. Female-dark brown. Cream crown.

Montagu's Harrier (M)

Montagu's Harrier (Imm.)

Related species:**Montagu's Harrier (Circus pygargus-191) 480 mm. Rare WM up to 1000 m. Male has grey head, back, breast, black primaries and underwing-coverts. Separated from Hen Harrier by the black band across secondaries, rufous streaks to pale belly. Female is like Palid Harrier but has dark trailing edges to secondaries, sharply barred underwings, paler head. Locust eater.*

Flight actions of the Marsh Harrier (F)
-soaring, hovering, predating
& scanning for prey..

Nictitating membrane is
protective during dives and
rapid movements.
(Short-toed Eagle's eye)

Prey like geckos, lizards, small birds are
a favourite of the harriers.

Marsh Harrier (F)

Habitats where harriers roost are often shared by blackbucks.

Pied Harrier (F)   **Kavdya Hareen**

Description: Flies tirelessly in search of
terrestrial prey. Glides, soars and ascends in
circles to considerable heights. Pairs rarely
seen together. Prefer to settle on ground. Male
has black head, upperparts, breast and
primaries. White below and on the forewing.
Female is brown with pale underwing and has
grey primaries, secondaries and barred tail.

Pale Harrier (F)   Hen Harrier (M)

Pale Harrier (F)

Related Species: *Pallid or Pale Harrier*
(C.macrourus-190) WM to scrub, agricultural,
marshy terrain. Male-grey above, pale below
with black wedge on primaries best seen in
flight. Female-larger, brown above, pale and
streaked below, pale collar, dark ear-coverts
and eye-stripe. In flight her wings are dark
above and patterned below with pale
primaries. Both have long yellow legs.
Near Threatened.

Pied Harrier

Accipitridae

Circus melanoleucos (192)

Patri (for harriers), **Shabal Patri**

Size: 460-490 mm.   M / F : Dimorphic   **WM**

Distribution : Throughout the region,up to
2100 m. in the Ghats. Rare in Malabar,Kokan.

Nest-site: In Assam, Manipur in India.
Commonly in China, Siberia, etc. *Apr. - Jul.*
Material: Leaves, grass, to make a rough
platform on ground in grasslands.
Parental care: Both.
Eggs: 4-6,white,red speckled. 43.6 x 34.5 mm.
Call: Silent in winter quarters. An occasional
cackle may be heard.
Ecological notes: Predates reptiles,
small mammals in grassland, farmland and
crabs, molluscs, frogs in marshes and coastal
areas. Undisturbed grasslands and reedy
marshes are required for their nesting, a
habitat currently under threat. At night, ground
roosting birds are known to be killed by
predators.
Cultural notes: In ancient Indian texts a
harrier was recognized as the less
accomplished hawk, capable of a sustained
flight but incapable of expertise in aerial
predation. Harriers are also known as
Field-Kites.
Status: Rare.

Montagu's Harrier (F)

Pied Harrier (M)   Pale Harrier (M)

Shikra

Accipitridae

Accipiter badius (137,138,139)

Sichana, Shikara

Size: 300-360 mm. M / F : Dimorphic **R, WM**

Distribution : The entire region u pto 1500 m.
Race *cenchroides* is WM to North Ghats, race
dussumieri all over, race *badius* in South.

Nest-site: 6-12 m. up in a leafy tree in a fork,
in towns, groves or forest edges. *Mar. - Jun.*
Material: Twigs, branches, lined with leaves.
Parental care: Both build the nest and feed
the chicks. Incubation is by the female.
Eggs: 3-4, blue-white,spotted.
38.8 x 31.1 mm.
Call: Loud screeches *Ti-uu, tii, tuii*. Silent
when looking for prey. Chicks are vocal.
Ecological notes: Predator. Preys upon
smaller birds. A pet Love-bird was once taken
by a Shikra after bending the thin wires of the
cage(Shyam Deshmukh).
Cultural notes: Though not as accomplished
as the Goshawk, Shikra was a favourite
amongst falconers since it trains easily. The
Goshawk,second only to the Peregrine, was
known as the gentle falcon in the art of
Falconry, and hence was named *A. gentilis*!
Status: Common.

Male.   **Shikra, Cheepak**

Description: Upperparts pale blue-grey in
male, brown in female. Wing tips are dark,
underparts have fine orange barring. Gular
stripe is seen in juveniles. Hidden in a leafy
branch, it scans the area and swiftly swoops
on the prey, surprising it. Passerines often
escape in the thickets and the Shikra pursues
the prey in the bush. Flies close to the ground
looking for prey and takes it unawares. Prey is
shredded with sharp talons and hooked beak.

Northern Goshawk (M), In flight.

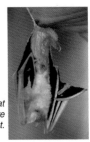

*The rare Painted Bat
may feature
in the Shikra's diet.*

Related species: *Crested Goshawk (A.trivir-
gatus-145) Kills partridges, green pigeons,
hare, smaller mammals by ambush. Name*
Chulankit *means crested (Chul-a crest). In the
Western Ghats, Goa south rarely to the north.
Not in Kokan. Builds a large nest in forest
trees. Call-a shrill cry.*
*Northern Goshawk (A.gentilis-136) Rare WM
to forests in the northern range. 500-610 mm.
Barred belly, banded long tail, rounded wings.*

Crested Goshawk.

Shikra often dresses the kill outside the nest and then feeds it to the chicks.

Shikra (F) taking a bath.

Shikra (F) with bat &
(M) with myna
held in
the talons.

Shikra - Eggs, chicks and Juvenile.

Eurasian Sparrowhawk

Sparrow Hawk.

Accipitridae

Accipiter nisus (147)

Kulingaka, Vasa, Vesar

Size: 310-360 mm.   M / F :Dimorphic   **WM**

Distribution : The entire region up to 800 m.

Nest-site: Ledge in a cliff, old corvid nests in forests. Resident race breeds in Himalayas and NE India. *Apr. - Jul.*
Material: Sticks, straw, etc.
Parental care: Female incubates. Other duties are shared.
Eggs: 4-6,pale red or white, brown spotted.
Call: Long whistling notes. A noisy hawk. It's Sanskrit name **Vasa** comes from the root word **Vacha**-word.
Ecological notes: Numbers declining probably due to habitat loss. Pesticides may also be responsible. Largely subsists on birds. Has an apt name **Kulingaka** -a sparrow-killer.
Cultural notes: Sparrowhawks were divided into three categories by falconers based on their colour, courage and flight. Female being a better predator, female chicks were removed from the nests for falconry. This may have added to their present decline.
Status: Rare.

Shyen

Description: Male is slaty-grey, female brown above. Male has rufous and female brown bars on the underparts. Black tail bands and white eyebrow are typical. Solitary. Keeps to trees from where it suddenly launches attacks on birds up to the size of pigeons, killing them by pursuit. Flies over a grassy patch with the hope of catching flushed birds.

Related Species:
*Besra Sparrowhawk (A.virgatus-151) R. In dense forests of the Western Ghats. Common Belgaum south. Speedy and bold short-winged hawk. Eats small birds, mammals, reptiles.
Look Alike: *Shikra (A.badius-139) is darker, has barred tail, underwings and longer legs. The Besra has a gular stripe and barred breast, which the Sparrowhawk lacks.

Besra Sparrowhawk.

Eurasian Sparrowhawk.

Tisa

White-eyed Buzzard-Eagle

Accipitridae

Butastur teesa (157)

Girishal

Size: 430 mm.   M / F : Alike   **R, LM**

Distribution : The entire region up to 600 m.

Nest-site: Fork of mango, tamarind tree in dense foliage, often overlooked. *Feb. - May.*
Material: Branches, twigs, straw.
Parental care: Female incubates. Both share other duties.
Eggs: 3, 46.4 x 38.4 mm.,greenish-white.
Call: Mewing calls uttered on the wing, vocal when breeding.
Ecological notes: Beneficial predator of drier biotope, since it eats rats and locusts. Also eats eggs and chicks of grassland birds.
Cultural notes: The name **Girishal** (**Giri**-mouse, **Shal**-to impale) is a generic name for buzzards.
Status: Ocassional.

Description: White nape-patch, dark gular stripe, black-tipped pale longish wings, rufous tail and yellow-white eyes. Underparts barred in adult, streaked in juvenile. Perch upright on a vantage point for prolonged periods. Take prey from ground. Soar at noon taking lift from thermals. Brown morph recorded(S.Pande).

*Related species: *****C**ommon or Eurasian Buzzard (Buteo buteo-156) Uncommon WM to our area. Pale black-tipped underwings, faintly barred tail. Head pale, flanks rufous, belly streaked.*
*Longlegged Buzzard (Buteo rufinus-153) Rare WM to the Deccan. Not Kerala. Unfeathered long legs, pale head, variable plumage.

Common Buzzard.

Longlegged Buzzard.

White-eyed Buzzard.

Crested Honey Buzzard protecting the killed female koel.

White-eyed Buzzard-Brown Morph.

White-eyed Buzzard-chicks with food in the nest.

Tawny Eagle on the nest.

Mating is observed throughout the nesting period.

With poultry for the chick.
Tawny Eagle.

Caring the chick.
Chick-4 weeks old.

Lizard features in their diet.

Tawny Eagle

Accipitridae

Aquila rapax (168)

Gomayu, Pichhabaan, Ragpakshi

Size: 630-710 mm.   M / F : Alike   **R**

Distribution : In the Deccan and eastern side of the Ghats up to Uttara Kannada. Not in Kerala.

Nest-site: Ficus etc. Spp. in cultivated area in the vicinity of a village. *Nov. - Apr.*
Material: Branches, sticks. Lined with leaves.
Parental care: Female incubates. Other duties are shared.
Eggs: 2-3, white, red spotted. 66 x 52.8 mm.
Call: Shrill screams *Ke-Ke-ke-ke* are uttered in flight.
Ecological notes: Scavenger and predator. Robs prey from falcons, hawks, kites. Also seen with vultures near a carcass, a rare sight in recent times. Devours rotting fish thrown up by the tide. A menace to poultry. Russell's viper, Common Sandpiper recorded in the stomach contents (Salim Ali,Handbook)
Cultural notes: Mentioned as one of the seven eagles of ancient India. Being vocal, it was offered in sacrifice to the Gandharvas-celestial musicians (Yajurveda,1500 BC). Its quills were used to feather arrows, hence the name *Pichhabaan* (Tail-Arrow).
Status: Uncommon.

Lesser Spotted Eagle (A. pomarina-171) 60-65 cm.
Flight feathers darker than wing-coverts.
In flight, two white carpal crescents on underwings.
Short billed, short tailed.
Vagrant.

Suparna

Description: A stately, bold and courageous eagle. Gape extends up to the eye and not distal as in the Steppe. Baggy trousers, oval nostril, yellow iris-typical. In flight head, neck are extended and a pale wedge is seen to the inner primaries. Gets excited when prey is seen in the possession of smaller predators, and robs it from them with much screaming. The Sanskrit name **Ragpakshi** means one who becomes jealous of others having a booty! Flocking of Tawny eagles is recorded.

Related Species:
*Greater Spotted Eagle (Aquila clanga-170)
Vociferous, sluggish. Does not resort to piracy.
Scavenger. Yelps like a dog. Perches in a lofty tree for hours waiting for prey. Round nostrils, short gape extending up to the eye, dark chin. Brown with a buff rump. Juvenile has a spotted look due to white tipped brown coverts. Rare WM to our region. Vulnerable.

Greater Spotted Eagles.

Garud

Steppe Eagle

Eastern Steppe Eagle

Accipitridae

Aquila nipalensis (169)

Suparna

Size: 760-800 mm.   M / F : Alike   **WM**

Distribution : Up to about Mumbai in winter, vagrant to the south.

Description: Yellow gape extends behind the eyes. A heavy eagle with big beak, rufous nape patch, baggy trousers, pale chin. Black trailing edge to the broad long wings, bars on underside of remiges, best seen in flight. Two pale bars are typically seen on outer wings. Ear-shaped nostril. Scavenger but also hunts small birds, reptiles. Like the Tawny Eagle it pirates prey from falcons, small eagles, hawks.

Nest-site: Breeds extralimitally from Altai to Mongolia, SE Siberia and probably in NW India. *Mar. - Jun.*
Material: Branches, sticks.
Parental care: Both.
Call: Silent in winter.
Ecological notes: The possible route of migration is over the hostile climate of the South Col pass in the Himalayas, from where a speciman has been collected for the BNHS (Salim Ali Handbook). The harsh climatic condition in the Himalayas takes a toll of migratory eagles.
Cultural notes: An unknown eagle in our area since it is rarely encountered.
Status: Rare.

Steppe Eagle.

Imperial Eagle.

Rufous-bellied Eagle (Juv.)

Rufous-bellied Eagle.

Related species:
*Rufous-bellied Eagle *(Hieraaetus kienerii-165) Forests of Western Ghats, Belgaum south. Soars with flat wings. Black streaked white breast, chestnut belly, black crested head, back. Barred tail is white below, grey above. Size-530-610 mm.*
*Imperial Eagle *(Aquila heliaca-167)WM. Vagrant to our area. Buff below, dark above, cream crown and white scapular patch. Gape extends behind the eyes. Primaries and secondaries dark and pale wedge to inner primaries. Vulnerable.*

Rufous-bellied eagle.

Steppe Eagle-Pale morph.

Imperial Eagle.

A rare flock of Steppe Eagles.

Steppe Eagles.

Naraach

Description: Long tailed, long winged slender eagle. Amber-brown above, pale, streaked below with a whitish mantle. Terminal black band on tail and underwing-coverts. Soars with flat wings. Vigilantly perches hidden in foliage. Ground prey is ambushed or pursued and killed by mutual pursuit by a pair.

Booted Eagle.

Changeable Hawk Eagle (S.c.limnaeetus) with Chital kill. This crestless eagle can be mistaken for Bonelli's Eagle.

Bonelli's Eagle

Bonelli's Hawk-Eagle

Accipitridae

Hieraaetus fasciatus (163)

Mayuraghni

Size: 680-720 mm.   M / F : Alike   **R**

Distribution : Seen patchily distributed in the entire region.

Nest-site: Tall, lofty trees, rarely on a ledge in a cliff-face. Large nests are built. *Dec. - Apr.*
Material: Sticks, branches, lined with leaves.
Parental care: Both.
Eggs: 2, white. 69.1 x 53.4 mm. Chicks leave the nest after about 3 months of hatching.
Call: Vocal during nesting. Pre-mating aerial displays with much screaming begin in late November. Call is *Kli, klu, kluii, kluii.*
Ecological notes: Predator. Lifter of poultry. Pigeons, doves, crows, shikra, blue-breasted quail, reptiles, young hare, rats are eaten (C.Bapat). Numbers are dwindling due to habitat loss. If undisturbed the nest is reused. A stolen chick from the nest was replaced and it was accepted by the eagle.(S.Pande, B.Pednekar)
Cultural notes: The Sanskrit name Mayuraghni means a killer of peafowl! Bonelli's eagle is capable of predating upon larger birds like Painted Storks, Geese (Salim Ali).
Status: Uncommon.

Bonelli's Eagle.

Related species:
*Booted Eagle (H.pennatus-164) Rare WM. A small, pale, kite-like eagle with feathered tarsus shows pale wedge on the dark inner primaries in flight. White shoulder patch constant. Long and square ended tail has dark terminal band and centre. White crescent on uppertail-coverts. Soars, suddenly stoops at prey.

The early life of the Bonelli's Eagle - a photographic depiction :
Both parents with two chicks on the nest. Food in the form of birds, lizards, hare is procured.

-Three week old chick shows black feather on white down..

*-Six weeks old chick is fully transformed from
the initial white plumage. (Incubation-43 days, Fledge-54 days.)*

*Mountain Hawk Eagle.
Juvenile Changeable Hawk & Bonelli's Eagles.*

Vyadh, Morghaar, Shendri Ghaar

Crested Hawk-Eagle

Accipitridae

Spizaetus cirrhatus (161)

Shashad Shyen, Shashaghni

Size: 720 mm   M / F : Alike

Distribution : The entire region.

Up to 1000 m. in the Ghats.

Mountain Hawk-Eagle.

Nest-site: Fork of a forest tree on a hill slope, at a considerable height. *Dec. - May.*

Material: Branches, sticks. Lined with leaves.

Parental care: Both. Owlet, koel, green pigeon, rats recorded to have been brought to the nest.

Eggs: 1, whitish, blotched at broad end. 64 x 50.5 mm.

Call: Scream, high pitched whistle *Ki-ki-ki* uttered in flight or when approaching the nest. Chick also calls from the nest.

Ecological notes: Predator. Poultry lifter from forest-side villages. Eucalyptus tree in a forest plantation has been used for nesting for consecutive years (Amit Pawashe).

Cultural notes: Mentioned as a hare-devouring eagle, **Shashad Shyen**, in the text Bruhadaranyaka(800 BC), as separate from the falcons. The two are also described as distinct entities by Charak (400 AD). It is the Shahbaaz of the falconers and was trained to chase prey.

Status: Uncommon.

Description: A graceful, slender eagle. Perches hidden in foliage, on the lookout for junglefowl, quails, reptiles, small mammals. Prey is taken in a swift attack. Female is larger. Terminal black tail band, crested head, white heavily brown streaked belly, coppery brown back. Plumage variable. A black morph has been seen to pair with a normal bird and raise a single normal chick in North Kokan.

(Sagar Mestri,S.Pande).

*Related species: *Hodgson's or Mountain Hawk-Eagle (Spizaetus nipalensis-158) R in forests of the Western Ghats, Goa south, vagrant northwards. Rufous barred underparts and black bands on tail. Crested white head is sreaked. Brown overall. Seen soaring over the forest canopy.*

Juvenile Changeable Hawk-Eagle.

Mountain Hawk-Eagles.

Osprey

Pandionidae

Pandion haliaetus (203)

Pankajit, Matsyarankashyen

Size: 560 mm.   M / F : Alike   **WM**

Distribution : In the entire region on marine and fresh water.

Nest-site: Tree, ledge, in or near water, 5-10 mt. up. Breeds in Europe. Patchily in India. *Apr. - Jun.*
Material: Sticks, twigs, branches, leaves.
Parental care: Both.
Eggs: 1-2. 61.6 x 46.3 mm.
Call: Silent in winter, when it visits us. Sharp musical whistle when breeding.
Ecological notes: A fish-eating hawk. Nests usually close to water. A bird ringed in Norway was recovered in Jamnagar(Salim Ali).
Cultural notes: The Osprey is mentioned in the works of Charak (400 AD). There is a story in the epic Mahabharat emphasising the virtue of vegetarianism, when the fish-eating Osprey was cursed by the Swan to be born as a Demoiselle crane, which was in those times thought not to eat flesh!
Status: Near Threatned.

The Osprey can often be seen on large fresh water lakes as on marine waters.

Kukri

Description: White underparts, brown breast band, brown above with a crested white head and a dark eye stripe. Black carpal patches and angled wings are seen in flight. Scans water in aerial beats, at times hovering where prey is suspected and then plunge-dives with closed wings often submerging. Fish is eaten piece-meal at a suitable perch. Likes to Sun dry the wings after a dive. Perches on a stake or a pole in water.

Lesser Kestrel in flight.

Common Kestrel accidentally snared in a flying kites twine. It was rescued & released..

Common Kestrel (M)

Common Kestrel (F)

Lesser Kestrel (M)

Common Kestrel

Kestrel

Falconidae

Falco tinnunculus (222)

Dishachakshu, Anal, Akash-Yogini

Size: 360 mm.   M / F : Dimorphic   **R, WM**

Distribution : The entire region up to 2500 m.

Race *objurgatus* is resident and *tinnunculus* is a winter migrant.

Nest-site: Ledge, a hole in a cliff, ruins of many of our hill forts or on trees. *Feb. - May.*
Material: Leaves, grass, sticks, rubbish.
Parental care: Both. The cliff-nest is approached in an ascending flight. The risk of the chicks falling from the high nest is thus reduced, since they do not see the parents approaching with food.
Eggs: 3-6, red, spotted. 38 x 30 mm. The red colour of the eggs has given it the Sanskrit name **Lohandi**.
Call: Shrill, high pitched *Ki-ki-ti-ti* is uttered in flight or from a perch. Chicks are noisy.
Ecological notes: Communal nesting of kestrels, falcons, vultures is often seen on the same cliff-face. Predates on crop-pests.
Cultural notes: The Kestrel is **Anal** (fire)due to its rufous colour,and **Dishachakshu** (*disha*-direction, *chakshu*-eyes) as it incessantly scans the locality, in the epic Mahabharat! It is called Windhover in England. The ability to stay suspended in the sky has accorded it another apt Sanskrit name **Akash-Yogini**.
Status: Occasional.

Common Kestrel (F)   **Kharuchi**

Description: Tail long, broad with rounded wing tips. Male-grey head, rufous back marked with arrowheads, moustachial stripe, grey tail with terminal dark band. Juvenile and female-streaked rufous head, neck, dark spotted rufous back, thin moustachial stripe. Seen perched in an upright stance. Intently inspects ground for reptiles, locusts, small birds, mammals, frogs and swiftly takes prey in its sharp talons. A victory call is often uttered after a successful catch. Also hovers.

Related Species:
Lesser Kestrel (F.naumanni-221) Rare WM. Takes crawling prey but prefers insects. Hovers less as compared to the Kestrel, swoops over grass in low sallies. Male lacks moustachial stripe, has an unmarked grey head, less spotted belly than the Kestrel. In flight the tail shows longer central feathers and black-white bands. **Vulnerable.*

Specialized beak of Common Kestrel.

Grass-hopper, prey.

Common Kestrel (M)

Laggad, Sasana

Description: Adults have ashy- brown back. Thighs and lower belly heavily streaked with brown, variable chest, prominent moustachial stripe and rufous crown. A sustained swift flight (up to 150 mph) is possible due to the long pointed wings. Turns and twists in air in pursuit of aerial prey, which it catches at extremely high speed. Ground prey is taken in a stoop with wings closed.

Shahin Falcon.

Related species:*Peregrine Falcon (Falco peregrinus japonensis-209) This Blue-Hawk is a rare WM to the hills. Black head, blue-grey back, white chin, barred belly. Accomplished aerial predator of aquatic and other birds. Was popular in the art of falconry. **Kshiprashyen-**fast hawk and **Amrutvaka-**singing hawk,are its apt names.
*Shahin Falcon (F.p. peregrinator-211) R in our hills. Ferruginous underparts typical. Perch on pinnacles, other vantage points in the hills.

Laggar

Laggar Falcon

Falconidae

Falco jugger (208)

Raghat, Laghad, Langan

Size: 430-460 mm.   M / F : Alike   **R**

Distribution : In the entire region, more so in the Deccan.

Nest-site: Trees, ruins, ledges in inaccessible cliffs. Old corvid nests are also used. *Jan. - Apr.*

Material: Sticks, twigs, lined with straw.

Parental care: Both. Parents teach juveniles to catch aerial prey by dropping an injured bird, caught earlier and inviting them to seize it in air. They are allowed to keep the catch.

Eggs: 3-5, pink, red blotched. 50 x 39.4 mm.

Call: Usually silent. A shrill prolonged whistle *Whii-ii* is rarely uttered on the wing.

Ecological notes: Dense humid forests are avoided. Rarely visits towns for the easy pigeons. Laggar chicks were removed on a large scale for falconry and their numbers are presently on the decline.

Cultural notes: Falconry was practiced since the Vedic period. The Rigveda mentions falcons and fire as precious guests of every tribe. Many texts like Mansollas, Shyenik-Shastra, mention falconry. Sanskrit name for the falcon-**Hapit**, derived from the root **Ha**, means a bird deployed from the hand (in falconry). Its connection with the word Hawk needs study. (K.N.Dave)

Status: Uncommon.

Peregrine Falcon, Juvenile.

Red-headed Falcon

Red-Necked Merlin

Falconidae

Falco chicquera (219)

Vegi, Turmati

Size: 310-360 mm.   M / F : Alike   **R, LM**

Distribution : Patchily in the entire region.

Avoids or absent on the coasts.

Nest-site: Branches close to the tree-top. Ruins ledges and poles are also used. *Jan. - May.*

Material: Sticks, branches, lined with leaves.

Parental care: Both. Ferociously guard the nest, even by attacking larger birds of prey.

Eggs: 3-4, pale red, brown blotched. 42.4 x 31.1 mm.

Call: High pitched *Ki, ki, ki* calls are given. Squeals are uttered from a perch or in flight.

Ecological notes: Predator, poultry lifter. Subsist on birds, small mammals, reptiles. Insects are fed to the chicks for the first few days. Avoid dense forests, prefer open country.

Cultural notes: The Sanskrit name **Vegi** indicates its swift flight. Falconers trained it to capture small birds like mynas. The flesh of the redlegged falcon was prescribed to patients with haemorrhoides, in ancient Indian medicine (Bhavprakash). Various such prescriptions were then rampant, but are without substance.

Status: Uncommon.

Turmati

Description: Rufous crown, nape, moustachial stripe and grey upper parts. Long tail has subterminal band, white belly finely barred with black. Yellow legs. Pointed wings enable swift flight. Dash speedily with rapid wing beats and birds are caught in air. Pairs hunt tactically, mutually and with persistence. Territorial. Local movement depends on food.

Related Species:
*Amur or Redlegged Falcon (Falco amurensis-220) A separate species from the extralimital Red-legged falcon (A.vespirtinus). Autumn passage migrant to our coasts on its way to S. Africa. Red-orange thigh, legs, vent, cere and white underwing-coverts. Sexes dimorphic. Grey above, streaked white chest and dark moustachial stripe. Juvenile brownish. Usually is seen in flight along the coastal area.
*Eurasian Hobby (Falco subbuteo-212) WM up to Belgaum. Call Tee, tee. Slaty grey above white belly streaked with black, thigh spared. Moustachial stripe, spotted underwings.

Eurasian Hobby (Imm.)

Redlegged Falcon (M)

Red-headed Falcon on the nest with chick.

Red-legged Falcon (F)

Falcon-headed ancient
Egyptian god - Horus.

Peregrine Falcons
(F.p.japonensis)

Red-headed Falcon.

Red-headed Falcon-threat display.

Shahin Falcon (Juv.)

Painted Francolin

Painted Partridge

Phasianidae

Francolinus pictus(241)

Chitrapaksha

Size: 310 mm.   M / F : Dimorphic   **R**

Distribution : Patchy. North Kokan, north Karnataka(Shimoga) and Western Ghats. Not in Malabar.

Nest-site: On bunds or embankments in paddy fields at the base of bushes well hidden in the grass. *Jun. - Sep.*
Material: Thinly lined with grass.
Parental care: Both.
Eggs: 4-8,in shades of brown, blotched. 35.7 x 29.5 mm.
Call: Loud *Chuck, chrik, cheek, chray, subhan teri kudrat.* Very vocal during the SW Monsoon (SWM) when fields resound with their calls!
Ecological notes: Though a grassland bird it is often seen near well watered sugarcane and paddy fields. Eats insects from the drying cowdung cakes near cattle sheds.
Cultural notes: Clandestine partridge fights are still conducted with spurs attached to their legs. The fighting cocks are highly prized and are bartered for dogs, other cocks or grain. Large sums are wagered on such fights in remote rural areas, though thankfully this sport is rapidly on the decline!
Status: Rare to occasional.

Painted Francolin, perching & incubating.

Chitur, Chittir

Description: The black undertail-coverts and chestnut wings are prominently seen in flight. The female has a whitish throat patch.
This partridge calls from a high perch, with its head held high and eyes closed. A closer approach can be made while the bird utters the call,and a good photograph can thus be obtained! It is more arboreal than the grey partridge and is seen in slightly wetter areas.

Related species: The chest pattern of the *Black Francolin (F.francolinus-238) is somewhat similar to that of the Painted Francolin, for which it may be mistaken, but the black face, chestnut neck and white ear-coverts of the former are distinct. It is not encountered in our range and resides further north.

Black Francolin.

Grey Francolins roosting on a tree.

Patridge fights are still conducted with high stakes.

Grey Francolin chicks.

Grey Francolins southern race-F.p.pondicerianus
on a water hole.

Albino Grey Francolin from
the BNHS museum.

Grey Francolin

Grey Partridge

Phasianidae

Francolinus pondicerianus(246)

Gaurtittir, Kapinjal

Size: 330 mm.   M / F : Dimorphic   **R**

Distribution : Patchily throughout the region up to 1000 m. Not seen on the coasts.

Nest-site: Eggs are laid in a ground scrape hidden in grass, scrub land, fallen walls of old wells, fallow fields. *Mar. - Jun.*
Material: Lined with dry grass.
Parental care: Both.
Eggs: 4-8,cream-coffee coloured. 34.5x26.1mm.
Call: *Pattetar,kattetar,pateela*, repetitive high pitched calls are intensified during breeding season. Female utters softer calls.
Ecological notes: Flocks congregate on freshly harvested fields at dawn and dusk to glean grains and seeds. A bird of drier areas avoids forests. The bird is still trapped for meat.
Cultural notes: As the ancient Yajurvedic story goes, the knowledge discarded by sage Yadnyavalkya in the form of a vomitus, was acquired by his desciples in the guise of the **Tittir**,the partridge. Hence the ancient Sanskrit text is known as the Taittiriya Upanishad or Krishna-Black Yajurved(1000 BC). Since this partridge nests during the spring, it was considered to be the deity of Vasantrutu, the spring season.
Status: Occasional.

Chitur, Tittir, Titur

Description: This stub-tailed secretive bird with a dark neck-ring and a buff throat is crepuscular. The cock has a spur on each leg. Though terrestrial, the coveys perch on thorny branches at noon and at night. They are often seen fighting. When surprised, they noisily scramble for cover. The chicks start walking in a day or two after hatching. North of Pune the southern race *pondicerianus* is replaced by the race *interpositus*. Throat is greyish in the latter.

Partridge Trapping: Partridges can be trained to answer the call of a special whistle.These caged birds are then deployed to trap wild partridges. The wild birds are very territorial and come out to fight rival cocks. The call of the caged cocks placed in their territory angers the wild birds,and they land in the traps. This ploy is still widely practised!

Juvenile.

Male.   Jungli Durlav

Description: One may encounter this quail as it rapidly scales a hill slope and vanishes in the undergrowth. Flocks explode with a 'whirr' from beneath the very feet of casual cross-country hikers, causing moments of anxiety. Coveys of quails roost in circular formations facing outwards. These wary birds are so well camouflaged that they are more often heard in the bush than seen.

Related species: *Painted Bush-Quail (P.erythrorhyncha-262) Male has a white Throat and head stripe(which the female lacks)both have red bill and legs. It is seen in forest clearings or open grassland patches above 600 m. Call- 'Kiri-kee'.
*Blue-breasted Quail (Coturnix chinensis-253) Rare in our area. Commoner in monsoon. Male has slaty blue underparts.
*Common or Grey Quail(C.coturnix-250) White eyebrow and buff streaks on back, spotted belly. Size-200mm. Resident up to about Satara and southwards in winter.

Jungle Bush-Quail

Phasianidae
Perdicula asiatica(255,257,257-a)
Gairik Laav
Size: 170 mm.   M / F : Dimorphic   **R**
Distribution : Lightly forested hilly areas of Kokan and the Western Ghats up to about 1500 m. Rare in south Malabar.

Nest-site: Eggs are laid on ground in a grass clump or at the base of a bush. *Jul. - Feb.*
Material: Lined with thin soft grass.
Parental care: Female incubates.
Eggs: 4-8, creamy white. 25.4 x 19.5 mm.
Call: Breeding males utter raucous aggressive calls. Softer notes *Whi, whi* are emitted by the group to keep in touch with one another.
Ecological notes: This quail prefers lightly wooded and stony hilly country. Avoids more arid land where the Rock Bush Quail is seen. Quails are popular game birds even today. Their eggs are easy prey for ground predators.
Cultural notes: Quails have been known since the Vedic times, when their flesh was considered a delicacy. Quail is cited in the famous Andhak-Vartak Nyaya of the Yajurveda, to illustrate the helplessness of the weak (**Vartak**-quail) against the strong(Falcon-**Shyen).**
 Status: Common.

Blue-breasted Quail (M)

Common Quail.
Barred primary.

Painted Bush-Quails.

Common Quail (F)

Small Quail (M)

Rock Bush-Quails roosting in a outward facing circle.

Common Quail (M, Imm.)

Common Quail (F)
without anchor marks.

Jungle Bush-Quail (P.asiatica punjaubi) is seen in Vidarbha.

Rain Quails (M & F)

Jungle Bush-Quails in a cage. They are deployed to trap other quails & to conduct fights.

Lava, Bater

Description: Male has barred chest, brick red throat and blunt leg spurs. It lacks the white moustachial stripe. Female has plain, buff-vinaceous underparts, head and a thin eyebrow. Their calls and the rustling in the bushes and undergrowth betray their presence. If noticed, they freeze and if approached, explode in flight. The dispersed flock soon re-unites.

*Related Species: *Rain Quail (Coturnix coromandelica-252) Male has black underparts. The primaries are unbarred. Breeds during SWM. Call is - 'Which,which'. Common in the area. Secretive like all other quails. *Japanese Quail(Coturnix japonica-251) Resident in Assam, Bhutan. It is bred in our region for culinary purposes. Escaped birds are encountered in the wild.*

Rain Quail (M)

Rain & Jungle Bush-Quails trapped for the table.

Rock Bush-Quail

Phasianidae

Perdicula argoondah(259,260)

Vartirak, Paansul Laav

Size: 170 mm.   M / F : Dimorphic   **R**

Distribution : North and Central Western Ghats, Kokan. Rare in Malabar and southern Ghats. Race *salimalii* in Wynaad.

Nest-site: On a bund or in a scrape on bare ground, hidden in grass or below a jutting rock. Mar. - Jun.

Material: Thinly lined with grass.

Parental care: Female incubates.

Eggs: 4-5,whitish. 25.6 x 20.1 mm.

Call: These noisy birds emit harsh calls when fighting each other. Feeding quails keep track of each other's whereabouts by emitting low pitched sounds.

Ecological notes: Quail eggs are removed and eaten by humans. Eggs and Quails are preyed upon by buzzards, falcons, shrikes, monitor lizards, mongooses, etc. A White-eyed Buzzard eating five eggs from a single nest has been observed (Amit Pawashe Per.Com.).

Cultural notes: Quail fights are still secretly conducted with heavy betting. Females are better fighters. Quail fighting was a royal sport in ancient south Indian kingdoms. An interesting story is narrated in the Sakunagdhi-Jatak, where a quail managed to escape from the attack of a falcon! Quails also feature in the Vedas, Mahabharat, Hitopadesh.

Status: Common.

Rock Bush-Quails (M)
Rain Quails (M, F)

Red Spurfowl

Phasianidae

Galloperdix spadicea(275,277)

Seerpad, Kukkutak

Size: 360 mm.   M / F : Dimorphic   **R**

Distribution : Patchy. Forests of the Western Ghats up to 1250 m. Rare in Kokan and Malabar. Race *stewarti* Palghat south.

Nest-site: On ground in bamboo clumps, under rocks, thorny thickets in stone strewn country. *Jan. - Jun.*
Material: Thinly lined with grass.
Parental care: Female incubates. To avoid attention, the nest is approached with caution.
Eggs: 3-5, buff. 40.4 x 29.5 mm.
Call: Cock emits a loud *Krr-rr-kwak-kwak*. Vocal at dawn and dusk. Contact calls are utterred when feeding with chicks.
Ecological notes: Endemic Indian resident. Trapped for fights and for the table. Spurfowls used to turn up in the market till a few years ago.
Cultural notes: The Sanskrit name **Seerpad** means one with a plough-share shaped leg spur. This appears to be a generic term applicable to many members of the Pheasant family. In the Vedic times, the spurfowl being a game bird was used as a sacrificial offering for the Yadnya (sacred fire).
Status: Uncommon. Patchily in the forests of the Western Ghats.

Female.   **Chakotri, Kakotri, Sakotri**

Description: Red legs, red bill and bare red peri-orbital patch are conspicuous. The tail is long. Male has grey head, neck and brown crown. The hen is rufous. The cock has 2-4 long pointed spurs while the hen has 1-2 short spurs. Extremely wary bird, runs for cover or flies to a branch at the slightest alarm. Rummages amongst fallen leaves in ravines, dry streams and forest clearings at dawn and late evening.

Related Species:
*****P**ainted Spurfowl (G.lunulata-278) Keeps to dry, stony, thorny ravines. Secretive. Hides in rock crevices. Male has greenish head, tail, chestnut upperparts and buff underparts. Fine barring and spotting on the body and the overall colour give a painted appearance. Female is brown, buff throated, with a rufous eye-brow. Both have dark, spurred legs. Rarely in the Ghats; not on coasts.*

Painted Spurfowls (M)

Red Spurfowl (F & M)

Red Spurfowl (Pair)

Male.   **Rakhi Rankombada**

Sonnerat's Junglefowl

Phasianidae

Gallus sonneratii(301)

Yavagreev, Badar Kukkut

Size: 600-800 mm.   M / F : Dimorphic   **R**

Distribution : Patchy. Forests of Kokan, Malabar and throughout the Western Ghats up to the highest altitude.

Description: Hen has a scaly white breast. The elegant sickle shaped dark tail and bright golden mantle and breast spots of the male are conspicuous. Cock has red legs with a single pointed spur. (Hence the Sanskrit name- **Charanayudh**, leg with a weapon!) Often seen in forest clearings and on jungle roads around dawn and at twilight. Confiding where not hunted, elsewhere they scramble for cover in the undergrowth or take flight and disappear.

Nest-site: Concealed in bamboo clumps or thickets. Eggs are laid on ground. *Feb. - May.*
Material: Lined with grass and thorny twigs.
Parental care: The hen incubates.
Eggs: 4-7, fawn coloured. 46.3 x 36.5 mm.
Call: *Kuk, ka, kak, kya, kya, kuk* in broken notes (Sanskrit name **Swarbhangi,** (**Swar-**note, **Bhang-**break). One of the first forest bird calls to be heard. Clucks when feeding.
Ecological notes: Their alarm calls *Khak, khagak* on sighting wild cats and predators alert other jungle fauna. These calls are joined by the langoor calls and a cacophony resounds in the forest. The junglefowl are preyed upon by larger birds of prey and carnivores.
Cultural notes: The Indo-Aryan Armies took cocks on expeditions to announce daybreak to the soldiers! In the Vedic period, the time of morning sacrifice, a ritual which granted wishes, was proclaimed by the cock and hence it was known as **Madhujivha-**sweet tongued. Cock-fighting was a popular ancient sport. The spurs of fighting cocks are even today adorned with a silver shoe.
Status: Common.

Related species:
*Red Junglefowl (Gallus gallus-299) Not a bird of our area, yet a few fowls continue to be found in the forests of northern Kokan. These probably are survivors of the captive fowls released in the wild from time to time.

Grey Junglefowl, subadult.

Grey Junglefowl (F)

Sri Lanka Junglefowl, Endemic to Sri Lanka.

Cock fights are still conducted in spite of being banned.

Indian Rock Art depicting a hunting party with two jungle fowls. These served the purpose of announcing the day break-live alarm clocks!

Grey Jungle Fowls (M)

A hunting scene depicted about 35,000 B.P.

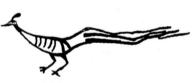

Peafowl on a seal & in rock art.

Feather of a peahen.

Peacock with several peahens.

Indian Peafowl

Common Peafowl

Phasianidae

Pavo cristatus(311)

Bhujangbhuk, Barhin, Meghanulasin

Size: 920-1220 mm. M / F : Dimorphic   R

Distribution : Kokan, Malabar and the Western Ghats up to 1800 m. Semi-feral populations exist patchily.

Nest-site: On high riverside rocks,boulders, in thorny thickets and dense undergrowth. Semiferal birds nest on rooftops. *Jan. - Oct.*
Material: Lined with twigs, sticks, leaves, etc.
Parental care: Hen incubates.
Eggs: 3-5,pale cream, glossy. 69.7 x 52.1mm.
Call: *Mee-aaw, Mee-aaw, kaan, kaan* repeated loudly. Often heard at night and on cloudy days.
Ecological notes: Tall leafy trees are required for roosting. Vegetation along rivers and ravines is a preferred habitat. Cause damage to standing crop, being fond of grain. Its alarm call is well known in forests. Features in the diet of large cats. Eggs when found can be easily incubated under a domestic hen, reared and later released.
Cultural notes: Our national bird since 1962. In many rural areas, 'People's Sanctuaries' have been setup to protect peacocks. Being associated with the goddess of learning, Saraswati and several other deities, it enjoys religious protection. Features in literature, works of art. Known since the Vedic times and has adorned flags, thrones, crowns. Jewels, ornaments are fashioned after peacock feathers. On the other hand, peacocks are also eaten.
Status: Common.

Indian rock painting-35000 Before Period.

Mor, Mayur, Anantchakshu

Description: Flocks, with a cock and several hens, feed in meadows, fields, on mountain slopes, etc. They are fleet of foot and move surprisingly well in impregnable thorny shrub land. Retire to tall trees if alarmed, being capable of strong short flight. Rains inspire them to dance and utter trumpeting calls. The cocks fan their train and perform a nuptial dance in forest clearings, in front of the hens, to lure them. This is an exquisite sight. Both sexes crested. The train of the male comprises elongated uppertail-coverts; moults after breeding.

Partial Albino Peacock.

Peahens.

Female.   **Darva, Durlav**

Description: Blue-grey legs, beak and yellow-white eyes are typical. Black barred chin, throat and upper breast of the cock are dark black in the hen. The latter is more colourful. The Buttonquails differ from others, being three-toed (hind toe absent), smaller sized and hens being polyandrous. They are not easily sighted and are seen only when flushed. Their flight is feeble and they prefer to scramble through the undergrowth.

(F)
Yellow-legged Buttonquails (M)

Common Buttonquails (M)

Common Buttonquail

Barred Button Quail, Blue legged Bustard Quail
Turnicidae

Turnix suscitator (318)

Laav, Lopa, Durva

Size: 130 mm.   M / F : Dimorphic   **R**

Distribution : Light deciduous forests and grasslands in Kokan. Malabar and the Western Ghats up to 2000 m.

Nest-site: A depression in a grass clump in lightly wooded areas or in scrub country.
Jun. - Oct.

Material: Lined thinly with grass and leaves.

Parental care: The polyandrous hen leaves the incubation of her successive clutches to the concerned males!

Eggs: 3-4, grey, speckled brown. 24.7x19.4mm.

Call: The hen is vocal during breeding, when loud resonant calls *Drrr-rr-rr* are given to attract the cocks. Fights with rival hens are vociferously conducted.

Ecological notes: If nests which are discovered in the standing crop are left undisturbed at the time of harvesting, they are not abandoned. These quails avoid dense jungles.

Cultural notes: The three-toed quails have been described as a separate class called **Lopa** since the time of the Yajurved (1400 BC). Charak (400AD) calls them **Laav.** This group was further classified. Barred Buttonquail was known as **Gundra** and **Durva**, after the kind of grass which it inhabits(Dhanwantarinighantu).

Status: Common.

Other Quails from the Turnicidae Family:
*Yellow-legged Buttonquail (Turnix tanki-314) Yellow bill, legs. Female has a rufous hind-collar. Call is a drumming 'Drr-rr'. Patchily throughout the area, but rare.
*Small Buttonquail or Little Bustard -Quail (T.sylvatica-313) Pointed tail is typical. The back looks scalloped. The male incubates. Patchy and rare in our area except in Kerala.

Small Buttonquail. Sadly, it still appears in the market.

Demoiselle Crane

Gruidae

Grus virgo (326)

Pankticharkurar, Neelaang-Saras

Size: 760 mms. M / F : Alike **WM**

Distribution : The Deccan, up to Belgaum.

Unrecorded in Kokan and Malabar.

Nest-site: Extralimital.
N&C Asia, S.Europe. *May - Jul.*
Parental care: Both.
Eggs: 2, blotched; grey, brown, yellow.
Call: Loud honking and trumpeting calls are heard, as the flocks graze on river beds or fields. *Kurr, kurr,* in a varied pitch. Deep calls while taking to flight. A noisy bird.
Ecological notes: Cause much damage to the standing winter crop(gram)and hence are occasionally killed by farmers.
Cultural notes: At Khichan, Rajasthan, these cranes are fed grain and protected by the local Bishnoi community! They are said to draw 'maps in the skies', owing to varied formations that the flocks make in flight! The Sanskrit name **Krishnakraunch** means a dark crane. They are seen standing in a long line-Pankti, and are called **Panktichar**!
Status: Rare.

Prey of Sarus Cranes.

Sarus Cranes.

Kandesar, Kandya-karkocha

Description: Black head, neck, breast, grey crown, white feather tuft behind eyes. Tertials are elongated and loosely hang like an arced tail. Common Crane lacks black breast. These cranes feed in the morning and evening when they appear quite bold.Resting flocks are seen at noon, standing almost motionless on banks of rivers, lakes all facing in one direction.

Common & Demoiselle Cranes.

Siberian Crane (G. leucogeranus)

***Related Species:** *Sarus Crane (Grus antigone-323) Vagrant in our area. Red head and neck. The similar sized Siberian Crane (G.leucogeranus-325) has pink restricted to face. Not recorded in our region. **Vulnerable.** *Common Crane (Grus grus-320) Rarely in the company of Demoiselle Cranes. Dull red neck patch and white band from the eyes along the sides of the neck aid identification.*

Sarus Cranes - pair.

Siberian Crane.

Commen Cranes - Adult and Juvenile.

Crane Flower (Strelitzia reginae)
Native South African flower resembles
the head of a crane.

A large flock of Demoiselle Cranes.

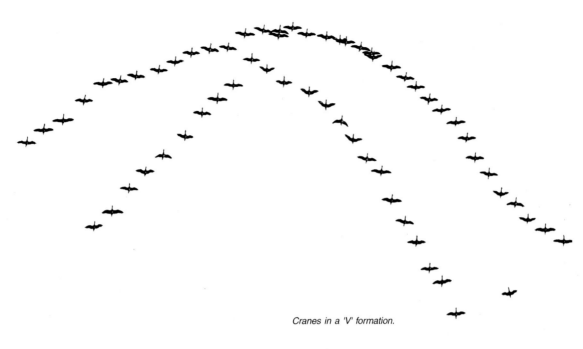

Cranes in a 'V' formation.

Demoiselle Cranes.

A typical habitat of rails, crakes & waterhens.

Little Crake.

Blue-breasted Rails-
In flight & preening.

Watercock (F)
White-breasted Waterhen.

Baillon's Crake.

White-breasted Waterhen.

Ruddy-breasted Crake

Ruddy Crake

Rallidae

Porzana fusca (340)

Ambukukkuti

Size: 220 mm.   M / F : Alike   **R, LM**

Distribution : The entire region, Mumbai south. Up to about 2000 m.

Nest-site: The nest platform is hidden in reedbeds, marshy or water clogged areas. *Jun. - Sep.*

Material: Leaves, aquatic weeds. Twigs are sometimes bent over the nest to form a canopy.   **Parental care:** Both. The nest is approached secretively and indirectly.

Eggs: 4-8. 22.3 x 30 mm.

Call: A mousy trilling call and a hoarse *Krek* are uttered if surprised or when confronted with some wader. *Chuck* when feeding.

Ecological notes: A wary wetland bird. Often in watered paddy-fields. Subject to local movements with the availability of water.

Cultural notes: These small water-hen like rails are known as **Ambukukkutika** in Sanskrit. Sushrut calls them **Chitraangak**, due to the banded body pattern.

Status: Uncommon.

Slaty-legged Crake.

Spotted Crake.

Lal Panlavi

Description: Red legs, brown back and chestnut underparts. Barred flanks, undertail-coverts indistinct. Keeps to the edges of marshes, momentarily coming in the open. It upends while feeding and submerges the head fully in murky water. Walks on delicate twigs when foraging. Vanishes if disturbed.

Little Crake.

Blue-breasted Rail.

Related Species: *****Baillon's Crake (P.pusilla-337) 190 mm. Grey breast, painted olive-brown back. WM to the region. Call 'Trrr-rr'.

***B**lue-breasted Rail (Gallirallus striatus-329) 270 mm. Long red-based bill, white-barred grey belly, chestnut crown, nape and grey legs. Rarely in our swamps.

***S**laty-legged Crake (Rallina eurizonoides-332) 250 mm. White throat, chestnut head, neck, chest, grey legs. Barred belly, brown unbarred wings. Ghats up to 1600 m.

***L**ittle Crake (P.parva-336) 200 mm. WM. Pale edged primaries. Breast-grey in male, rufous in female. Flanks barred. ***S**potted Crake(P.porzana-338)230 mm. WM. Not in Kerala. Spots on head, neck, belly and barred flanks. Brown above, grey below. Red based yellow bill, whitish vent.
All the rails are sighted with difficulty.

Juvenile **Lajari Pankombadi, Kuku Kombadi**

Description: Grey back, white neck, belly and rufous undertail-coverts. These shy birds are confiding if undisturbed, and freely enter gardens around houses in swampy areas. Continually search for insects and vegetable matter. Move with agility over floating vegetation and climb reeds to command a view. Fly over compound walls and jump from bush to bush. Take cover if disturbed.

White-breasted Waterhen, chick.

Brown Crake.

Related species: *Watercock, Kora (Gallicrex cinerea-346)This pugnacious, big swamp bird rightly deserves its name. Plumage variation with age and breeding status. A red fleshy horn at the beak-base when breeding. Red legs, eyes. Rare, crepuscular. Loud call booms 'Utumb, utumb'.Size-430 mm. Hen is smaller.*
*Brown Crake(Amourornis akool-342) Well watered paddy, sugarcane fields and marshes. Pinkish legs critical to identification. The similar juvenile white-breasted waterhen has a rufous rump and juvenile common moorhen has a white rump. Both have green legs.

White-breasted Waterhen

Rallidae

Amaurornis phoenicurus (343)

Meghrao, Datyuha

Size: 320 mm.   M / F : Alike   **R, LM**

Distribution : Kokan, Malabar and the Western Ghats up to 2000 m.

Nest-site: The broad cup is hidden in a tree or a creeper near a water-body. *Jun. - Oct.*
Material: Twigs, leaves, stems, reeds are woven to make the shallow platform-cup.
Parental care: Both. The nest is never directly approached. Chicks follow the parents secretively, one after the other.
Eggs: 6-7, creamy-white,red blotched. 40.5 x 29.7 mm. The chicks are jet black and the juveniles are brown.
Call: Grunts, chuckles, harsh cries are uttered. *Kurr, kwa-kwa-kwak* call is repeated. Low pitched calls when communicating with the chicks.
Ecological notes: Local movement during SW Monsoon, when the waterhen is seen on inland water-logged fields and swamps.
Cultural notes: The name ***Meghrao*** found in Charak and Sushrut-Samhita's relates the sound of the bird to the noise of the clouds.
Status: Common.

Watercock (M)

Common Coot

Coot

Rallidae

Fulica atra (350)

Karandav, Marul, Shakatavii

Size: 420 mm.   M / F : Alike   **R**

Distribution : On water bodies in the entire region. Uncommon in Kokan.

Nest-site: Small islands in water bodies, near the edge of tanks, lakes and rivers in dense reed beds. Well concealed. *May - Aug.*

Material: Water-weeds, dry rootlets, twigs of water-plants, leaves are used to make the nest.   **Parental care:** Both.

Eggs: 6-10, yellow-brown, blotched. 53.1 x 35.6 mm.

Call: A croak. Loud trumpeting is heard during the breeding season at night and before dawn. **Ecological notes:** Coots keep to clean and fresh water. Prefer proximity to human habitations. Rarely in partially polluted water. Cause damage to young paddy, being fond of shoots.

Cultural notes: Commonly featured in ancient literature mentioned as **Karandav**. The vernacular names Nama and Warkari draw a similarity with the white sandalwood-paste or Nam, applied on the forehead by devotees (Warkaris), making their annual pilgrimage to Pandharpur.

Status: Common.

Chandwa, Warkari, Nama, Karandav

Description: Deep grey-black, duck-like aquatic rail. Runs on water, vigorously flapping the wings, when forced to fly. Does not soar high in the air but flies low. Enjoy chasing each other. Territorial. Diver, surface feeder. Flight is typically rail-like, with legs trailing behind, differing from the flying duck.

The vernacular name Nama for the Coot is derived from the custom of the Varkaris-devotees of applying white sandalwood paste to the forehead.

Look Alike: *The immature Coot has a red crown. It can be mistaken for an adult breeding *Little Grebe (T.ruficollis-5) if not observed carefully. Both birds often share the same water bodies. Coot is darker than the Grebe.*

Little Grebe-Breeding.

Coot, chick.

Ducks were sacrificed to Gods in ancient Egypt.

New born Coot chicks.

Coot feeding the chick.

A congregation of Coots.

Purple Moorhen

Purple Swamphen

Rallidae

Porphyrio porphyrio (349)

Kamal, Manjul-datyuha

Size: 430 mm.   M / F : Alike   **R, LM**

Distribution : Patchy. Kokan, Malabar and the Western Ghats.

Nest-site: In reed-beds near lakes, rivers, etc. *Jun. - Sep.*

Material: A platform is made from matted reed, aquatic weeds, just above the water level.   **Parental care:** Both.

Eggs: 3-7, yellowish brown, blotched. 50.5 x 35.7 mm.

Call: A noisy rail. Honks, cackles, harsh notes are uttered. Loud calls are given when breeding. Call is *Crek, crek.*

Ecological notes: Harmful to standing paddy crop, which it devours. Presence of dense cover around perrenial water is essential. Hence marshes should be preserved.

Cultural notes: The Sanskrit name ***Kamal*** indicates the birds bluish-purple colour and preference for lotus plants. It is mentioned in the text Matsyapuran.

Status: Uncommon.

Purple Moorhens.

Common Moorhen.

Jambhali Pankombadi

Description: A beautiful shy rail. Flocks grazing on the edges of lakes take to cover when human presence is sensed and scramble up the reeds to look around. Playfully chase each-other. White rump flashes as the tail is flicked now and again. The low flight is clumsy, laboured, with much effort and wing fluttering. Walk on floating vegetation, eating tubers and insects.

Purple Moorhen.

Related species: ******Common, Indian Moorhen (Gallinula chloropus-347) Red based yellowish bill. Closed wings show a white edge. Adult is dark slaty-brown and juvenile brown, both have green legs. More aquatic than purple moorhen; also likes to swim. Head bobbing and tail flicking typical. Throughout our range up to 2000 m in Ghats.*

*Look Alike:

*Brown Crake(A.akool-342) Like juvenile common moorhen; former has pinkish legs vs. Green, and greyish underparts vs. black.

Tanmor

Description: Male-white collar, throat, wing patches. Vermiculated grey above, otherwise jet-black. Magnificent, long, spatulate tipped plumes spring from the sides of the head, up to five on each side. Female-darkish crown, back, other wise rufous brown. Move with rainfall.

Great Indian Bustards (F)

*Related species: *Great Indian Bustard (Ardeotis nigriceps-354) Formerly up to eastern fringes of Western Ghats, but due to hunting, now restricted to Nannaj, Solapur and around Dharwad, Karnataka, in our range. Breeds with SW Monsoon. Displaying male puffs neck feathers and pouch. Female is smaller. Locusts, lizards, grain and occasionally chicks of larks are eaten. Beneficial to farmers. Status-Threatened.*

Lesser Florican

Leekh

Otididae

Sypheotides indica (357)

Varat, Anjalikarna

Size: M-460,F-510 mm. M / F : Dimorphic **WM**

Distribution : Very patchily, locally and now rarely in the Deccan and Western Ghats up to 1000 m. Not in Kokan and Malabar.

Nest-site: In a secluded grassland, fallow field, on ground in a grass clump. *Jul. - Oct.*
Material: Eggs are laid amidst a few pebbles.
Parental care: Incubation and chick rearing by the female.
Eggs: 3-4, olive-brown,mottled. 49.1 x 41.3 mm.
Call: Wing clapping sounds are produced by the dancing, breeding male to attract the females. The springing male jumps up to 600 times in a single session. Whistles are uttered.
Ecological notes: Its present range has shrunk to a few scattered patches. Injured birds are found once in a while, proving their presence in the locality.
Cultural notes: Was extensively hunted for the table. The beautiful male has attracted the attention of many artists. The name Likh is derived from the Sanskrit name **Khilkhilla**, a playful grassland bird. Sushrut, the great surgeon of ancient India, had designed a surgical forceps *(Anjalikarna Shalaka-Yantra)*, after the shape of the Florican's beak!
Status: Threatened.

Great Indian Bustard's Feather. Are we destined to see only this in the future ?

*Houbara or MacQueen's Bustard(Chlymydotis undulata-355)550-650 mm. Dark linear neck stripe. Once in Kerala. Status-Near Threatened.
Houbara Bustard.

Great Indian Bustard (F)

Lesser Florican (M) with the spatulate tipped plumes.

Lesser Florican (F)

Male puffing the throat feathers during breeding.

Great Indian Bustards eggs are often predated by the Jackal.

Lesser Florican (M) springs in the air to attract the female.

Kamalpakshi

Description: Fly with legs dangling clumsily. Non-breeding adults have a black breast necklace and golden nape. Wings have spurs. Female is slightly larger.

Related Species: *Bronze-winged Jacana (Metopidius indicus-359) Lack the saber-tail. (It is shed in the non-breeding H.chirurgus also) White eyebrow is conspicuous. Partial to hydrilla, waterlily, singara-Trapa and water hyacinth. Call is 'Seek, seeek'. Jacanas are swimmers and divers. Long toes help them trot on floating plants by wider weight distribution. M.indicus is rare in Kokan.*

Bronze-winged Jacana, Nest & Chick.

Non-breeding Pheasant-tailed Jacana.

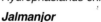

Pheasant-tailed Jacana

Lily Trotter

Jacanidae

Hydrophasianus chirurgus (358)

Jalmanjor

Size: 310 mm.   M / F : Alike   **R**

Distribution : Kokan, Malabar and on the Deccan. Rarely on water bodies in the Western Ghats.

Nest-site: Lakes, pond shores close to semi-aquatic vegetation. Nest is a floating affair. Marshes important for breeding. *May - Sep.*

Material: Aquatic weeds, rootlets, leaves.

Parental care: Male. Female polyandrous but sometimes visits the nest.

Eggs: 4-5, brownish, glossy, peg top shaped. 37.4 x 27.6 mm.

Call: Loud notes *Ki, ki, tie, teon, teon* especially in flight and while taking off. Vociferously chases away coots, ducks if active nest is approached.

Ecological notes: Seldom seen on rivers. The chicks move quickly after hatching. On sensing danger they remain submerged with beaks jutting outside the water! Predated by snakes, birds of prey. Parents carry chicks under their wings.

Cultural notes: Generate awe due to the polyandrous nature of the female.

Status: Occasional.

Pheasant-tailed Jacana on nest. Chick & Peg-top Egg.

Greater Painted-Snipe

Painted Snipe

Rostratulidae

Rostratula benghalensis (429)

Kunal, Jalvartika (For snipes)

Size: 250 mm.   M / F : Dimorphic   **R, LM**

Distribution: Kokan, Malabar and wetlands in the Western Ghats, up to 1400m.

Nest-site: Marshes, mangroves and bunds in watered fields, concealed in reed beds.
May - Sep.

Material: Stems, aquatic weeds and twigs.

Parental care: The female is polyandrous and dominating. Parenting is left to the male.

Eggs: 3-4, grey-yellow, streaked.
35.9 x 25.5 mm.

Call: A resonant deep *Ook, hoook* is uttered at night and it is a familiar sound of the swamps.

Ecological notes: Snipes blinded by vehicle lights get injured when flying across roads at night. Head-on collision results in a fracture of the long bill. If other injuries are not fatal, the broken bill can be fixed. Care should be taken not to block the nostril with adhesives.

Cultural notes: This snipe with a weak flight is easy to shoot, but it was not a favourite amongst the hunters due to the poor table value! Still persecuted by tribals in Kokan.

Status: Uncommon.

Female.   **Rangit Panlava**

Description: Typically long down-curved bill, white 'Rucksack' shoulder-straps and goggle eyes. This rail like wader, grazes in the solitude of swamps and well watered fields. Freezes when approached and suddenly takes flight. Females fight each other for cocks. Males are duller in colour.

Broken beak.

Look Alike: The male ***P**ainted Snipe can be confused with other snipes, but the buff-white shoulder straps of the former are diagnostic.

Male approaching the nest.

Eurasian Oystercatcher

Sea-Pie

Haematopodidae

Haematopus ostralegus (360)

No Sanskrit name recorded

Size: 420 mm   M / F : Alike   **WM**

Distribution : Coastal Kokan and Malabar.

Breeding.   **Shimple-phodya**

Description: Long red bill, eyes, white wing bar seen in flight. White collar in non-breeding plumage. Crepuscular. Wary. Flocks roost on rocks at high tide. Seen with waders like plovers, dunlins, curlews. Food is obtained by probing sand.

Nest-site: Extralimital. Depression on coasts above the tidal mark. *May - Jun.*
Material: Grass, weed.
Parental care: Both.
Eggs: 3-4, buff brown, black spotted. 53.6x39.7 mm.
Call: Whistles, grunts, piping calls.
Ecological notes: Marine bird. Bill enables it to extract limpets off marine rocks. Also seen on coral reefs. Molluscs, sandworms, aquatic insects are also eaten.
Cultural notes: None in particular.
Status: Rare.

Non-Breeding.

Pacific Golden-Plover

Eastern Golden Plover

Charadriidae

Pluvialis fulva (373)

Khanjanika (For plovers)

Size: 230-260 mm.   M / F : Alike   **WM**

Distribution : Mainly in coastal Kokan and Malabar. Rarely near inland waters.

Nest-site: In N. Siberia. Ground scrape on a mud bank is used for nesting. *May - Jun.*
Material: Unlined.
Call: A high pitched *Tuii-chi, wii, chii-vit* notes are uttered as the flock rises with unison and twists and turns in the air.
Ecological notes: This typical plover keeps to the vicinity of coastal and inland waters and marshy areas.
Cultural notes: Popular amongst sportsmen, being palatable. Wary and capable of erratic speedy flight, making a difficult shot.
Status: Occasional.

Asian Dowitcher (Limnodromus semipalmatus) is not seen on the West coast. It may be mistaken for the Pacific Golden-Plover.

Eurasian Dotterel
(Charadrius morinellus)
Size-21 cm.
May occur as
a winter vagrant.

Caspian Plover.

Related Species: **Greater Sand Plover (Charadrius leschenaultii -374) Size-220-250 mm. WM to coasts. The dark bill is longer and pointed than that of the Lesser Sand, legs yellowish. Back brownish, the white eyebrow and gonys are striking. *Caspian Plover (C.asiaticus-376) Size-180-200 mm. Like Greater Sand but the broad band extends across the breast. Legs green. Vagrant.*

Soneri Chikhlya

Description: Stout thick-billed plover with dark wings, gold spangled back and brownish-grey streaked whitish underparts. Species *dominica* is now the extralimital American Golden Plover. Often encountered in two's and three's on stubble fields and grassy marshes, close to the coasts. Feeds by making short spurts in plover fashion. Less active as compared to other plovers.

Pacific Golden-Plover.

Greater Sand Plover.

Breeding. **Ardhakanthi Chikhlya**

Description: White outer tail feathers, wingbars distinctive in flight. Greyish patches on either side of breast, white hind collar. Dark legs, absence of dark ear-coverts and white forehead help differentiation from Little Ringed Plover. Crown rufous when breeding.

Lesser Sand Plover (Br)
Kentish Plover (Non-Br)   *Sociable Lapwing Juvenile.*

Related species: **Grey Plover(Pluvialis squatarola-371) Size-270-300 mm.*
Uncommon WM to our coasts. White wing bars and rump prominent in flight. Grey upper parts of this stout plover become spangled and streaked belly assumes black when breeding.
*Lesser Sand or Mongolian Plover (Charadrius mongolus-384) Like kentish but larger and lacks white hind collar. Short bill, green legs. When breeding, the crown and breast band become rufous and a black eye-mask is seen.

Kentish Plover

Charadriidae.

Charadrius alexandrinus (381)

Khanjanika (For plovers)

Size: 170 mm.   M / F : Alike.   **WM**

Distribution : Winter visitor to Kokan, Malabar, Western Ghats and the adjacent Deccan. Resident in Gujarat, Vidarbha & Bangalore.

Nest-site: Usually amidst foliage, well above the water line on a mud bank. Breeds within our range. *May - Jul.*
Material: Tiny pebbles. Unlined.
Parental care: Both. Intruders near nest are distracted with the broken wing display.
Eggs: 2-4, sandy-green,brown blotched.
Call: A pipping *Pi-pi* or a trilling *Trii-trr.*
Ecological notes: A widespread winter visitor to saline and fresh waters alike. Mixes freely with other plovers.
Cultural notes: The vernacular name Chikhlya is descriptive of the habit of keeping to mudflats. Ardhakanthi means half collared.
Status: Common.

Grey Plover (Non-Br)
Grey Plover (Br)

Little Ringed Plover

Charadriidae

Charadrius dubius (380)

Rajputrika, Sarshapi

Size: 170 mm.   M / F : Alike   **R, WM**

Distribution : The entire region. Race *jerdoni* is resident and *curonicus* is winter migrant.

Nest-site: Dry mud-flats, sandbanks, scrape or sun backed cattle-hoofprint. Often with terns, pratincoles. *Mar. - Jun.*

Material: Few tiny pebbles are arranged around the nest. Unlined.

Parental care: Both.

Eggs: 3-4, stone-grey, brown spotted. Typically peg-top shaped. 27.5 x 20.7 mm.

Call: A high pitched whistle *Phiu, phee-weep, too-lee* is uttered on the wing.

Ecological notes: The lack of hind toe makes them terrestrial birds. Live near water. Tiny aquatic invertebrates and insects eaten. Migrate locally if the water-body dries up.

Cultural notes: In Sanskrit, this graceful plover is called **Rajputrika**, a little princess. The name **Sarshapi** (Mustard seed) indicates its tiny size and the tripping, rolling movement, the black neck-ring appearing like a rolling mustard seed!

Status: Common.

Related Species: **Common Ringed Plover* *(Charadrius hiaticula-378) Like Little Ringed but has orange base to the black tipped bill and orange legs, when breeding. Lacks eye-ring. White wing bar seen in flight. White supercilium prominent when not breeding. Rare.*

Kantheri Chikhlya

Description: Golden eye-ring prominent. Black forecrown, ear-coverts and white forehead. Legs yellow and bill uniformly black when breeding. Runs in short spurts, pausing to eat and dashes again. Takes flight when disturbed, otherwise confiding. Resident populations are augmented by winter migrants. Movement alone betrays the presence of this obliteratively coloured plover.

Chick-Little Ringed Plover.

Common Ringed Plover.

Little Ringed Plover (Non-Br.)

Yellow-wattled Lapwing on the nest & tail feather.
Chick resembling Lepidagatus cristata-local vegetation.

Grey-headed Lapwing.

Yellow-wattled Lapwing chick panting in the sun.

Yellow-wattled Lapwing parent with two chicks.

Yellow-wattled Lapwing (Juv.)

Yellow-wattled Lapwing

Charadriidae

Vanellus malarbaricus (370)

Sadaluta, Peetpaad

Size: 270 mm.   M / F : Alike   **R, LM**

Distribution : Patchily throughout the region but keeps to the dry and open country. Avoids forests.

Nest-site: Bare soil in fallow or scrub country. *Apr. - Jul.*
Material: Scrape is lined with small pebbles.
Parental care: Both. Hysterical alarm-calls are given when the nest is approached. Mock attacks are launched. Chicks remain immobile and hence invisible, till the intruder is around and an all-clear sound is uttered by the parents. If the chick is taken in hand it feigns death!
Eggs: 3-4,stone-brown, blotched, obliterative. 36.4 x 26.9 mm.
Call: Silent and hence overlooked. Harsh melancholy *Tweet, cheet, ti-ee* when disturbed. Calls from the ground and also on the wing.
Ecological notes: A lapwing of the arid wasteland, less dependent on water. Chicks resemble the locally found vegetation *Lepidagatus cristata* (S.Ingalhalikar).
Cultural notes: The Sanskrit name **Sadaluta** means one who has a permanent scar (The scar-like wattle). **Peetpaad**, yellow-legged, would be applicable to any yellow legged bird.
Status: Uncommon.

*Related Species: *Sociable Lapwing (V.gregarius-363) Rare WM to our arid country. Has dark cap, white supercilia fusing on the nape, black bill, legs and streaked breast (Maroon in breeding adult). It's Sanskrit name **Katukwan**, appearing in the ancient text Matsyapuran, describes the mewing quality of its call. (See page 106)*
*Grey-headed Lapwing (V.cinereus-365) WM, vagrant. Grey head, neck, breast and black chest and tail band.

Maaltitwi

Description: Dark cap, white eye brow, yellow wattle and legs. Prefers to run and reluctantly takes to wings. Lazily forages on the ground for insects, interspersed with bursts of sudden dashes. Flocks are seen in winter. They often come to a preferred water body, leaking tap or canal at noon, to quench thirst. Cap is brown in the juvenile. A limping bird with missing toes of one leg is recorded to have successfully raised a brood (Amit Pawashe).

*Northern Lapwing(V.vanellus-364) Vagrant to wetlands in our northern area.

Titwi

Description: Lanky bird with a red wattle, incapable of perching. Confiding and unlike its Yellow-wattled cousin is always found near water. The alarmed chick crouching on the scrub land becomes invisible, for such is its camouflage.

Red-wattled Lapwing.

White-tailed Lapwing.

White-tailed Lapwing.

Red-wattled Lapwing with juvenile.

Red-wattled Lapwing

Charadriidae

Vanellus indicus (366)

Utpadshayan, Tittibh

Size: 330 mm.　　M / F : Alike　　**R**

Distribution : Kokan, Malabar and the Western Ghats up to 2000 m. Uncommon in forests.

Nest-site: Open ground, ploughed fields, roofs of tall buildings.
Material: Small stones, pebbles, gravel. Up to a thousand pebbles were seen around a single nest on the terrace.(S.Nalawade)
Parental care: Both. Broken wing display may be given to distract the intruders from the nest.
Eggs: 3-5, brownish, blotched. Unhatched eggs are often seen. 42 x 30 mm.
Call: A low pitched *Tit, tit*, but when alarmed *Ti, ti, ti, tivit* getting louder with the approach of the intruder near the nest. A familiar night call.
Ecological notes: A bird well known by its alarm call.
Cultural notes: Call mistakenly believed to be an ill omen! Features in texts Panchtantra and Dnyaneshwari. According to folklore, the lapwing vowed to drink and empty the ocean, to find its eggs stolen by a toad, that had hidden in the sea. An impossible task!
The name **Utpadshayan** means standing on one leg.
Status: Common.

These innumerable pebbles were brought to the terrace of a four storied building in Pune city by a Lapwing, to make a nest.
This is a vivid story of habitat loss & adaptation.

*Related species: *White-tailed Lapwing (V.leucurus-362) Rare WM around Mumbai and recently from Kerala (JBNHS 98(2),Aug 2001, 280) Frequents marshes. Long yellow legs, black bill and unbanded pale brown head. Black outer wings are seen in flight.*

Common Snipe

Fantail Snipe

Scolopacidae

Gallinago gallinago (409)

Jalvartika, Pankakeer, Gobhandir

Size: 270 mm.   M / F : Alike   **WM**

Distribution : Kokan, Malabar and the Western

Ghats from August end onwards.

Nest-site: Marshy areas of north Uttar
Pradesh, Kashmir and the Himalayas.
May - Jun.
Material: The floating platform in a bog is lined
with weeds and grass.
Parental care: Both. Incubation is 19-20 days.
Eggs: 3-5, yellow-green, mottled.
38.3 x 28.5 mm.
Call: A nasal call *Scape, pench, pench* is
uttered in flight. Silent when feeding.
Ecological notes: Essentially haunt marshes,
intertidal weed infested zones, mangroves,
flooded fields. Largely insectivorous.
Albino birds are known.
Cultural notes: Male utters a drumming and
bleating sound during the aerial nuptial display.
It was therefore known as the 'Goat Of The
Sky' in Germany! (Salim Ali,Handbook) Its vocal
aerial display is indicated in the Sanskrit name
Gobhandir. As it scatters mud while feeding it
is called **Pankakeer** (Panka-mud). The snipe
and the woodcock were popular game birds
and were considered a table delicacy.
Status: Uncommon.

Wood Snipe.

Pintail Snipe.

Paan-Laava

Description: White trailing edge to pointed
wings, white bars on underwings, white-tipped
rufous fan like tail and long straight beak.
Longitudinal buff lines on dark back, white
belly and buff streaked chest. When flushed
takes a zig-zag flight, hence was considered to
be a prize shot by snipe shooters. Feeds in
squelching mud by probing with the sensory
bill.

Eurasian Woodcock.

Related Species: *Wood Snipe (G.nemoricola
-405) Barred below. Mumbai south.
Vulnerable.*
*Swinhoe's Snipe(G.megala-407)Vagrant.
Mumbai south.*
*Pintail Snipe (G. Stenura-406) Flight less zig-
zag than others, dark wings lack white trailing
edge. Supercilium bulky anterior to eyes.
Central tail feathers rufous, outer pied.*
*Jack Snipe (Lymnocryptes minimus-410)
Lacks pale crown stripe. Supercilium is
divided. Our smallest, short billed snipe.*
*Eurasian Woodcock(Scolopax rusticola-411)
Black bands on crown and nape, brown
unstriped back, barred breast. In dense Ghat
forests above 600 m. ALL THESE SNIPES ARE
WINTER MIGRANTS.*

Tail Patterns of Snipes.

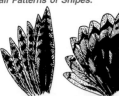 

Jack Snipe.   Fantail Snipe.   Swinhoe's Snipe.   Great Snipe.   Wood Snipe.   Solitary Snipe.   Pintail Snipe.

Jack Snipe.

Common Snipe.

*A typical
Snipe silhouette.*

Swinhoe's Snipe.

A typical habitat where snipes can be seen.

Black-tailed Godwit

Scolpacidae

Limosa limosa (389)

Aaramukhi

Size: 410-500 mm.   M / F : Dimorphic   **WM**

Distribution : Patchy. Kokan, Malabar and rarely on shallow water bodies and seashore. Rarer in the hills.

Nest-site: Breeds outside India. *Apr. - Jun.*
Material: Weeds, twigs.
Parental care: Both.
Eggs: 3-4.
Call: *Kek, kek, kek* is uttered when the disturbed bird takes to wings.
Ecological notes: This semiaquatic winter migrant is seen on the coasts and inland water bodies alike. It has not escaped the shotgun of the bird hunters.
Cultural notes: The local name Pankaj derives from the bird's habit of keeping to muddy areas. In Nepalese, Godwit is called Malguza, meaning one who probes mud with a bamboo-pin like bill.
Status: Occasional.

Kalya-shepticha Pankaj

Description: White wing bars, black banded white tail, dark lanky legs and uniform brownish colour in the non-breeding plumage. Breeding male is chestnut-red with black bands on flanks. Probes beneath the turbid water with the long bill. Long legs enable deep water wading. Longevity at least 17 years.

Bar-tailed Godwit.

Snipe-billed Godwit (Limnodromus semipalmatus) -of the East coast.

Black-tailed Godwits.

Related species: ******Bar-tailed Godwit* (Limosa lapponica-391) *Breeding birds are dimorphic; underparts of male chestnut-red, female cinnamon. Non-breeding birds brown, streaked. Barring on the tail is seen in flight. Seashore bird, unrecorded inland. Up curved bill distinct. Compared to L.limosa, legs are shorter and white wing bar is absent.*

Nakshidar Lahan Kudalya

Description: Dark crown with white central stripe and two lateral superciliary stripes, latter visible from a distance. Long curved bill and grey legs. Underwing coverts and lower back are white in race *phaeopus* vs. *variegatus*. Feeds from the surface. Wary. Flight swift.

Broadbilled Sandpiper.

Other Waders: **Great or Eastern Knot (Calidris tenuirostris-413) WM.* Grey-brown, streaked above. Few streaks on underparts. Size-270 mm. Rarely on coasts.
*Red Knot (C.canutus-412) Vagrant. Straight dark bill, stocky, grey back, streaked whitish belly. Red when breeding. Gregarious.
*Broadbilled Sandpiper(Limicola falcinellus-424) Bill downcurved, posteriorly placed legs, prominent white supercilium, dark eye-stripe, streaked breast and dark patch on wing-shoulder. On intertidal zones. WM. Flocks speedily fly along the coastline.
*Spoonbill Sandpiper(Calidris pygmaeus 423) Rare WM, coasts. 140 mm. Grey above, white belly. Face orange when breeding.

Whimbrel

Scolopacidae

Numenius phaeopus (385)

Tishti

Size: 430 mm.   M / F : Alike   **WM**

Distribution : Patchy. On coastal Kokan and Malabar on mangroves and inland creeks.

Nest-site: Nesting is outside India in Northern Europe, etc. Nest is concealed in reed beds. *May - Jul.*
Material: Leaves, aquatic weeds, twigs.
Eggs: 3-4, olive-green, blotched.
Call: A peculiar laughter like *He, he, he, tetti, tet-tii, tee* is repeatedly uttered in flight.
Ecological notes: A coastal winter migrant, but is also recorded inland during passage. Crabs, snails, other molluscs and crustaceans form a major part of their diet. Migrate during night, when calls are heard from the sky.
Cultural notes: Hunted for flesh in the coastal areas. The Sanskrit name *Tishti* is onomatopoeic. In England, the Whimbrel is known as 'Titterel' (Call- *Titti, titti*).
Status: Uncommon.

Red Knot (Br.)

Spoonbill Sandpiper.

Whimbrel & Godwit.

Red-necked Stint (Non-Br.)

Whimbrel wading in shallow water.

Great Knot in partial breeding plumage (top & bottom) and non-breeding plumage (left).

A flock of Whimbrels.

Kural, Kuree

Curlew

Scolopacidae

Numenius arquata (388)

Utkroshi-Kurari

Size: 600 mm.   M / F : Alike   **WM**

Distribution : Kokan, Malabar. Not inland.

Description: Breeding and non-breeding plumage differs, being darker when breeding. Dull, sand-brown, streaked with lighter rump and lower back in winter. Upper tail-coverts whitish. Down curved bill visible from afar. Shy. Searches intertidal zones, mud flats for grub. Hectic activity is noted when food is found. Probes mud singly and slowly, wading with a deliberate step, in a hunchback posture.

Common Greenshanks.

Other Waders: *****C**ommon Greenshank (T.nebularia-396) Green legs, grey underparts. White breast streaked in winter. Long bill. Call 'Too, too'. Size-360 mm. *****D****unlin (C.alpina-420) Faint eyebrow, dark centred rump, short curved bill, grey-brown above with rufous scapulars, mantle. Call -'Tue-ep, wee-et'. On coasts, inland in winter. Eats by probing. *****G****reen Sandpiper (Tringa ochropus-397) Green legs, olive streaked back. Rump band white. 210 mm.*

Curlews are rarely seen in flocks in our region.

Nest-site: In Europe to the east of Siberia, from northern temperate to arctic regions. Nest is placed on ground often in agricultural areas. *Apr. - Jun.*

Material: Twigs, weeds.

Parental care: Both. Broken wing display by parents to distract intruders from the nest site.

Eggs: 4.

Call: The mournful high pitched call *Coor-lew, Ooor-rii-ooor* is usually uttered in flight.

Ecological notes: The burrows of polychaete worms and mollusc siphons in the marshes are probed with the long bill. Their bills have tactile sensitive Herbst corpuscles. Vision is also used for finding food. Fruits of aquatic vegetation are also eaten.

Cultural notes: Mentioned in the epic Ramayan. Curlew's wailing call has inspired poets like Kalidas, Bhavbhuti, to draw similes when describing mournful situations! Similar use is seen in Australian Aboriginal folklore (Suruchi Pande,Per.Com.). These heavy birds are extensively hunted Their populations are now on the decline.

Status: Uncommon.

Green Sandpiper.

Curlews.

Curlew-feather.

Dunlin (Non-Br.)

Dunlin (Br.)

Mixed flock-Curlews, Crab Plovers, Oystercatchers.

Mangroves-A threatened habitat of the waders.

Reeves.

A flock of Common Redshanks & Curlew sandpipers.

Common Redshank (Br.)

Blackwinged Stilt on the nest.

Spotted Redshank (Br.)

Common Redshank

Redshank

Scolopacidae

Tringa totanus (394)

Bhat, Pinga

Size: 280 mm.   M / F : Alike   **WM**

Distribution : Kokan, Malabar and eastern margins of the Western Ghats on water bodies where wading is possible.

Nest-site: Kashmir and Ladakh. The ground scrape is concealed in a bog. *May - Jul.*
Material: Leaves, grass, twigs.
Parental care: Both.
Eggs: 4, drab coloured but richly spotted. 46.1 x 31.8 mm.
Call: *Teuu-tuu-uu.* Noisy bird. The piping call is shrill and loud. It serves as an alarm call for the mixed company of waders.
Ecological notes: Inhabit fresh and coastal waters alike. Mainly subsist on worms, Molluscs, crustaceans and small insects.
Cultural notes: The Sanskrit name **Pink** means one with a piping call and **Bhat** means a fighter, the spirit well exhibited on the breeding grounds! Vernacular name Surma derives from **Shoormanya** (brave fighter).
Status: Common.

Other Tringa Sandpipers: **Wood, Spotted Sandpiper(T.glareola-398) White, barred rump. Wings black above, white below. Yellow legs, upperparts woody brown. Call 'Chiff-iff-iff'. Size-210 mm.*
*Marsh Sandpiper (T.stagnatilis-395) Delicate, long legged, long necked wader. White below in winter. Call 'Pep, pep'. Size-250 mm.
*Spotted Redshank (T.erythropus-392) Differs from redshank by longer bill, absence of white in wings and less grey on breast. Call 'Tuick'. Size-330 mm. All are winter migrants.

Marsh Sandpiper.

Rakta-Surma

Description: Wings have a white band on trailing edge. Black tipped bill has a red base. Legs orange red, breast grey. Noisy in their winter quarters, indulging in fights during the breeding season. They feed on water's edge along with other waders and fly when flushed.

Spotted Redshank.

Wood Sandpiper.

Marsh Sandpiper.

Tiwala, Tutwar

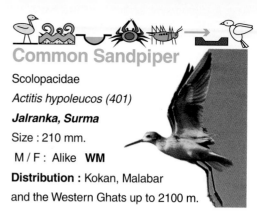

Scolopacidae

Actitis hypoleucos (401)

Jalranka, Surma

Size : 210 mm.

M / F : Alike **WM**

Distribution : Kokan, Malabar and the Western Ghats up to 2100 m.

Description: Head bobbing motion of this wader is distinct. Wings are flicked in low flight close to the water. Grey-brown upper parts, breast, yellow legs, white supercilium and belly. White wing-bar and rump best seen in flight. Unlike snipes, these waders keep to open areas of the wetlands.

Ruff (M, Br.)

Terek Sandpiper.

Related species: *Terek Sandpiper (Xenus cinereus-400) Yellow-orange legs, up curved bill, white trailing edges to secondaries. On coasts. 240 mm.*
Ruff (Philomachus pugnax-426) On coasts and inland wetlands. Designed scaly upper parts, yellow legs and drab underparts. Has a horizontal disposition. Breeding male gets the dark chestnut ruff. Non-breeding female or Reeve is similar to male, but smaller.310 mm.

Nest-site: In North India in Kumaon, Kashmir. On riverbanks, islands in hill-streams. *May - Jul.*
Material: Leaves, rubbish, aquatic weeds.
Parental care: Both.
Eggs: 3-4,stone-grey, pink spotted. 35.6 x 26.2 mm.
Call: Shrill *Teeit, teeit, trii,trii, tee, tee, wee, wee.* The piping call is given in flight or when feeding on sand bars-hence the name Sandpiper.
Ecological notes: Earliest winter migrant on fresh and marine waters alike. Non-breeding birds often stay back on the winter grounds throughout the year.
Cultural notes: The sandpipers were called **Surma** in Sanskrit, due to their habit of indulging in wars and fights when breeding. No wonder that alcoholic beverages are named after these spirited waders!
Status: Common.

Ruddy Turnstone (Arenaria interpres-402) On rocky coasts. Dark breast pattern, orangish stumpy legs, black-banded tail. White wing-bars in flight. 220 mm. Flocks.
Ruddy Turnstone (Br.)

Curlew Sandpiper (Non-Br.)

Ruddy Turnstones (Non-Br.)

Temminck's Stint (Non-Br.)

Terek Sandpiper (Non-Br.)

Ruff (Br.)

Curlew Sandpiper (Partial Br.)

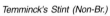

Red-necked Stint (Non-Br.)

Sanderling in mixed flock. **Safed Tutwar**

Description: Active flocks keep to intertidal zones. Can be confused with stints and sandpipers in mixed flocks. Grey above, white below has a short black bill and black legs. Narrow white wing bar is seen in flight. Black shoulder patch is prominent.

Curlew Sandpiper.

Related Species: **Rufous-necked, Red-necked Stint (Calidris ruficollis-415) WM to Goa. Rufous head, neck when breeding. Grey back, flanks when nonbreeding.*
*Little Stint (C.minuta-416) Black legs, bill, dusky brown above white below. Runs in search of little insects, near water. Active.
*Temminck's Stint (C.temminckii-417) Olive-green legs, white outer tail-feathers. Similar to Common Sandpiper but smaller-150 mm. WM. Commonly on sand bars and mud flats.

Sanderling

Scolopacidae

Little Stint.

Calidris alba (414)

Jalalobhin, Jalranka

Size: 190 mm.   M / F : Alike   **WM**

Distribution : The entire sea-coast of Kokan, Malabar. Not in the Western Ghats.

Nest-site: On ground, in grasslands in NE Europe and Siberia. *Jun. - Jul.*
Material: Willow leaves, twigs, etc.
Parental care: Probably both.
Eggs: 4,greenish, red blotched.
Call: Sharp *Wick, wick, wit, wit, kiwiri, rri.*
Ecological notes: A bird ringed in the former USSR at Alma-Ata in 1977, was taken at Point Calimere in India in 1990(SA Handbook).
Cultural notes: The sandpipers and the stints are loosely called 'Snippets',and they were hunted by less accomplished snipe shooters since their flight is poorer than the snipes! The Little Stint is called Panloha in Hindi, which means, one greedy for water.
Status: Uncommon.   *Spoonbill Sandpiper(Br)*

*Curlew Sandpiper (C.ferruginea-422) WM to coasts and inland waters. White rump, long legs, down curved long bill. Grey white plumage turns to chestnut when breeding.

Kentish Plovers & Little Stints.

Black-winged Stilt

Recurvirostridae

Himantopus himantopus (430)

Yashtik

Size: 250 mm.   M / F : Dimorphic   **WM**

Distribution: Throughout the region, on lakes, rivers, mud flats and coasts. Commoner in the plains than in the hills.

Nest-site: On ground in a scrape, raised platform in shallow water. Extralimital. Rarely in the region. *Apr. - Aug.*
Material: Twigs, water-weeds, pebbles.
Parental care: Both. Breeding is colonial.
Eggs: 3-4,pale, blotched. 44 x 31 mm.
Call: Sharp *Ki, ki, ki* or *Kenk, kenk, tewn, tewn* notes uttered while quarreling or taking to wings.
Ecological notes: One of the first winter migrants to arrive and the last to leave. An indicator species of polluted water. Long legs enable them to wade in deeper water, as compared to the other waders of the same size, hence there is no competition for food.
Cultural notes: The name Shekatya decribes the lanky legs, an apt depiction of this long legged bird. It also carries the distinction of possessing the longest legs with respect to the body size, in the avian world!
Status: Common.

In the avian world,the stilt has the longest legs with respect to body length.

Mating stilts. They rarely breed in our region.

Shekatya

Description: Black wings, (browner in female) long black bill, white underparts and long pink legs. Gregarious. Probes the bottom mud with head fully immersed. Also takes floating organic matter. Partial to polluted and sewage water. Often stands on one leg or rests on the belly. Flies with lanky legs clumsily dangling. Can swim for some time. A stilt with one leg amputated, probably from a bite has been recorded to be wading in the company of other waders(Satish Pande).

Immature.

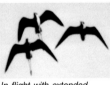

In flight with extended & folded legs.

Black-winged Stilt with chick.

Pied Avocet

Recurvirostridae

Recurvirostra avosetta (432)

Kushitak, Kashikani

Size: 460 mm.   M / F : Alike   **WM, PM**

Distribution: Coasts of Kokan and Malabar. Rarely on inland waters in passage. Not in the Ghats.

Tilwa

Description: This majestic, pied waterbird carries itself with elegance. Also swims with the partially webbed toes. The unique up curved bill is used for churning mud with swaying movement of the head. Enters knee deep squelch in search of food running in short spurts. The bright contrasting plumage and the unique bill are not easily forgotten. In flight seven black patches are seen on the upper body .

Pied Avocets.

Red-Necked Phalarope (Phalaropus lobatus) Non-Br. can be mistaken for Pied Avocet. Not seen in our region.

Pied Avocets.

Nest-site: Shallow ground scrape in large nesting colonies. *Apr. - May.*
Material: Unlined.
Parental care: Both.
Eggs: 4, spotted, drab coloured. 49.2 x 35 mm.
Call: The high pitched call *Kleet, kleet* is uttered on the wing. Usually silent.
Ecological notes: The only known breeding site in India is in the Great Rann of Kutch. Also wades in sewage water. Recently recorded near Shirwal on the banks of the Neera river, district Satara in the company of Spoonbills and smaller waders (Saleel Tambe).
Cultural notes: The Sanskrit name *Kashikani,* of the Charak-Samhita, means a waterbird with a plough-share like beak. The name *Kushitak* is derived from its habit of extracting worms from the mud !
Status: Uncommon.

Note the seven black patches in flight.

Crab-Plover

Crab-plover
Dromadidae
Dromas ardeola (434)
Vailatak, Veladhar
Size: 410 mm.   M / F : Alike   **WM**

Distribution : On coasts. Recently recorded on tidal mud flat, Agardanda, Dighi, Dist. Raigad. (JBNHS,vol.98(2),280-281). Rarely on inland waters.

Nest-site: On marine islands in burrows dug in sand-dunes. Nearest place is around Sri-Lanka. *Apr. - May.*
Material: Unlined nest hole.
Parental care: Chick is fed by the parents even after it is capable of flight.
Eggs: 1, white. 65.4 x 45.9 mm.
Call: *Twell-tak, twell-twell* is uttered in flight when disturbed.
Ecological notes: This littoral wader is not a true plover and is the only member of the subfamily Dromadinae. Apart from Crabs (Locally known as Chimbori), small Crustaceans and mudskippers are also eaten.
Cultural notes: The bird was known to ancient Indians due to its peculiar nesting habit. Sage Vasishtha had imposed a ban on Hindus around Rameshwaram against eating the Crab-Plovers and their eggs. The birds nesting in the burrows were otherwise easy to catch. This highlights the humanitarian spirit, love and respect for nature and the power of observation of the ancient Indian law-makers! The Sanskrit names **Veladhar** and **Vailatak** mean those concealing the egg in a hole near the coast.
Status: Rare.

Chimbori-khaoo

Description: Though pied, looks more white when standing with wings closed. Dark biconvex bill, long grey legs. Black in the wings best seen in flight. Follows the receding tide and catches the exposed crabs by chasing them. Catch is hammered on the legs or on the ground and then swallowed as a whole or in chunks. Feeding activity becomes frantic as the twilight becomes dimmer and dimmer.

Crab Plovers rest at noon.
They are active near dawn & dusk.

Flocks of Crab Plovers are seen on islands in the Gulf of Kutchh.
In our region this bird is a vagrant.

Great Stone-Plover

Great Thick-knee

Burhinidae

Esacus recurvirostris (437)

Kalpaanvik, Karvaan

Size: 500 mm.   M / F : Alike   **R, LM**

Distribution : Patchily in the entire region.

Nest-site: Ground scrape in a rocky area near water. Also on a flat surface. *Feb. - jul.*
Material: Usually none. Rubbish may be piled.
Parental care: Both.
Eggs: 2-3,sandy, grey-green. 54.4 x 41 mm.
Call: Loud creaking and drumming call often after dusk, *Kree-kre-kre*. Vocal on moonlit nights.
Ecological notes: Specially shaped bill aids the plover to handle hard shelled crabs and molluscs, which feature in its diet.
Cultural notes: Mentioned in the works of Sushrut. The name **Kalpaanvik** means a drummer, for so is the sound of this bird.
Status: Uncommon.

Related species: **Stone-Curlew or Eurasian Thick-knee (Burhinus oedicnemus-436) Like Great Thick-knee, smaller, lacks black on the head, has less bulky bill. Wing bar is white. Call-'Curlew'. Crepuscular, nocturnal, in arid habitat. When disturbed, flattens on ground with neck extended. Patchily in our area. Recorded in North Kokan.*

Karvanak

Description: Bulky upturned yellow-black bill, goggle eyes, stout yellow legs and black wing bar help instant spotting. Spend the day lazily basking in the sun, standing on a rock or sand bar. Keep to the chosen territory for years. Nesting overlaps with terns and nests may be seen in a tern colony. Local movement is with the availability of water. Legs have three toes.

Great Stone-Plovers.
Stone-Curlew on the nest.

Stone-Curlew Chick.

Indian Courser

Glareolidae

Cursorius coromandelicus (440)

Shwetcharan, Kshiprachala

Size: 260 mm. M / F : Alike **R, LM**

Distribution : Patchily in the Deccan and along the eastern fringes of the Western Ghats in arid habitat. Not on coasts.

Nest-site: On ground in fallow land and fields.
Material: A few pebbles are arranged around the ground scrape. *Mar. - Jul.*
Parental care: Both.
Eggs: 2-3,stony brown, black-blotched.
Call: A silent bird. Low pitched *Wit, twit* is uttered in flight, if disturbed.
Ecological notes: Bird of arid and dry land. Also seen near fallow fields migrating when ploughing begins. Many habitats are lost due to human encroachment and urbanisation. Courser populations are declining.
Cultural notes: If pursued, the courser runs, stops and surveys. Chasing it is a favourite past-time of wanton boys, hence the bird's rural name-Porechalawani-one who stimulates children! It's Sanskrit name **Shwetcharan** indicates its white legs, **Kshiprachala** means swift-footed.
Status: Rare.

Indian Courser on nest.

Dhawik, Porechalawani

Description: Bright red crown and peri-orbital white stripes aid identification. Underparts of adults are plain rufous and black, which are lightly barred in juveniles. Family flocks with juveniles are seen in August. Show local movement. Night roosting is in loose circular formations. These wary birds allow close approach in four wheelers, offering opportunity for photography. Their habitat is shared by lapwings, sandgrouses, larks, pipits, doves.

Chick.

Indian Courser-incubating, broken wing display & Juvenile.

Panbhingari

Description: Swallow-like bird with pointed black edged wings, black lore and a forked squarish black tipped white tail. Numbers often underestimated due to the drab colour and the flock size is apparent only when the birds take to wings. Winged insects are hawked over water. Chicks are often fed moistened grub. Small fish are taken from the water surface. Parents cool their eggs on hot summer noons by wetting belly feathers.

Collared Pratincole.

Related species: **Collared or Large Indian Pratincole (G.pratincola-442) Rare. Pointed wings extend beyond the forked tail at rest. Collar of the breeding bird becomes duller later. Size-240 mm.*
*Oriental Pratincole (G.maldivarum-443) At rest, tail shorter than wings, that lack white edge. The neck gorget of the breeding bird becomes very faint in the non-breeding plumage. Rarely and patchily in our area.

Small Pratincole

Swallow-Plover

Glareolidae

Glareola lactea (444)

Haputrika, Vishuvi, Vishwaka

Size: 170 mm.   M / F : Alike   **R, LM**

Distribution : Seen patchily distributed in Kokan, Malabar and the Western Ghats.

Common around water bodies on the plains.

Nest-site: Colonial nesting, often with terns, Little ringed plovers, on dry mud flats, ground scrape. *Mar. - May.*
Material: None. Unlined.
Parental care: Both.
Eggs: 2-3,matching perfectly with the ground colour. Eggs with multiple hues, muddy-brown to dull green, are often seen in the same colony. 25.9 x 20.5 mm.
Call: A whistle *Tri, tit, tri, tit, tit* is given in flight. Noisy in the evening, when flocks vocally communicate with one another.
Ecological notes: Local migration is seen from one percolation tank to another, depending on the availability of water. Eggs in the ground-nests are sometimes trampled under the hooves of cattle or feet of herdsmen.
Cultural notes: Names *Vishuwi* and *Vishwaka* mean one flying everywhere, in all directions. The typical broken wing display of the pratincoles, gives it the name *Haputrika*, meaning one who is bemoaning his son by helplessly prostrating on the ground!
Status: Unommon.

Oriental Pratincole.

Collared Pratincole.

The nesting habit of Small Pratincole - Incubating.

At noon temperature is controlled by wetting the belly feathers in water before incubating the eggs.

A newly hatched Small Pratincole chick (egg-tooth is seen) & an unhatched egg. Feeding the chick.

Samudrakak

Description: Wide based bill with culmen bent over lower mandible. Nostrils excrete saline water. Broad based wings, long tail with spatulate bulbous ends due to twisting. Pale and dark morphs, former with barred chest, flanks and under-tail-coverts. Webbed feet with sharp claws. Pirate prey from gulls.

Brown Skua

***Related Species:** *B*rown or Great Skua (Catheracta antarctica-446) 610 mm. Largest skua, pelagic hunter, vagrant to Kerala, Ratnagiri. Breeds on subantarctic islands. ***S*outh Polar Skua (C.maccormicki-446a) 530 mm. Bird ringed in 1961 in Antarctic peninsula was retrieved at Udipi, Karnataka in 1964* (Handbook,Salim Ali).

Parasitic Jaegers.

Pomarine Jaeger

Pomatornine Skua

Stercorariidae

Stercorarius pomarinus (447)

Samudrakak

Size: 530 mm   M / F : Alike   **MM, PM**

Distribution : Marine waters off the West coast, Vengurla Rocks,also near Mumbai.

Nest-site: Extralimital. Arctic coasts, Siberian islands. *Mar. - Jun.*
Parental care: Both.
Call: Silent in winter.
Ecological notes: Pelagic. Predator of eggs, chicks of colonial nesters like terns, gulls. In our region they are seen on passage to Australia.
Cultural notes: Fishermen know them as sea-crows which appear briefly during winter.
Status: Vagrant.

South Polar Skua

**Parasitic Jaeger* (Stercorarius parasiticus-448) 480 mm. Wedge like tail feathers are faintly barred. Brown to pale morphs. Flocks rarely on our marine waters in monsoon, seen by the authors in 2000 near Burnt Island, Kokan.*

South Polar Skua

Breeding.

Kira, Tapkiri Dokyacha Kurav

Non-Br.

Laridae

Larus brunnicephalus (454)

Vichikak (Wave-crow), **Kharshabdakurar, Damar**

Size: 460 mm.   M / F : Alike   **WM, PM**

Distribution : On the coasts of Kokan and Malabar. Vagrant on inland waters.

Description: White wing-tip mirrors, yellow iris, grey-black underwings. Head pale brown when breeding. These masters of wind pick food from the water surface, manoeuvering with their broad wings and square tail. Swim well with webbed toes, elegantly riding the waves in the roughest weather. Clams and crustaceans are reportedly dropped on rocks and on hard ground to break them open.

Slender-billed Gull.

Nest-site: Colonial nesting in marshes, bogs in Ladakh, Tibet and near Manasarovar, Rhamtso, etc. high altitude lakes. *Jun. - Jul.*

Material: Aquatic weeds, stems.

Parental care: Both.

Eggs: 2-3,bluish green, blotched. 61.3 x 42.6 mm.

Call: Loud calls and croaks. A quarrelsome bird. *Kraa, kree, kek, keeah* calls are uttered when roosting on the shores or on the wing.

Ecological notes: Omnivorous scavenger. Visits seaside garbage dumps, fishing hamlets, harbours for offal, dead fish and other discarded organic material. Eats tender shoots and small reptiles when inland.

Cultural notes: Featured in the epic Mahabharata. Sanskrit names **Damar** and **Kharshabdakurar** indicate their noisy nature. Statue of the gulls is erected at Salt Lake City, Utah,in appreciation of their help during the famine when they devoured locusts and saved the remaining crop. The aesthetic smile design, popular amongst contestants in beauty pagents, is said to have a Gull Wing Appearance!

Status: Common.

Yellow-legged Gull.

Other Gulls: All Are Winter Migrants

*Black-headed (L.ridibundus-455) No wing mirrors, wings white, breeding birds have dark brown head. Size-430mm. Common.

*Heuglin's (L.heuglini-450)Our large, dark coloured gull with streaked head. Solitary.

*Yellow-legged or Herring (L.cachinnans-451) Pale grey upperparts, mirrors on primaries, red patch on lower mandible. On coasts. Size-600 mm. Rare.

*Pallas's or Great Blackheaded (L.ichtyaetus-453) Incomplete mask on sloping head, white eye-crescent, black-tipped bill. Largely white.

*Little (L.minutus-457) White with grey under-wing. Rear crown, bill, posterior eye spot black.

*Sooty Gull (L.hemprichii-449) 480 mm. Brownhood. Blacktipped yellow bill, yellow legs. Vagrant near Mumbai marine waters.

*Slender-billed Gull (L.genei-456)430mm. White longish head, dark ear spot variable, white iris, scarlet legs, beak. Marine.

Mixed flock of Brown-headed &
Black-headed Gulls.

Heuglin's Gull.
Heuglin's Gull (Imm.)

Little Gull.

Yellow-legged Gull (Non-Br.)

Black-headed Gulls (Non-Br.)

River Tern

Indian River Tern

Laridae

Sterna aurantia (463)

Kurari

Size: 410 mm.   M / F : Alike   **R, LM**

Distribution : Throughout the region, on large water bodies.

Nest-site: Bare ground in a shallow scrape. Colonial nesting on secluded shores, islands. *Feb. - May.*

Material: Lined with pebbles and small sticks.

Parental care: Both. Fearlessly launch attack on intruders with pointed bill and may cause minor abrasions on forehead(Satish Pande).

Eggs: 2-4, greenish, brown blotched. 42 x 31.4 mm.

Call: Harsh calls *Tewn, tewn, kiuk-keeuk* are usually uttered in flight. The tern islands become frantically noisy if approached.

Ecological notes: Eggs on ground are vulnerable to mammalian and avian predators. Nesting is near water bodies rich with fish. Chicks are known to die, if the days are very hot.

Cultural notes: Name **Kurari** means a vociferous and noisy bird!

Status: Common.

Nadi-Suray

Description: Graceful, sublime fliers and swimmers. Webbed red legs, orange-yellow bill striking. Black crown, long pointed wings and forked tail. Fly above water with bill pointing below. Dive when fish is sighted, submerging beneath the water. Fish is flicked in the air and eaten in flight. Breeding pairs feed each other and mating is seen even after egg laying. The flight displays are spectacular.

With chick.

Juvenile taking the first swim.

Courtship feeding.

Arctic Tern (Sterna paradisaea) is not seen in our region.

Caspian Tern & Heuglin's Gull.

Black-Naped Tern & Common Tern.
These archival photographs are taken by Lok Wan Tho in 1951 & 1958.(From the BNHS Collection)

River Tern nest-New born chick, hatching & unhatched eggs. *White Tern (Gygis alba) Can be encountered as a vagrant.*

Black-bellied Tern on the nest.

Gull-billed Tern

Laridae

Gelochelidon nilotica (460)

Kurari

Size: 380 mm.   M / F : Alike   **WM**

Distribution : In the entire region.

Hiwali Suray

Nest-site: Shallow sand pits in mixed tern colonies in Pakistan. In India near lake Chilka, Orissa and in West Bengal. *Apr. - Jun.*
Material: Scrape is lined with debris.
Parental care: Both.
Eggs: 2-3, dull yellow, blotched in various hues. 47.9 x 34.2 mm.
Call: Very vocal like all terns. Call is a high pitched *Feek-fik-fik, kit-vit.*
Ecological notes: On inland and coastal water alike. Scavenge from garbage.
Cultural notes: The arrival of gulls and winter are closely and inseparably linked in the minds of people living along the coasts.
Status: Uncommon.

Description: Grey-white tern. Flies slowly keeping close to the surface scanning for fish. Distinguished by the thick black bill, broad wings and heavy body. In non- breeding plumage black cap replaced by a black patch on ear-covert,grey underparts become buff.

Indian Skimmer.

Black-bellied Terns, Resting & Taking off.

Sandwich Tern

Related Species:

*Sandwich Tern (S.sandvicensis-480) Like breeding Gullbilled Tern, but the yellow tipped bill is longer and thinner. On our coasts. Gregarious.
*Black-bellied Tern (S.acuticauda-470) Black underparts and crown when breeding, turning ashy grey later. Tail-deeply forked. Nests patchily on freshwater islands in March-April.
Near Threatened.
*Indian Skimmer(Rynchops albicollis-484) Size-400 mm. Vagrant. Black cap, mantle, wings. White below. Orange red bill, legs. Catches fish on wings as it skims with sensory long lower mandible immersed in water.
Vulnerable.

Large Crested Tern 1st Week Chick on Burnt Island..

Large Crested Tern with fish for the chick.

Large Crested Tern 3rd week chick on Burnt Island.

Large Crested Terns on the Burnt island. This is their largest nesting colony off the west coast.

Little Tern

Ternlet

Laridae.

Sterna albifrons (475,476)

Laghukurari

Size: 230 mm.   M / F : Alike   **WM**

Distribution : On the coasts of Kokan, Malabar, adjacent inland water bodies.

Chota Suray

Lesser Crested Tern.

Nest-site: Patchily in coastal Kokan on isolated sandbars or rocky surfaces.
Mar. - Aug.
Material: Small pebbles line the nest scrape.
Parental care: Both.
Eggs: 2-3,grey-green. 32 x 24 mm.
Call: The call is similar to that of the Pied Kingfisher's, *Kee, kriee-krieek, kik, kik.*
Ecological notes: Our smallest marine tern, rarely venturing inland. Breeds locally and patchily. Exhausted migratory terns if found are often nursed by local bird watchers.
Cultural notes: A well known bird amongst fishermen. Visits fishing hamlets and jetties in search of fish.
Status: Rare.

Description: Black bill. Brownish legs Become yellow-orange when breeding. Cap, nape become black and forehead remains white. Shafts of first three primaries brown (race *albifrons*), white*(pusilla)*. Hovers before plunging. Tirelessly scans the water surface, looking for fish, throughout the day. Roost at night on sandbars in mixed company.

Lesser Crested Tern.

Large Crested Tern.

Large Crested Tern with chick on Burnt Island.

*Related Species: *Lesser Crested Tern*
(S.bengalensis-479) Rare WM to our coasts.
Marine tern. Like the Gullbilled but with an
orange beak. Pale grey above with a small
crest when not breeding.
**Large Crested Tern*(S.bergii-478) Rare WM.*
Seen at considerable distances from the coast
on the sea. Nuchal crest prominent. Head
becomes dark black when breeding.
Both breed on Vengurla Rocks, former to a lesser extent.

Black-Naped Tern.

Roseate Tern on the Burnt Island.

Sooty Tern can be mistaken for the Bridled Tern.

Common Tern. The tail and wings are of the same length.
Lesser Crested Terns.

Whiskered Tern

Laridae

Childonias hybridus (458)

Pushkarsad, Padmapatrashayika

Size: 250 mm.   M / F : Alike   **WM**

Distribution : Entire region, on coastal and inland waters.

Nest-site: Lakes, rivers and large water bodies in Kashmir and Assam. *May - Aug.*

Material: The floating nest is constructed with reeds, aquatic vegetation, often on lotus leaves (Sanskrit name-**Padmapatrashayika**).

Parental care: Both.

Eggs: 2-3,spotted brownish-blue. 36.9 x 27.4 mm.

Call: Harsh *Creerp, eerk, eerk* uttered in flight.

Ecological notes: Being a marsh tern it is largely a surface feeder, but unlike the other marsh terns it also eats fish.

Cultural notes: The Whiskered Tern builds a beautiful nest, a fact known since the Vedic times. This bird-architect was sacrificed to Twashtra, the artisan of gods!(Yajurved)

Status: Rare.

Look Alikes: For Common Tern:
*Non-breeding Roseate Tern-upper wings paler. *Arctic Tern-shorter,pointed black tipped wings. **For Caspian Tern:** *Great Crested Tern-Yellow bill. (Red of Caspian) *Blackheaded Gull-smaller size,narrow bill.

Shirwa Suray

Description: Takes food from surface, hawks insects in air and dives in water for fish. The flight is deliberate, slow. Breeding birds have a black cap, grey underparts, red bill, legs and prominent whiskers. In non-breeding plumage the crown is streaked, bill and legs black and whiskers inconspicuous.

Related species:*White-winged Black Tern (C.leucopterus-459) Marsh tern with rounded wing tips. More delicately built than Whiskered. Non-breeding adults-white rump, black ear-coverts drop below the eye, nape line is black. WM to coastal and inland waters.*
STERNA TERNS:
**Common Tern* (Sterna hirundo-465) Greyish, with dark cap, red legs and bill. Non-breeding birds-grey streaked cap. Wing tips do not project beyond the tail.
**Caspian Tern* (S.caspia -462) Our largest tern. Thick red bill, short black legs, short tail. Breeding birds have a crested black head which is streaked in winter. On coastal, inland waters. A large breeding colony at Jamnagar.

Common Tern.   White-winged Black Terns on nest.

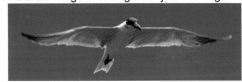

Caspian Tern.

Gull-billed Tern.

Bridled Terns in various wing positions.

Bridled Terns during courtship. Bridled Terns are marine terns & come to land for breeding.

White-cheeked Tern

Laridae

Sterna repressa (467)

Kurari

Size: 350 mm.   M / F : Alike   **R, LM**

Distribution : Seen patchily distributed in Kokan along the coasts, tidal creeks and on inland lakes and rivers in winter.

Nest-site: On high marine rocks and islands. *Feb. - May.*

Material: Lined scantily with gravel.

Parental care: Pre-mating aerial displays follow egg laying. Both attend to the nest. Incubating female is fed by the male.

Eggs: 2-4. 40.8 x 30.4 mm.

Call: Plaintive calls uttered on the wing. Harsh calls when breeding colony is approached.

Ecological notes: These terns breed on the Vengurla rocks. Brown Noddy breeds in Lakshadweep, Maldives.

Cultural notes: The activity of the terns over water, as they chase shoals of fish, guides fishermen to possible sites where nets can get a good yield of edible fish!

Status: Uncommon.

Safedgalya Suray

Description: Head grey in winter, black when breeding and underparts grey. Dark, thin, long bill. White cheek prominent. Rump, tail concolorous with grey back. Smaller than Common Tern. Underwing has a dark edge. Birds launch buoyant sallies over water.

Bridled Tern.

Brown Noddy.

Lesser Noddy.

Related species:

Prior two breed On Vengurla Rocks

*Roseate Tern (S.dougalli-466) On coasts and marine islands. Yellow bill becomes red with black tip, when breeding and underparts get a rosy tinge. Grey above, tail long. 310 mm.

*Bridled Tern (S.anaethetus-471) Grey-brown upper parts. White eye brow is longer than Sooty tern when breeding. Call-Kek, kuroo. Marine tern. 360 mm.

*Sooty Tern (S.fuscata-474) Black above, white neck, belly, forehead. Black trailing under edges of primaries. Juvenile all black except whitish belly. Marine tern-400 mm. Call is shrill-Kurr-kak.

*Saunder's Tern (S.saundersi-476) Marine. Yellow tipped black bill, black cap, white frontal patch not extending behind eyes. Vagrant.

*Brown Noddy (Anous stolidus-481) 420 mm. Brown with pale coverts, curved bill. Juvenile-brown crown. Vagrant.

*Lesser Noddy (Anous tenuirostris-N) Pelagic vagrant to West coast.

Brown Noddy with nesting material.

Caspian Tern (Non. Br.)

Black-headed Gull.

Bridled Terns during courtship feeding.

Sea Snakes sometimes predate on juvenile marine terns. In their own turn they often fall prey to Sea Eagles.

Bridled Terns (Juv.).

Brown-headed Gull. (Non. Br.)

Yellow-legged Gull.

Brown-headed Gull (Br.)

Black-headed Gull (Non-Br. & Br.)

Pallas's Gull.

Heuglin's Gull in a mixed flock.

Male.   **Pakurdi, Bhat-Tittir**

Description: Yellowish brown bird with a
pointed tail and dark belly. Female-streaked,
spotted, barred body, the chin is spared. Black
chest band. Flocks gather at water hole,
morning and evening, at the same time
everyday. Walk with chest close to the ground,
picking tiny grit with the small bill. Pugnacious
nature of these partridge like birds has
conferred upon them the name Bhat-Tittir
(Fighting-Partridge). Chicks are well
camouflaged and impossible to locate, if they
remain motionless on the ground. Fall prey to
harriers, shrikes, etc..

Chestnut-bellied Sandgrouse-Chick.

Painted Sandgrouse (M)

Pair with two chicks.
Male is in front.

Chestnut-bellied Sandgrouse

Indian Sandgrouse

Pteroclididae

Pterocles exustus (487)

Krukal, Bhatkukkut, Vikakar (Painted Sandgrouse)

Size: 280 mm.   M / F : Dimorphic   **R.**

Distribution : Patchily in the adjacent dry
and arid areas to the east of the Western
Ghats. Not Kokan and Malabar.

Nest-site: Ground scrape or a gentle
depression on the ground in stone studded
areas. *Feb. - May.*

Material: Unlined. Few pebbles may be
randomly gathered around the ground-nest.

Parental care: Both.

Eggs: 2-3, spotted, stone coloured.
36.8 x 26.2 mm.

Call: High pitched sharp *Kuturr-rro, Katarr-ka*,
uttered on wings, when coming to a water
hole.

Ecological notes: Essentially a bird of arid
habitat. They require water, at least twice a
day, to quench their thirst, a fact which has led
to extensive trapping of the sandgrouses.

Cultural notes: Sandgrouses have been
popular game birds! They are still hunted and
secretly sold in the market. The bird is known
since the Vedic times. In the text Kakkar-Jatak,
a decoy technique for catching sandgrouse
(Kakkar) is mentioned.

Status: Uncommon.

Painted Sandgrouse (M)

Related species:
***P**ainted Sandgrouse (P.indicus-492)
*Lacks tail pin-feathers and is smaller
than Indian Sandgrouse.
Male-black bar on white forehead,
gorgeted chest, barred belly.
Female- uniformly barred and buff brown.*

Pigeon in Sri Lankan Rock Art.

Yellow-legged Green Pigeon incubating.

Nilgiri Wood Pigeon.

Sri-Lanka
Wood Pigeon.

Chick taking the regurgitated food from Blue Rock Pigeon.

Yellow-legged Green Pigeon-feather and adult.

Parva

Description: Two black wing bars, dark terminal band on the grey tail and shiny Multi-hued sheen on neck and chest. Flocks alight in unison with much wing clapping, on hearing a loud sound. Confiding, first to feed at bird-seed platforms. Aggressive and selfish when feeding. Wild birds appear stouter than semi-feral cousin. Eggs laid on narrow ledges are prone to fall and to predation. Litters urban building and pigeon repelling chemicals are available to lessen their nuisance. A partial albinos is seen on Vengurla Rocks.

Blue Rock Pigeon Egg & Chick.

Blue Rock Pigeon with Chicks.

Related Species: *Nilgiri Wood-Pigeon (Columba elphinstonii-521) Local name-Taam. Uncommon. **Endemic** to the Western Ghats, Mumbai south up to 2000 m, also Nandi Hills. Maroon brown back, chessboard on the neck, pale grey head, underparts, dark grey tail. Call 'Whoo-who-who'. Keep to foliage. Eat berries, figs, drupes on trees or from the forest floor. Flight swift. Nest-a flimsy cup. Lays one egg. **Vulnerable.**

Blue Rock Pigeon

Rock Pigeon

Columbidae

Columba livia (517)

Neelkapot, Paraavat, Gruhakapot

Size: 330 mm.   M / F : Alike   **R**

Distribution : Kokan, Malabar and the Western Ghats up to 2000 m. In and outside the cities.

Nest-site: Wild birds nest on cliff ledges, ruins, niches in well walls. Semi-feral birds on buildings, rafters, roofs. *Jan. - Dec.*
Material: Twigs, small sticks, pebbles, rubbish.
Parental care: Both. Chicks suck regurgitated pigeons milk from the mouths of their parents.
Eggs: 2-3, white. 36.9 x 27.8 mm.
Call: *Gooturr-goo* is repeated. Familiar urban sound. Birds are vocal at noon.
Ecological notes: Well adapted to urban life. Cause much littering in terraces, verandas. Glean cereals, nuts, grain kept for sun-drying. Infection can be spread to humans via this route from soiling of grain with bird excreta.
Cultural notes: Smoky plumage and glistening reddish neck have drawn simile with smoke emerging from sacrificial fire, in the epic Ramayana. In the Rigveda (1500 BC)the pigeons were called **Paraavat**, since they came to hamlets from distant hills and returned at dusk. In Matsyapuran (400 AD) there are descriptions of temple walls adorned with these pigeons. This may throw some light as to when they may have settled in cities.
Status: Common.

Nilgiri Wood-Pigeon (Sub-adult)

Spotted Dove

Columbidae

Streptopelia chinensis (537)

Chitrapaksha-Kapot

Size: 300 mm.   M / F : Alike   **R, LM**

Distribution : Western Ghats up to 1500 m.
Common in Kokan, Malabar. Rare in Deccan.

Nest-site: The flimsy nest platform is placed in a bush, creeper, on roofs or rafters. *Jan. - Jun.*
Material: Twigs, straw, stems.
Parental care: Both.
Eggs: 2-3, white. 27.2 x 21.8 mm. Chicks are born with down.
Call: *Kroo, kroo, kruckroo* and similar notes.
Ecological notes: Nests are prone to predation. A common dove in Kokan where it replaces the Little Brown Dove, Ratnagiri south. Avoids dry, arid country. Altitudinal and local migrant to the wetter areas in summer.
Cultural notes: In the Mahabharata, spotted dove is glorified for fidelity and compassion. A similar belief exists in the West. The red, bare orbital patch had prompted the ancient Indians of the Vedic period to give this dove the Sanskrit name-**Chitratanuruha**! It is caged as a song bird in the eastern range.
Status: Common.

Kavda, Thipkya Kavda

Description: A graceful dove with a chess-board pattern on nape. Brown above, spotted back, vinous underparts. White tail band is seen in flight. Gleans seeds from harvested fields, road-side, village fringes. Freely enters gardens if unmolested. Usually perch in foliage.

Oriental Turtle-Dove.

Spotted Dove.

Spotted Dove-Nest.

Related species:
*Oriental Turtle or Rufous Turtle-Dove
(S.orientalis-531,532,533) North Kokan up to about Mumbai in winter. Race meena in Western Ghats north of Goa and erythrocephala southwards. Fly with a flutter, returning to the same spot after intrusion ends. Call 'Gurr, gurrgoo'.

Little Brown Dove

Laughing Dove

Columbidae

Streptopelia senegalensis (541)

Dhoosarkapot, Kumkum-Dhoomrakapot

Size: 270 mm.   M / F : Alike   **R**

Distribution : Western Ghats up to 1500 m, Malabar, North Kokan. Absent or scarce in Central and South Kokan.

Nest-site: Bushes, ledges, hanging flower pots in terraces, rafters in houses. Rarely on ground. *Jan. - Jun.*

Material: Leaves, grass, stems to make a very flimsy platform.

Parental care: Both.

Eggs: 2-3, white, ovoid. 25.3 x 19.3 mm.

Call: Resonant *Coo-coo-croo-koo-kroo*. A familiar call in scrub and agricultural habitat.

Ecological notes: Adapted well to urbanization. Freely nests in houses. The rare ground nests have been observed to have burnt in forest fires and predated (Mongoose, lizard).

Cultural notes: As per a popular story in the Mahabharata, in order to test king Shibi, Indra the king of gods took the form of a falcon and Fire that of a dove. The latter alighted on the thigh of Shibi, on being pursued by the falcon, and asked for mercy. King Shibi in order to save the dove, offered flesh of his own thigh to the falcon, proving his moral integrity!

Status: Common.

Hola, Nagari-Hola

Description: Delicate, slim, long-tailed dove. Pink-brown head, back and grey-brown underparts. Black spots on upper breast, black primaries. Pairs perch on bushes or wires cooing at each other. Glean seeds from stubble fields hurriedly filling their crops. Tame.

Eurasian Collared Dove on nest.
Eurasian Collared Dove

Little Brown Dove's nest in a pot !

Related Species: *Eurasian Collared, Ringed- Dove (S.decaocto-534) Arid, scrub habitat. Large sandy brown dove with black half collar on nape. Size-320 mm. Call 'Coo-cook'. Eggs are known to be predated by shrike, buzzard.* ***R**ed Collared or Red Turtle-Dove 230 mm.(S.tranquebarica-535) Blue grey head, dark half collar on nape, pinkish below, maroon above. Upperparts brownish in female. In scrub country. Call 'Groo, gurr, guu'. Rare. mm.*

Red Collared-Dove (Male)

Yellow-legged Green-Pigeon

Common or Yellow-footed Green-Pigeon

Columbidae

Treron phoenicoptera (504)

Haritaang, Haarit, Haritaal

Size: 330 mm.   M / F : Alike   **R, LM**

Distribution : Kokan, Malabar and the Western Ghats often on fruiting trees.

Nest-site: Tree fork 1.5-10m.up. *Mar. - Jun.*
Material: Twigs, straw, grass, leaves.
Parental care: Both.
Eggs: 2, shiny pearly-white. 31.8 x 24.6 mm.
Call: Loud pleasant whistle *Oo-oo-r, ooo-roo.* Wing clapping is heard during take-off. Flock alights with commotion.
Ecological notes: Frugivorous forest bird associated with barbets, mynas, parakeets. Plums are swallowed whole hence help seed dispersal. Cause damage to groves. Movement is with fruiting trees. A tree bearing an active-nest being indiscriminately felled for fire-wood, during the Holi festival is recorded(Vishwas Joshi).
Cultural notes: This pigeon is the State Bird of Maharashtra. Was considered to be auspicious in Buddhist times. Features in the Puranas and graphic descriptions are seen in Raghuvansha by Kalidas. Was much sought after as a cage bird due to it's musical call. Still sadly extensively hunted for flesh.
Status: Common.

Pusava, Hariyal

Description: Green body, grey head, faint mauve shoulder patch and yellow legs. Arrive in Deccan in winter when flocks are seen in groves, on fruiting ficuses. Eat fruits in acrobatic manner. Take to flight when disturbed and their numbers are astonishingly deceptive since they camouflage perfectly in the green foliage. Night roosting is communal, on leafless trees on hill slopes, from where they command a good view. It is at this time that their beauty is best appreciated, in the tender golden rays of the setting sun.

Orange-breasted Green-Pigeon.

Pompadour Green-Pigeon.

Related species: *Pompadour or Grey-fronted Green-Pigeon* (T.pompadora-496) Red Legs, thin grey bill and crown, green tail, maroon mantle(absent in female). In Ghats up to 1200 m.
Orange-breasted Green-Pigeon (T.bicincta-501)In Ghats, up to 1200 m.,Belgaum south. Red legs, grey tail with subterminal dark band. Breast lilac and orange banded in male, yellow in female. Keeps to the canopy.

Female.　　　　**Bhil Kavda, Pachoo Kavda**

Description: Shiny bronze-green above, vineous brown below with a dark stubby tail. Male- white shoulder patch, forehead and crown; black and white bands on rump. Chestnut underwings with dark primaries helpful when seen dashing through the forest at great speed. Flocks gather at water-holes at dusk. The beauty of this dove, the emerald of the Ghats, is seen at its best, when rays of the sun reflect from it's glistening back!

Mountain Imperial-Pigeon.

Related Species:
*Green Imperial-Pigeon (Ducula aenea-507) Size-430-470 mm. Glossy bronze-green upperparts, tail, otherwise grey. Keeps to canopy of well wooded areas of the Ghats and foothills of Kokan. Rare.
*Mountain Imperial-Pigeon (Ducula badia-511) Western Ghats, Dandeli south up to 2000 m. A rare canopy pigeon. Size-430-510 mm. Brownish above with dark tail band at base. Head, neck and underparts are grey.

Emerald Dove

Bronzewinged Dove

Columbidae

Chalcophaps indica (542)

Raktakantha, Shukachavi (Green like a parrot)

Size: 270 mm.　　M / F : Dimorphic　　**R, LM**

Distribution : Western Ghats up to 1800 m.

Nest-site: Shrubs, trees or in bamboo clumps.
Material: Leaves, grass, twigs.
Parental care: Both.
Eggs: 2,pale yellowish to muddy. 27 x 21 mm.
Call: Low, mournful, deep *Khoon, khoon.*
Ecological notes: Frugivorous forest dove. Gleans fruits, berries, seeds from ground.
Cultural notes: In the Vedic times this dove was considered inauspicious, since it was thought to be a messenger of Yama, the lord of death. Like fire, it symbolized destruction (Shiv-Puran 500AD). It was linked with fire probably due to it's vinous breast and chestnut underwings, which are usually seen in flight.
Status: Rare.

Emerald Dove (Male)
Green Imperial-Pigeon.

Rose-ringed Parakeets near the nest. Much aggression is seen prior to acquisition of the nest hole.

A pair of Rose-ringed Parakeets.

Plum-headed Parakeet (M)

Rose-ringed Parakeets at the night roost.

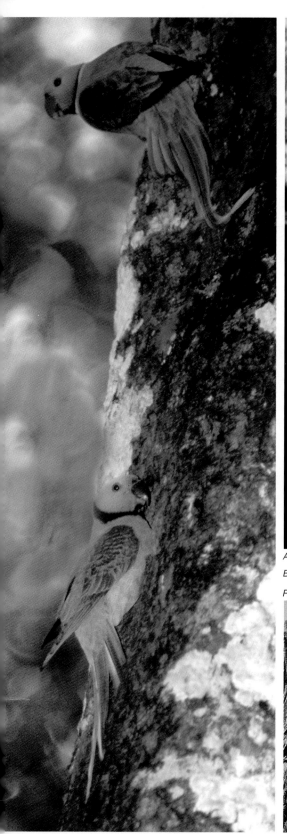

Alexandrine Parakeet.

Blue-winged Parakeets near the nest (F & M)

Rose-ringed Parakeets mating.

Rose-ringed Parakeet

Parakeet

Psittacidae

Psittacula krameri (550)

Kashthashuka, Chiri

Size: 420 mm.   M / F : Dimorphic   **R**

Distribution : Kokan, Malabar and the Western Ghats, in deciduous forests and near human habitations.

Nest-site: In secondary or self-excavated holes in live or dead tree trunks, wall-holes, crevices under roofs, holes in rocky cliffs. *Feb. -May.*

Material: Tail-feathers, saw-dust, excreta. The saw-dust falls in the nest-hole while boring, and a handful of it is seen in a single nest.

Parental care: Both. Parents guard the nests from marauding crows, hornbills, etc.

Eggs: 4-6,white. 29.3 x 24 mm.

Call: Loud calls *Keeak, keea, keeak* uttered in flight. Mumbling sounds are produced. Can recite words taught to it.

Ecological notes: Highly destructive to fruit groves and crop. Wastes and damages more than what is actually eaten.

Cultural notes: Still a popular cage bird, though banned. Can be trained to talk and perform tricks. It is used by certain Jyotishis (astrologers), to pick the right card, from which future is predicted! In poetry and literature, similies are drawn with the caged life of the parrot, when describing human confinement!

Status: Common.

Popat, Raghu, Keer

Description: Flies with agility and swiftness, hurtling through the air, between trees and houses. Female lacks the rose-pink, black collar of the male. Perches on narrow ledges on vertical surfaces with the tail as a strut, often upside down. Flocks roost in large leafy trees at night. Male feeds the female prior to pair formation.

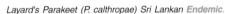

Layard's Parakeet (P. calthropae) Sri Lankan Endemic.

Such intentional nest holes were offered by our ancestors as a part of service to birds and nature.

Plum-headed Parakeet

Tooiya, Rangit-Popat

Description: Male: plum-red head, Female: grey head, no neck collar. Tail white- tipped bluish green and upper mandible yellow in both. Flocks are seen on flowering, fruiting forest trees. Pods, flowers and grain are devoured. These slender birds fly at great speed zigzagging through the forest canopy.

Plum-headed Parakeet male attracted to a caged female. This female was then successfully released.

Alexandrine Parakeet eating soil.

Alexandrine Parakeet.

Blossomheaded Parakeet

Psittacidae

Psittacula cyanocephala (558)

Krushnaangshuka, Krushnottamaang

Size: 360 mm.　**M / F :** Dimorphic　**R, LM**

Distribution : Throughout the region up to about 1300 m.

Nest-site: In self excavated or secondary tree-hole in *Acacia* spp. *Tamarindus indica, Mangifera indica* etc. The entrance hole is surprisingly small in diameter. The egg chamber is bigger. *Jan. - May.*
Material: Wood shavings, saw-dust, parrots tail-feathers, droppings.
Parental care: Both.
Eggs: 4-6, white, round. 24.9 x 20.2 mm.
Call: The typical interrogative Tooi, tooi ? is uttered by the speedily flying flock.
Ecological notes: Partial to forested areas and groves amidst farmland. The smaller nest entrance hole prevents insertion of the smallest human hand, hence the nest-bearing branch is broken to collect the chicks.
Cultural notes: Parakeets were once trained to recite Buddhist maxims and such birds were called **Shakya-shasan-kushal** ! Training parakeets was an acclaimed trade. Many ancient literary works like Kathasaritsagar sing praises of these beautiful birds. It is still a popular cage bird.
Status: Uncommon.

Alexandrine Parakeet in nest.

Related Species: *Alexandrine Parakeet (P.eupatria-545, 546) Like roseringed but larger, has maroon shoulder patch. Uncommon. Female lacks collar. Loud call echoes in well wooded areas where it dwells. Race nipalensis is in North, Central Ghats. Not in Malabar.*
Look Alike: *P.himalayana-562 and P.finschii-563, have grey head and red bill, but female P.cyanocephala has a yellow bill. The former two are not seen in our range.*

Blue-winged Parakeet

Malabar or Western Ghats Parakeet

Psittacidae

Psittacula columboides (564)

Vakratunda (For all parrots)

Size: 380 mm.   M / F : Dimorphic   **R**

Distribution : Entire Western Ghats, Mumbai south, from 500 m up to 1500 m.

Nest-site: A secondary tree hole. Usually a barbet or woodpecker nest is used. *Jan. - Mar.*
Material: Tail-feathers, droppings, saw-dust are seen in the nest-hole.
Parental care: Both.
Eggs: 3-5, white (As in all hole-nesters). 28.3 x 24.5 mm.
Call: Noisy forest bird. Screams uttered on the wings. Noisy flocks speedily navigate through tree-tops, in frantic bursts of activity.
Ecological notes: Endemic. In our area it is seen in the Western Ghats and also in foothills.
Cultural notes: It is kept as a pet by tribals residing in the Ghats, and can be seen in cages hanging outside the hamlets of hill men! This parakeet is not an accomplished talker.
Status: Uncommon in our northern range, common in the Southern Western Ghats.

Neelpankhi Popat

Description: Glossy blue-green. Black collar in male, absent in female. Both blue winged, yellow tipped blue tailed. Large trees near old forest clearings are used for roosting. Juveniles are seen in the flocks from June onwards. Flowering *Erythrina, Salmalia, Butea* are their favourite haunts and flower petals are also eaten. This frugivorous bird feeds insects to the chicks.

Blue-winged Parakeet chicks.

Blue-winged Parakeet.

Related species: *Indian Hanging-Parrot or Lorikeet (Loriculus vernalis-566)Male has a blue throat patch. Crimson rump is absent in immatures. In Kokan, Malabar, Ghats up to 1800 m. This sparrow sized parrot is called **Patrashuka** and **Parnashuka** (Leaf Parrot) in Sanskrit. It breaks open hard nuts and hence is called **Bhedashi**! Call 'Chee, chee'gives away the presence of this cryptically coloured bird which is otherwise seen only in flight. They roost at night, inverted like a bat. Nest in holes.*

Indian Hanging-Parrot.

Chatak, Sarang-Chatak

Description: White wing patch and white tipped black tail are seen in flight. Parasitic cuckoo; shows zygodactyly, with first and fourth toes facing front, second and third pointing behind. Courtship display is seen after their arrival on surprisingly exact dates in successive years. Keeps to trees but also descends to the ground. Contrary to the popular belief that they drink only rainwater, ground water from puddles is taken like any other bird.

Related Species: *Red-winged Cuckoo (Clamator coromandus -569) Patchily in winter to our forests. Crested head and chestnut wings. Long tail, white collar and belly. Size-470 mm. *Red-winged Cuckoo.*

Pied Crested Cuckoo

Pied Cuckoo

Cuculidae

Clamator jacobinus (570,571)

Chatak, Divaukas

Size: 330 mm.   M / F : Alike   **MM**

Red-winged Cuckoo.

Distribution : The entire region up to 2000 m.

Race *serratus* in Monsoon to North Ghats.

Nest-site: Nests of the foster parents. *Turdoides* babblers are parasitized. Such nests are in sparsely forested, scrub Country, usually away from habitation. *May - Sep.*

Material: No extra material is added except its own eggs!

Parental care: Eggs of the fosterers are often removed by the cuckoo, prior to egg laying.

Eggs: Sky blue, round to oval, matching well with the fosterer's eggs. 23.9 x 18.6 mm.

Call: A well known pleasant call *Piu, piu, pi, pi, pee, peeu* immortalized by poets! Linked with the SW monsoon.

Ecological notes: Arrives in the Northern Ghats with SW Monsoon and departs southwards and presumably outside India in September. Chicks are fed in and outside the nests by fosters. The foster chicks are sometimes evicted by the larger cuckoo chicks. In our area fosters are jungle babblers. *(Turdoides striatus)*

Cultural notes: Features in poetry and religious texts where this bird is wrongly thought to sustain itself only on water from the falling raindrops. Symbolized as eternally awaiting the thirst quenching rain. A Philosopher's quest for enlightenment is compared to the eternal expectation of the *Chatak* for rainfall.

Status: Common.

Red-winged Cuckoo.

Brainfever Bird

Common Hawk-Cuckoo

Cuculidae

Hierococcyx varius (573)

Varshapriya

Size: 340 mm.   M / F : Alike   **R, LM**

Distribution : Kokan, Malabar and the Western Ghats up to 1400 m.

Nest-site: In our region, in the nests of large grey babblers. *Mar. - Jun.*
Material: None is added.
Parental care: Transferred to the fosters.
Eggs: 1, blue, like *Turdoides* babblers eggs. 26 x 20 mm.
Call: In late summer, loud calls *Pee-ah, peerte-whaa, peerte-whaa* in a rising crescendo repeated monotonously. Silent in winter. *Pee-kahan*, a hindi syllabalisation, meaning where is my love?
Ecological notes: Parasitic cuckoo. Causes damage to fruit orchards. Insects eaten.
Cultural notes: This is the famous Papiha, the love lorn bird of Hindi literature. Marathi name Pawsha deciphers the bird's call Peerte-wha, uttered before SW monsoon, as if telling the farmers to sow seeds in time.
Status: Uncommon.

Brainfever Bird Juvenile.
Large Grey Babbler has been observed
feeding Juvenile Brainfever Bird on the ground.

***L**arge Hawk-Cuckoo*
(Hierococcycx sparverioides-572)
Like Brainfever Bird, larger, browner
back and broad barred tail.
Patchily in the region.

Large Hawk-Cuckoo

Pawsha

Description: Yellowish beak is straight and not curved like the Shikra, which it otherwise resembles. Calls are heard from before dawn till midnight. Encountered singly, stealthily eating fruits in orchards. Also enters urban gardens. Hawk-cuckoo chick being reared by foster Large Grey Babblers is documented. The entire babbler sisterhood was seen feeding the chick on the branches of trees and also on the ground (Amit Pawashe, Satish Pande.).

Brainfever Bird

Male Koels-sparring.

Female's Feather.

Kokil, Kokila

Description: Male: black, crimson eyes. Female: spotted, barred, brown, pied. Female chick, when in the foster crow's nest, is more black than brown, to avoid recognition, turning brown later. Female chick being fed by crows, even after getting brown plumage has been observed. So also, a blackish-brown female chick being evicted from the nest and abandoned by the foster crows has been noticed. It may have a bearing on the intelligence of the crows. Initially the voice of the Koel chicks is similar to that of the crow. It assumes the musical note later.

Related species:*Common Cuckoo (C.canorus-578,579) WM from August to May. A bird has been seen in January,in the company of C.jacobinus. Call is 'Cuckoo,cuckoo'. *Banded Bay Cuckoo(C.sonneratii-582) Often overlooked. Call is 'Titeeti, titee, ti, ti'. Brood parasitic on Ioras, smaller babblers.(Salim Ali- Handbook)
*Indian or Eurasian Cuckoo(C.micropterus- 576) WM. Call 'Orange-pekoe' is rarely heard. All are seen in our entire range.
*Lesser Cuckoo(C.poliocephalus-581)Size- 260 mm. Like Common Cuckoo, smaller. Dark rump. *Asian Emerald Cuckoo(Chrysococcyx maculatus-586) Size-180 mm. Glossy green head, back, chin. White barred below. Female rufous head, throat, tail. *Oriental Cuckoo(C. saturatus-580) Like Common Cuckoo but buff barred belly and paler head. Winter vagrant.

Eurasian Cuckoo.

Asian Koel

Koel

Cuculidae

Eudynamys scolopacea (590)

Pik, Kuhumukh, Kuhukashtha

Size: 430 mm.   M / F : Dimorphic   **R, LM .**

Distribution : Kokan, Malabar and the Western Ghats up to 1000 m, being partial to fruiting trees and groves.

Nest-site: Parasitic. In house crow's nest. *Apr. - Sep.*
Material: None. Crow's eggs are evicted and their eggs are laid!
Parental care: Shuns the act of parenting.
Eggs: Up to 13 in a single crow's nest are known. Pale green, blotched. 31 x 23.6 mm.
Call: Silent in winter. In summer, calls from dawn to dusk and well till midnight. Male sings *Ku-oo, ku-ooo* in cresendo. Female-harsh *Kik.*
Ecological notes: Causes damage to orchards by partially eating fruits. The half-eaten fruits are later eaten by the White-eyes, Sunbirds, Flowerpeckers, Bulbuls, Ioras, etc. Altitudinal migration is seen in winter.
Cultural notes: Features commonly in poetry and literature. A custom called 'Kokilavrat', when a fast is not broken till a Koel is seen or it's call heard, has led to extensive trapping of these birds! Contrary to the common belief, female Koel does not sing.   **Status:** Common.

Asian Koel (M)

Lesser Cuckoo (Juvenile)

Asian Koel (F)

Large Hawk-Cuckoo (Juv.)

Male.   Aangara, Karunya-Kokila

Indian Plaintive Cuckoo

Grey-Bellied Cuckoo

Cuculidae

Cuculus passerinus (584)

Varshapriya, Trishankh

Size: 230 mm. M / F : Dimorphic **R, LM**

Distribution : Kokan, Malabar   *Drongo-Cuckoo*

and the Western Ghats up to 1800 m.

Nest-site: Nests of Prinias, Cisticolas and Orthotomus genera are parasitized. *May - Sep.*
Material: No new nest material is added.
Parental care: The parental role ends at the point when eggs are deposited in the host's nest. This is by no means an easy task!
Eggs: Oval. 19.9 x 14 mm.
Call: Silent in the non-breeding season and hence overlooked. Call *Pi, pi, pee, pi, piee,* is given by the male with tail depressed. The silent female lurks near a host's nest, awaiting an opportunity to lay the eggs clandestinely! Another ventriloquist call is *Piteer, piteer.* Calls incessantly during the breeding season, at night and on cloudy days.
Ecological notes: Fosters observed are Tailor bird, Purple Sunbird, Ashy Wren-warbler. These smaller birds feed the larger sized chick rather fearfully, as the chick aggressively demands food! Foster eggs, chicks are evicted.

Description: White under-wing spots are seen in flight. Female is dimorphic; dark chestnut above with fine bars on white underparts in hepatic phase. More heard than seen. The patience of this bird is noteworthy. A female has been observed to stay concealed in a dense bush near a Tailor- birds nest from noon till late evening, awaiting the right opportunity to lay the egg, which never came!

Tailor Bird feeding the Indian Plaintive Cuckoo.
Note the aggressive demand of the large sized chick.

Drongo-Cuckoo.

Related Species: **Drongo-Cuckoo (Surniculus lugubris-588) More like Drongo but with a shallow fork to its tail. Underside of tail is white-barred and undertail-coverts are white, but these are difficult to observe on the field, hence is passed for the Drongo. Beak is cuckoo like. Uncommon in our region.* **MM.** *Secretive. Call is 'pip-pip-pip'. Brood parasitic on Drongos, Babblers, Minivets, etc. Black Drongo feeding the Drongo-Cuckoo is observed (*S.Pande, A.Pawashe*).*

Eurasian Cuckoo (M)

Lesser Cuckoo (F)

Grey-bellied Cuckoo (Juv.)

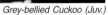
Eurasian Cuckoo (M & F)

This juvenile male Koel was evicted from the nest by crows. It was nursed & released.

Asian Emerald Cuckoo (M) is a vagrant to our region.

Female Sunbird is feeding a Plaintive Cuckoo chick. The chick is larger than the adult foster parent.

Violet Cuckoo
Chrysococcyx xanthorhynchus.

Crow fostering a female Koel.

Male Purple-rumped Sunbird feeding a foster Plaintive Cuckoo.

Drongo-Cuckoo.

Small Green-billed Malkoha

Bluefaced Malkoha

Cuculidae

Phoenicophaeus viridirostris (595)

Kairat, Gelat

Size: 390 mm.   M / F : Alike   **R, LM**

Distribution : Kokan and Malabar foothills

profuse with *lantana* and *euphorbia* bushes.

Patchily in the Western Ghats up to 1000 m.

Nest-site: Bamboo clumps, *euphorbia* and *lantana* thickets, 1-3 m. above the ground. *Mar. - Jun.*

Material: The unlined untidy shallow nest is almost entirely made of thorny twigs.

Parental care: Both.

Eggs: 2-3, chalky-white. 29.4 x 24.8 mm.

Call: This silent bird is often overlooked. A low croak like sound *Kraa* is occasionally uttered.

Ecological notes: Non-parasitic cuckoo. Preys upon eggs, chicks of smaller birds.

Cultural notes: The skulking habit of this bird is described in the old texts like 'Ratnavali'. The name **Gelat** features on the fifth pillar edict of Ashok, at Delhi-Tapora.

Status: Uncommon. Rare in north Kokan.

Malkoha, Mungshya (For Sirkeer Cuckoo)

Description: Warty blue patch around the eyes, green bill and white tipped tail help identification. Usually climbs branches and flies only from one tree to the other in short sallies. Keeps to shrubbery and seldom descends to the ground. Rarely sighted.

Sirkeer Malkoha.

Related species:

*****S**irkeer Malkoha (P.leshenaultii-598) Terrestrial, skulking, non-parasitic cuckoo. Reluctantly takes flight, prefers thorny country. Bright red beak is conspicuous. Harsh 'Kek, kek, kit, kit' Uttered. Nests during the SW monsoon. 2-3 eggs are laid. The vernacular name Mungshya means mongoose like. It predates upon lizards, eggs & small birds. Usually keeps to bushes & small trees. Often overlooked due to the habit of remaining silent.*

***R**ed-faced Malkoha (P.pyrrhocephalus-599) Rarely in south Kerala hills; no recent records. Red face patch, green graduated tail.*

Small Green-billed Malkoha.

Bhardwaj, Sonkawla, Kukkudkumbha

Description: Adults have red iris, black chestnut wings and long tail. Immature birds have a black iris. Commonly encountered in urban gardens and groves.

An injured Coucal. This is a common urban sight. Such birds can often be rescued.

Greater Coucal.

Lesser Coucal.

Greater Coucal

Crow pheasant, Coucal

Cuculidae

Centropus sinensis (602)

Kulalkukkut, Kubbha

Size: 480 mm.   M / F : Alike   **R**

Distribution : Seen throughout the region up to about 2200 m.

Nest-site: *Azadirachta indica, Tamarindus indicus, Thevetia nerifolia, Loranthaceae* members*, Mangifera indica, Bambusa arundinaceae* clumps at 10-15 ft. *May - Oct.*

Material: Twigs, palm leaf-shreds, grass, plastic.

Parental care: Both.

Eggs: 3-4, chalky white. 36.2 x 26.3 mm.

Call: A resonant deep *coop, coop, hoop, hoop.* Often in duet. Also utters croaks and chuckles.

Ecological notes: Adult birds predate the chicks and eggs of smaller birds. Immature birds are prone to collide with vehicles and are found injured on streets due to clumsy flight.

Cultural notes: The sighting of the coucal is considered auspicious, specially on the 'Padwa' day. The mythical goldmaking philosopher's stone, 'Parees', is thought to be present in the active nest of the coucal, according to a superstition in Kokan. Many active nests are therefore destroyed in it's futile search!

Status: Common.

Greater Coucal (Juv.) Iris is not red.

*Green-billed Coucal (C. chlororhynchus) Sri Lankan **Endangered** Endemic.*

Related Species: **Lesser Coucal (C.benghalensis-605) Rarely in southern Ghats up to 900m. Tail is white- tipped, scapulars are buff streaked. Size-350 mm. Call -'Whoot, kurook'.*

Barn Owl

Tytonidae

Tyto alba(606)

Ghargharak, Chandrak, Shwetolook

Size: 360 mm.   M / F : Alike   **R**

Distribution : Kokan, Malabar, Western Ghats up to 1000 m. One of our most familiar owls.

Nest-site: In natural tree holes, old fortifications, unused lofts, bell towers, behind arches, old temples, wells, ruins, barns, etc. *Feb. - Jun.*
Material: None. Nest is filthy due to the droppings and pellets of the chicks, containing bones, hairballs, etc. of the devoured prey.
Parental care: Both.
Eggs: 3-4, white, round. 40.7 x 32.5 mm.
Call: The screeching quality of its call rightly gives the names screech-owl and **Ghargharak**. Hungry chicks incessantly screech, begging for food all the night, and this becomes a major nuisance, when the nest is in residential localities.
Ecological notes: In urban areas, these owls are seen close to the garbage depots infested with rats, their staple diet. Immature barn owls, either exhausted or injured by crows, dogs are often encountered and respond to nursing.
Cultural notes: It features in the epic 'Mahabharat', as the destroyer of rats, a correct ecological observation! **Chandrak** links it to the moon; **Shwetoluk**- a pale owl.
Status: Common.

Gawhani Ghubad

Description: The heart-shaped flat face, peculiar head gyrations and golden-buff colour are characteristic. Chicks are white balls of feathers. Inspite of their cute appearance, when found, are cruelly killed in rural areas even today, due to the rampant superstitions. A screech owl being attacked by a pair of Brahminy Kites has been reported. (Bombay Gazetteer vol.10,1880, reprinted,1996)

Barn Owl adult & chicks.

Barn Owl juveniles in the nest.

Oriental Bay-Owl.

Related Species: *G*rass Owl (T.capensis-608) Prominent black medial eye spots, cross barred tail, feathered legs longer than Barn Owl's, brown colour and buff chest. In Nilgiris. Vagrant in northern area. One record from Chiplun*(V.Katdare, Per. Com.).* ***O***riental Bay-Owl (Phodilus badius-610a) 290 mm. Eared chestnut owl, spotted, rufous around eyes, pinkish belly and barred tail. Seen in Anaimalais and Nelliampathy Hills.*

Collared Scops-Owl

Indian Scops Owl

Strigidae

Eurasian Scops-Owl

Ottus bakkamoena (623)

Urdhwakarnolook (Owl with upward ears)

Size: 230 mm.　　M / F : Alike　　**R**

Distribution : Kokan, Malabar, Western Ghats up to about 1200 m.

Nest-site: Hollows in ancient secluded trees, under sloping roofs, lofts in unused or dilapidated buildings.
Material: None. Nest typically filthy.
Parental care: Both.
Eggs: 3-5, white, round. 31.8 x 27 mm.
Call: Usually silent in the day time. At night the presence of this secretive bird is betrayed by a hooting call *Wut, wut, wut.*
Ecological notes: This owl is a devourer of reptiles, rodents and insects.
Cultural notes: Owl feathers were used for lining the shoes and clothes by the ancient Buddhist monks, as a protection against cold, when they traveled to the colder areas in the Northern Himalayas! Many owls were killed to meet this demand and hence people were aware of various kinds of owls, their dwellings and calls. In the Indo-Aryan and Epic periods, owls were considered to be birds of ill omen.
Status: Uncommon.

Shingla Ghubad, Kutruz, Kuta

Description: Delicate 'ears' or horns with a pale collar at the base of the hind-neck, upper back. Grey-rufous brown, timid owl with white barred throat. Tarsus feathered up to the toes.

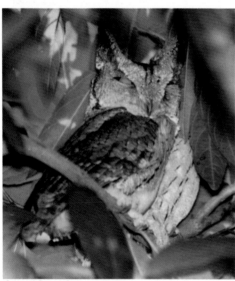

Oriental Scops-Owl.

Related species: **Oriental Scops-Owl (O.sunia-616) Call-'Wut,-chu-chraii'. Lacks collar. Slim-eared owl. Has grey, rufous-cinnamon and bay phases. Entire range.
Pallid Scops- Owl (O.brucei-614) Call-'Whoop,whoop'. Fewer bars on the tail than the
*Eurasian Scops-Owl (O.scops-615). Latter has prominent white scapular spots. Seen in the North Ghats.

Pallid Scops-Owls (Juveniles & Adult)

Eurasian Eagle-Owl

Indian Great Horned Owl

Strigidae

Bubo bubo (627)

Huhu

Size: 560 mm. M / F : Alike **R, LM**

Zimbabwe

Distribution : Kokan, Malabar and the Western Ghats in rocky terrain, cliffs, ravines and in the well wooded areas near waterfalls.

Nest-site: Wall recesses in discarded wells, ledges on cliffs, old stone quarries, earth cuttings, under thorny bushes, cactii near waterfalls. *Dec. - Apr.*

Material: Bones, hair-balls from the vomitus. No intentional nest lining.

Parental care: Both.

Eggs: 3-4, white, oval, laid on the bare ground. 53.6 x 43.8 mm.

Call: The scary, resonant call *Bu-boo,bu-boo* is heard over long distances at night.

Ecological notes: An extremely useful bird to agriculture, being a devourer of vermin.

Cultural notes: The very sighting of this bird is considered to be an ill omen and generates fear, even amongst literates! Wrongly thought to kill the attacker by holding and rubbing the missile like a stone thrown at it, by slowly rubbing it on the ground, thereby slowly draining assailant's strength. The belief has conferred immunity to the bird! Features in religious literature.

Status: Rare.

Owl attacking a kite.   *Prey of the Eagle-Owls.*

Related Species: *Forest Eagle-Owl (B. nipalensis-628)* Western Ghats, Goa south Up to 2100m. The night hoots, screams are terrifying, hence called the 'Devil Bird'. This large owl can prey on peafowl, jackal and hare. *Dusky Eagle-Owl (B.coromandus-630)* Northern Ghats. A resonant 'Wo, wo, oo' is heard day and night. Has yellow eyes and is greyer than Bubo bubo.

Shrungi Ghubad

Description: Feathered legs and orange iris. Noiseless flight. Ear-like tufts on the head are made of feathers. Often attacks intruders near the nest. Usually not seen till it takes flight often from a close quarter, surprising the observer. Comes back to the same spot when the intrusion ends. A lethal attack on a peafowl is reported.(Bombay Gazetteer,1880)

Eurasian Eagle-Owl & Forest Eagle-Owl (Juv.)

Dusky Eagle-Owl in an open nest.

Hooman, Matsya Ghubad

Description: A large owl with tufted head, vertically streaked underparts, unfeathered legs and yellow iris. By day it roosts in ancient trees on riverbanks, near graveyards, in dense waterside vegetation or ground holes. Hunts by night, rarely in daylight. Not disturbed easily. Pair starts hooting with sundown.

Brown Hawk-Owl.   *Brown Fish-Owl,Juv.*

Brown Fish-Owl.

Brown Fish-Owl

Strigidae

Ketupa zeylonensis (631)

Konth, Kotth

Size: 560 mm.   M / F : Alike   **R**

Distribution : Patchy. Kokan, Malabar Western Ghats up to 1400 m. Prefers vicinity of water.

Nest-site: Holes or ledges in steep cliffs or in a tree hollow or in an unused nest of a raptor. *Dec. - Mar.*
Material: Lined with sticks.
Parental care: Both.
Eggs: 2, white. 58.4 x 48.9 mm.
Call: *Hum, hoom, hoom, boom* is uttered before dawn and after twilight. Has a loud, resonant and ventriloquist quality.
Ecological notes: Keeps to thick foliage near water bodies. Not entirely dependent on aquatic food; also takes reptiles and birds.
Cultural notes: The large hooting owls are known to us since ancient times.They were called **Konth, Ghook** (due to their hooting sound), **Kaushik** (one residing in holes), **Pechak** (brown coloured, like baked mud).
Status: Uncommon.

Brown Hawk-Owl.

Related Species:*Brown Hawk-Owl (Ninox scutulata-644) Size-320 mm.This hawk like dark brown owl has red-brown spotted white belly, yellow iris and barred tail. Keeps to the canopy of the Ghat forests. Call- 'Oo-uk, oo-uk' is repeated. Vocal when breeding. Rare. Race lugubris is seen Mumbai north, hirsuta southwards.*

Brown Wood-Owl

Strigidae

Strix leptogrammica (659)

Kumbholuk, Ghordarshan

Size: 530 mm.   M / F : Alike.   **R**

Distribution : Patchy. Western Ghats up to 1800 m. Mahabaleshwar south up to Kerala.

Nest-site: Rock ledges, cliffs or rarely on ground under a boulder or hidden in the exposed roots of a large forest tree. *Jan. - Apr.*
Material: Sticks, twigs.
Parental care: Both.
Eggs: 2, white. 49.9 x 44.1 mm.
Call: Loud screeches and shrieks. Chuckles are uttered. Call *Hoo-hoo-hooo, tok, to.* Beak snapping noise is emitted. Shrieks of the owl echoing in forest ravines in the dead of night is a chilling experience.
Ecological notes: Controls the rodent and snake population. Keeps to the forested tracts of the Ghats.
Cultural notes: Glimpse of this owl generate fear in the minds of people, hence the name *Ghordarshan*! The sanskrit name **Kumbholuk** appears in the epic Mahabharat, meaning an owl with an arcuate, elephant like head.
Status: Uncommon.

Mottled Wood-Owls.

Mottled Wood-Owls (Juv.)

Laltondya Ghubad

Description: Barred abdomen and white half neck collar. The facial disc in our species is rufous and lacks concentric rings. Crown is unmarked. Adult facies is attained after a few months. The chick is white. An orphaned chick has been successfully rescued and reared in captivity. It was first fed minced meat, later dead squirrels. Live rats and frogs were given to develop the hunting habit (Subhash Puranik, Per. Com.). A shy owl.

Mottled Wood-Owl.

Related Species: **Mottled Wood-Owl (S.ocellata-657) Facial disc shows concentric dark rings. Brown upperparts and crown are mottled with rufous. Call is an eerie 'Whaa-aa-aah, Chi-huaa'. It is seen in old groves and around cultivation in the entire region. Prefers well wooded areas in the plains. Eats rodents, pests and is a farmers friend.*

Brown Wood-Owls (Juv. & Adult)

Pingala

Description: Yellow-eyed, white-spotted, grey-brown owl. Clownish head movements while staring. Capable of considerable neck gyrations. Has forward facing eyes. Night hunting is aided by vision, noiseless flight and by the capability of picking up the slightest new auditory signals from the familiar locality. Often seen during the day, hiding in the thick foliage. Readily accepts man made nest sites.

Jungle Owlet.

Related species: *Jungle Owlet (Glaucidium radiatum-636,637) Mixed forests of Ghats and Kokan, race malabaricum Goa south. Active during the day. Call 'Kao, kao, kuk'.
*Short- eared Owl (Asio flammeus-664) WM to scrub land and Ghats up to 1400 m.,singly or in small groups. Takes a short flight when flushed, hides in bushes or grass.

Spotted Owlet

Strigidae

Athene brama (652)

Pingachakshu, Krukalika

Size: 210 mm.   M / F : Alike   **R**

Distribution : Uniformly distributed in Kokan, Malabar and the Western Ghats.

Nest-site: Holes and crevices in walls, under roofs, bridges, trees like *Tamarindus indicus, Mangifera indica, Ficus* spp. *Palm* spp. *Nov. - May.*

Material: Fibres and bark are occasionally found in a few nest holes. Often unlined.

Parental care: Both.

Eggs: 3-4, white, roundish. 31.6 x 27.4 mm.

Call: Bursts of chuckling duets *Chirr, chirr, Chirr, chluk, chuk*. Commonly heard at night near old river-side temples and ruins .

Ecological notes: Rodent controller. Often chased by crows. Eggs in shallow nest cavities are sometimes eaten by birds of prey. Injured owls are occasionally found and respond well to nursing and medical aid.

Cultural notes: This familiar owl does not excite fear. The time of dawn and dusk, when the owls are commonly seen, is called the Pingala-Vel. (The time of the Spotted Owl).The prayer-chanting holy man visiting during this time is called 'Pingala'. The Sanskrit name **Krukalika** means one who eats lizards.

Status: Common.

Forest Owlet (Heteroglaux blewitti) can be mistken for the Spotted Owlet. It was recently rediscovered. It is not recorded in our region. Status-Critical.

Spotted Owlet nesting in a unique nest.

Forest Eagle-Owl (Adult)

Forest Owlet (H.blewittti) is not seen in our region.

Jungle Owlets are often seen during the day time.

Death Head Moth.
Moths have developed
such designs with false
eyes to scare predators like owls.

Barn Owl roosting in a Palm tree.

Spotted Owlet chick being nursed prior to release.
(A.brama brama)

Short-Eared Owl.

Spotted Owlet (A.brama indica)

Chestnut-Backed Owlet.
Sri Lankan Endemic.

Malayan Eagle-Owl is not seen our region.

Grass Owl (Juvenile)
Eurasian Eagle-Owl.

Draco (Draco dussumieri). This gliding lizard uses camouflage colouration as a protection against predation.

Dusky Horned-Owl kept as a pet by poachers.

Ceylon Frogmouth

Indian Frogmouth

Batrachostomidae

Batrachostomus moniliger (666)

Dardur

Size: 230 mm.   M / F : Dimorphic   **R**

Distribution : Patchy. Western Ghats along the heavy rainfall zones up to 1200 m. Not recorded north of Sindhudurga district, Kokan.

Nest-site: Tree fork or on a horizontal branch.
Material: Leaves, grass, twigs, moss, lichen, bark. The inner nest lining is of the bird's down feathers from the under-plumage. *Jan. - May.*
Parental care: Both.
Eggs: Single, white, elliptical. 29.9 x 20.6 mm.
Call: Monotonous *Kroo, kooroo, whoo, whooo.* More vocal during the SW monsoon, when the nesting areas become almost unapproachable.
Ecological notes: Inhabit undisturbed moist evergreen forests, which are disappearing.
Cultural notes: A little known bird. The Sanskrit name ***Dardur*** literally means a frog, and probably also indicates the frogmouth. Interestingly though, in England the Nightjar is known as the 'Flying Toad'!
Status: Rare. ?Threatened. Recorded in Goa, Top Slip, Amboli and Radhanagari range (Varad Giri, per.com.); near Belgaum in the Mahadei forest (Niranjan Sant, Per. Com.).

Beduktondya

Description: Obliteratively coloured. Wide gape and swollen bill, reminiscent of a frog. Adult male- grey-brown, mottled with white, buff, brown, black. Adult female- dull rufous, upper wing coverts spotted with black bordered spots. Perch motionless on chosen branches, with the tail pointing downwards mistaken for a stump. On closer approach stretches the neck upwards, keeps a vigil by rolling the large eyes and slowly screwing the neck towards the observer! Readily answers recorded calls, a ploy used by spotters.

Ceylon Frogmouths (M,F)

Ceylon Frogmouth roosting & on the nest with chick.

Common Indian Nightjar

Indian Nightjar

Caprimulgidae

Caprimulgus asiaticus (680)

Naptruka

Size: 240 mm.　M / F : Alike　**R, LM**

Distribution : Seen patchily distributed in Kokan ,Malabar and the Western Ghats up to about 1500 m.

Nest-site: On the ground in a dried stream, on a barren hill-slope abounding in stones. *Mar. - Jun.*

Material: None. A few pebbles are sometimes seen around the eggs.

Parental care: Both.

Eggs: 2, buff, creamy-salmon, speckled. 26.5 x 19.9 mm.

Call: *Chuk, chuk, chuk-rr-rrr.* The loud resonant call is repeated from a branch or ground.

Ecological notes: Moths, beetles, crickets, hemipteran bugs, winged insects are taken in flight in the wide gape. Our commonest Nightjar. Seen in scrub areas and deciduous forests.

Cultural notes: Though well known in the countryside, unknown in urban areas. Due to the wide gape, it was falsely thought to be capable of sucking milk of grazing goats and was called 'Goatsucker' in Europe! Its Sanskrit name **Naptruka** is aptly descriptive of the dipping flight while landing on the ground. The name literally means one who falls like a stone, only to arise!

Status: Common.

Ratwa, Ratrinchar, Chakwa

Description: A small nightjar with white side-throat patches, black streaked crown, buff collar, boldly marked scapulars, white wing patches and white distal outer tail feathers. Often seen on forest paths at night, their eyes shine red like that of a hare, in the headlights of a vehicle, hence called Chakwa-one who misleads! A collision with the vehicle is miraculously avoided due to the birds ability of suddenly changing the course of its flight! The nightjar is capable of flapping the wings in the reverse manner. The flight is noiseless.

Indian Nightjar's nest with salmon coloured eggs.

Jerdon's Nightjar (F)

Jerdon's Nightjar (M)

Related species: **Jerdon's or Longtailed Nightjar (C. atripennis-676) Medium sized, large-headed,long and dark tailed bird. Rufous nape and central white throat patch. Feathered legs. Three primaries spotted. Call 'Chukoo'. Entire range.*

Indian Jungle Nightjar

Grey Nightjar

Caprimulgidae

Caprimulgus indicus(671)

Sichapu

Size: 290 mm.   M / F : Dimorphic   **R, LM**

Distribution : Throughout the region and up to about 2300 m in the Western Ghats.

Nest-site: On a stone strewn hill slope or forest clearing on bare ground. *Mar. - Jun.*
Material: None.
Parental care: Both.
Eggs: 2, buff, spotted with brown. 30.4 x 21.3 mm.
Call: First nocturnal bird call at the same time each evening. *Kapoo, kapoo, chuckoo, chuckoo, chunk,* repeated for a few minutes at a stretch. Call is uttered on the wing or from a perch.
Ecological notes: Nocturnal feeder chiefly taking winged insects. A forest nightjar.
Cultural notes: Nightjar was offered to the deity of the Night at the Ashwamedh sacrifice in the Epic period. This ritual is also mentioned in the Yajurved.
Status: Common.

Eggs of Grey Nightjar.

Great Eared Nightjar.

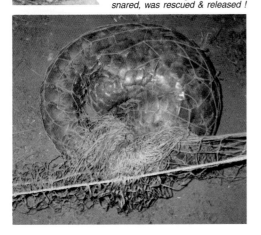

Bird watching at night can give an opportunity of helping other wild life in distress. This Pangolin was found snared, was rescued & released !

Kapoo

Description: Feathered tarsus. Three to four outer tail feathers are white spotted in male but not in female. Primaries unspotted in female. Streaked crown, scapulars. Large white central throat patch. Wide gape and long tactile bristles aid catch flying insects at night in a zig-zag flight. By day, they rest on a branch or on ground among leaves, horizontally (hence the Sanskrit name **Sichapu**). When flushed, they hurriedly alight, and settle a few meters away.

Eurasian Nightjar.

Large-Tailed Nightjar. Not recorded in our region.

Related Species: *European or Eurasian Nightjar (C.europaeus-673) WM, vagrant. White throat crescent, white distal outer tail. Call -'Quoit, quoit'.*
Great Eared-Nightjar (Eurostopodus macrotis-668-669) Family=Eurostopodidae. Resident in the Western Ghats south of Palghat Gap. Our largest Nightjar.

Pandhrya Shepticha Ratwa

Description: Unfeathered tarsus. White outer tail feathers in male, not in female, centers of primaries white spotted in male and buff in female. Both have white lateral neck patches. Also seen in scrub, deciduous country along with the savannas. During day, they roost in parties on chosen hill slopes in dried fallen leaves or on favourite branches if undisturbed.

Franklin's Nightjar chicks.

Sykes's Nightjar

Franklin's Nightjar

Savanna or Allied Nightjar

Caprimulgidae

Caprimulgus affinis (682)

Vishkir

Size: 250 mm.   M / F : Dimorphic   **R**

Distribution : Kokan, Malabar and the Western Ghats up to 1800 m.

Nest-site: On ground near a dry nullah, stony country, hill slopes, base of bamboo thickets. *Mar. - Aug.*

Material: None.

Parental care: Both.

Eggs: 2,salmon, red spotted. 30.2 x 22.1 mm.

Call: Penetrating *Chwees, chwees, sweesh, Sweesh, dheet, dheet*, more so during courtship. A relatively silent bird.

Ecological notes: Several nests are seen in the same locality. These may be inadvertently trampled under the hooves of grazing cattle or eaten by snake, mongoose, monitor lizard.

Cultural notes: In the ancient text Vishnusmruti, a cloth thief is said to be born as a nightjar! In Tamil and Telugu, the Nightjar is known as the Frogbird, due to its wide gape and the habit of sitting horizontally on the ground.

Status: Common.

Related species:
Sykes's Nightjar (C.mahrattensis-674) Small, light sandy brown Nightjar, with white distal outertail feathers, white spotted primaries and white throat crescent. Inverted anchor shaped marks on scapulars, streaked crown. Call- 'Cluk, cluk, prrr, prrr'. Rare WM, Pune south (NLBW vol.41,2,March-April,2001).

House Swift

Indian House Swift

Apodidae

Apus affinis (703,704)

Vatashin, Chirilli

Size: 130 mm.   M / F : Alike   **R**

Distribution : Throughout the region up to about 1000 m. Also on the coasts.

Pakoli, Pangli

Nest-site: In the angles of wall and roof of bridges, arches, temples, mosques, ruins, caves. Nest colonies are called nest villages. Undisturbed nests are reused. *Feb. - Sep.*

Material: Nest material is collected in the air! Feathers, flimsy grass strands are cemented by bird's own saliva to make the nest. Feathers of common myna, sparrow, parakeet, hen have been seen. Nest chamber is lined with down feathers.

Parental care: Both. Incubation-3 weeks.

Eggs: 2-4, white, oval. 22.2 x 14.2 mm.

Call: Pleasant high pitched vibrant chorus of *Sik, sik, chi, chi, chirr, chirr* uttered in flight, morning and evening and when approaching the nest.

Ecological notes: Food is captured in mid-air(winged ants, midgets, bugs, beetles, flies). In winter, due to lesser numbers of insects, birds migrate locally to warmer regions. The pest insect 'Mango-hopper' has been found in the stomach content of the Edible-nest Swift. The nests were sold during 1867-78 in Goa. Recent trade has also been exposed.

Cultural notes: The onomatopoeic name ***Chirilli*** is due to the bird call Chirr, chirr.

Status: Patchily common.

Edible-nest Swiftlet

Description: Sickle shaped wings extend behind the tail. White throat, rump. The first toe is reversible. Swift flight explains the name. They enjoy sporting glides, swings, aerobatic sallies in air as they call joyously. Copulate in mid-air. Unable to perch due to the peculiar toe structure and a grounded bird cannot rise, hence the local name 'Pangli' (lame bird). Birds cling to vertical surfaces or to each other.

White-rumped Needletail-Swift.

Alpine Swift.

Forktailed Swift.   *Brown-backed Needletail-Swift.*

Related Species: **Alpine Swift (Tachymarptis melba-694) Sahyadri cliffs. *Edible-nest Swiftlet(Collocalia unicolor-685) Nest colonies at the Vengurla Rocks. The canned Edible-nest drinks are readily available in Far East in departmental stores. *Pacific or Forktailed Swift (A.pacificus-700) Vagrant. *Brown-backed Needletail-Swift (Hirundapus giganteus-691) Western Ghats Goa south. White undertail coverts. *White-rumped Needletail-Swift (Zoonavena sylvatica-692) A small forest swift seen Goa south.*

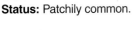

Indian Swiftlets nesting in a cave on the Burnt Island :

The egg is adherent to the outer side of the sticky nest & Chick in the nest. All four toes face anteriorly.

The reason behind nest poaching. Indian Swiftlet on the nest. The nest is more useful to the bird than to us.

House Swift

Fracture means death to this aerial bird.

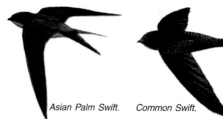

Asian Palm Swift. Common Swift.

Nest colony of the Indian Swiftlet on the Burnt Island.

Crested Tree-Swift

Hemiprocnidae

Hemiprocne coronata(709)

H.longipennis is the Grey-rumped Tree Swift

Ekputrak

Size: 230 mm.   M / F : Dimorphic   **R**

Distribution : Kokan, Malabar and the Western Ghats up to 1200 m.

Nest-site: On a bare branch of a leafless tree, and yet very difficult to notice, passed off as a wood-knot! On trees like *Tectona grandis, Anogeissus latifolia*. The astonishingly small cup is 50x30 mm. deep, and is stuck sideways to a branch about 4 to 18 m. above the ground. *Dec. - Mar.*

Material: Thin bark flakes, lichen and feathers gummed with bird's saliva.

Parental care: Both. When on the nest, the birds puff out the breast and abdominal feathers to cover the egg /chick. Female is bolder on the nest than the male, the latter abandoning it hastily on the slightest alarm!

Eggs: Single with obtuse ends. White, elongated and filling the nest completely. 23.7 x 17.1 mm.

Call: Harsh *chia, chia, chia, ti-chuk, t-chuk, whit-tuck, whit-tuck* uttered in flight or from a perch.

Ecological notes: Flying insects like *hemiptera, coleoptera* are taken on the wing.

Cultural notes: The Sanskrit name ***Ekputrak*** meaning one child is apt since the bird lays a single egg and hence has a single chick!

Status: Uncommon.

Asian Palm Swift.

Crested Tree-Swift (F) on the nest.

Male.   **Turebaaz Pangli**

Description: Non-reversible first toe. Can perch, unlike other swifts, which can only cling to the vertical surfaces. Female lacks chestnut colour on the face and throat. Erectile crest. Sickle shaped wings do not extend behind the tail which appears pointed in flight. Perch bolt upright. Perform aerobatic aerial sallies in the pursuit of winged insects. Drink water by a swift dipping flight.

Crested Tree-Swift (F) on the nest with chick.

Related species: *Asian Palm-Swift (Cypsiurus balasiensis-707) Family-Apodidae. Partial to palmyra palms. The tail is forked . H.coronata is sometimes seen with palm swifts, with whom it may be confused. It is distinguished by the more deeply forked tail than that of the latter.*

Male.   **Karna**

Description: Underparts of the female are dull orange-brown and not crimson-pink as in the male. This secretive bird keeps to dense foliage and catches winged insects in the air in an aerobatic manner. Returns to cover on catching the prey. The bright breast is seldom revealed. Crepuscular, hence infrequent sightings. Occasionally flits to the ground or clings to a vertical branch in search of food.

Malabar Trogon (F)

Malabar Trogon

Indian Trogon   *Malabar Banded Peacock.*
Trogonidae

Harpactes fasciatus(710,711)

Lohaprushtha

Size: 310 mm.   M / F : Dimorphic   **R**

Distribution : Patchy. Western Ghats up to 1500 m. Rare in Kokan and Malabar.

Nest-site: Natural holes in low forest trees. Occasionally a flimsy nest platform may be built. *Feb. - May.*
Material: Unlined. Nest is built of sticks, twigs.
Parental care: Both.
Eggs: 2-4, ivory-white, round. 26.7 x 23.4 mm.
Call: A plaintive low *too, too, too* or a whisper like *Mew, mew* or *Cue, cue*. When alarmed emits sudden *Hrr-rr-r* and then takes to wings. Moves the squarish tail in a screwing manner.
Ecological notes: Insectivorous. This forest bird is generally overlooked due to its shy and sluggish nature.
Cultural notes: In the ancient epic Markandeya Purana, a prince is advised to acquire the qualities of the Trogon (keeping to forest cover, talk in low tone, suddenly ambush the enemy and retire to cover),when on an expedition! The name ***Lohaprushtha*** meaning one with a rust-red back, is more apt for the Red-headed trogon of north India.

Malabar Trogon (M)

Small Blue Kingfisher

Common kingfisher

Alcedinidae

Alcedo atthis(723,724)

Matsyaranka, Suchitrak, Zampashi

Size: 180 mm.   M / F : Alike   **R**

Distribution : Entire region up to 1800 m.

Race *bengalenesis* in Gujarat and *taprobana*

in the Ghats.

Nest-site: Hole in an earth cutting, close to a stream, riverbank, pool or a creek. An old wellwall is rarely used. *Jan. - Aug.*

Material: None. Shed feathers, fish bones left over from the past meals make the nest stink.

Parental care: Both.

Eggs: 5-7, white,oval. 20.9 x 17.6 mm.

Call: A familiar call *Ki,ki,ki*, or *chi, chee, chi, chee* is uttered on wings as it flies close to the water surface or when hovering.

Ecological notes: Partial to clean water abounding in fish. But also frequents polluted water bodies, tidal zones and mangroves.

Cultural notes: Popular subject for many a painting. Depicted on postal stamps.

Status: Common.

Khandya, Dhiwar

Description: Jerky body and tail movements. White cheek spot. Plunges in water from a favourite perch, submerges fully, emerges with a fish held laterally in the beak. All dives are not successful. It has to work hard to get its quota of fish. Covers its favourite beat over the stream day after day flying close to the water surface. Territorial.

Small Blue Kingfisher.

Blue-eared Kingfisher.

Oriental Dwarf Kingfisher.

Related Species: **Oriental Dwarf Kingfisher (Ceyx erithacus-727) Western Ghats up to 1000 m. LM. Breeds during the SW Monsoon. One self excavated tunnel nest was 1 m. Long, with a terminal egg chamber. Another was in a hollow drain pipe in a stone wall. Often collide with glass and get injured.*
*Blue-eared Kingfisher (Alcedo meninting-725, 726) Rarely in the forests of the Western Ghats; Kokan and Goa south. Keeps near water.

Oriental Dwarf Kingfisher.
White-throated Kingfisher.

Oriental Dwarf Kingfisher also takes non-aquatic prey.
Their juveniles often collide against glass panes of
windows at night and are found injured.
They can be successfully rescued if not fatally injured.

Ruddy Kingfisher is seen along the Eastern Coast

Small Blue Kingfisher takes only aquatic prey
and hence is dependent on water.

White-breasted Kingfisher

Whitethroated Kingfisher

Alcedinidae

Halycon smyrnensis (735)

Chandrakant, Kikideewi

Size: 280 mm.   M / F : Alike   **R**

Distribution : The entire region up to 2000 m.

Nest-site: Self excavated horizontal tunnel on a vertical earth bank often away from water. Tunneling is done by tirelessly pecking at the earth bank with the stout beak. Excavated earth is pushed out with feet and the beak. *Mar. - Jul.*

Material: Littered with fish bones, droppings, shells, hence very smelly.

Parental care: Both.

Eggs: 4-7, white, round. 29.4 x 26.2 mm.

Call: Loud cackle *Kil, kil, kil, kil*, giving it the apt name Kilkilya! A harsh chatter is also given from a treetop, antenna or other perch.

Ecological notes: Often roost in the nest tunnel at night and are vulnerable to nocturnal predators like snake, mongoose, jungle-cat, etc. A *Naja naja* was recently reported to have choked over a swallowed *H.smyrnensis*. The large bill of the latter obstructed the snake's throat! (R.Whitaker,Hornbill,Misc.Notes,Apr.-June,2000)

Cultural notes: The Sanskrit name **Chandrakant** means white like the moon, an impression one gets when the bird is seen from front.

Status: Common.

Khandya, Kilkilya

Description: The white throat and wing patch are seen in flight. Turquoise wings shine brilliantly in the sunlight. Apart from the usual aquatic food, the bird has adapted well to a diet of non-aquatic creatures like lizards, mice, snails, insects, birds, eggs, hence it is seen away from water bodies.

Adult.

Immature.

The Immature Kingfisher: *Black bill, as against the red one of the adult. The white on the throat and the breast is not uniform. Immature birds are fed for a long time by the parents, even after they start acquiring food independently. ***T**wo adult birds involved in a fight, with one bird getting injured and falling temporarily on the ground, has been recorded. It flew after drinking the offered water.(C.Shete)*

Stork-billed Kingfisher.

Jalmadgu, Meenaranka

Description: Large coral red bill, black cap, white collar and rusty underparts. White wing patch is seen in flight. Fish, crabs, prawns & other aquatic animals feature in their diet.

Collared Kingfisher.

Related Species:*Stork-billed Kingfisher (H.capensis-730) On inland water bodies and coastal Kokan, Malabar, Western Ghats up to 1000 m. Large red bill is striking.
*Collared Kingfisher (Todiramphus chloris-740) Twice at Donawali tidal mangrove forest near Chiplun in Kokan (V.Katdare.Per.Com.). Feeds at low tide and is seen perched on mangroves overhanging the tidal channel. Readily flies on human approach. A colony was found in the mangrove swamp near Kelshi near Dapoli, a century ago (Gazetteer). No recent records here.

Stork-billed Kingfisher.

Alcedinidae
Halcyon pileata (739)
Dhiwar (For all kingfishers)
Size: 300 mm.   M / F : Alike   **R, LM**

Distribution : Kokan Mumbai south, Malabar on coasts, mangroves, estuaries, tidal creeks and rivers. Patchily inland. Not in the Ghats.

Nest-site: Tunnel in an earth-bank near coasts and tidal zones. *May - Jul.*
Material: Unlined terminal egg chamber at the end of about 1.5 m. long tunnel.
Parental care: Both.
Eggs: 4-5, white, round. 29.6 x 26.3 mm.
Call: A relatively silent kingfisher. A shrill *Ki, ki, ki, ki* is uttered, usually from a perch.
Ecological notes: Mainly dependent on salty and brackish waters. Sometimes seen in forests around rivers above the tidal zones.
Status: Rare.

Prawns-food.

Stork-billed Kingfisher.

Lesser Pied Kingfisher

Pied Kingfisher

Alcedinidae

Ceryle rudis (719,720)

Kshatrak, Kachaksha, Matsyarank

Size: 310 mm.   M / F : Alike   **R**

Distribution : The enrire region up to 1500 m.

Race *travancorensis* occurs Calicut south.

Nest-site: Self excavated tunnel in an earth bank, close to jheels, canals, pools, rivers, etc. *May - Jul.*

Material: Unlined but littered with fish bones, excreta and hence smelly.

Parental care: Both excavate the tunnel and feed the chicks.

Eggs: 5-6,white,round to oval. 29.9 x 21.4 mm.

Call: A harsh, high pitched *Chirruk, chirr-chirr* is uttered on the wings. Very vocal during breeding. Also, calls joyously when a fish is taken.

Ecological notes: Roost in nests and may fall prey to nocturnal predators. Nests near the water-line are known to get drowned in floods. Nest-bearing walls sometimes suffer land slides thereby burying the nestlings, eggs. Ground leeches attacking chicks is recorded (JBNHS 41(1):173,1939).

Cultural notes: The Sanskrit name *Matsyrank* means one dancing in the air for fish! *Kshatrak* means one suspended in the air, and *Kachaksha* means one with brown eyes; all accurately descriptive.

Status: Common.

Kawdya-Dhiwar, Bandya, Disa

Description: Male has double, female single gorget on breast. When fishing it hovers over the water, bill pointing downwards, shuffling in the air, till a fish is sighted, when it plunges headlong with wings and eyes closed. If successful, it emerges with the prey held in the beak. Fish and tadpole are eaten on the wing. Larger prey is taken to the favourite perch, battered and swallowed head first. About every fifth dive is successful. Bobs while perching. Flight speed is about 49 km/hr.

Male.

Female.   *Male.*

Related Species: *Greater Pied Kingfisher (*C. lugubris-717*) is seen in the Himalayas and NE India. It is much larger and lacks the white supercilium of the Lesser Pied Kingfisher.

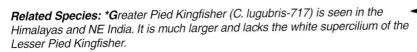

Veda Raghu, Bahira Popat

Description: Blue throat, rusty head, neck and long tail pins (lost in the moult). Chick has a yellowish neck. Flight is an aerobatic sally on outstretched wings. Captures apparently invisible bees from the air, in its curved beak, comes to a favourite perch, batters the bee and then devours it. The bee-eater with a colourful butterfly held in the beak is an interesting sight. Roost in large congregations huddled on a favourite perch.

Small Bee-eater chicks. Such chicks are exposed when earth banks with nest holes are flattened in road widening operations during summer.

Apart from the bees other insects are also eaten.

Small Bee-eater

Green Bee-eater

Meropidae

Merops orientalis (750)

Divyak, Sharg

Size: 210 mm.   M / F : Alike   **R, LM**

Distribution : The entire region up to about 2000 m.

Nest-site: Self excavated holes in earth-cuttings on hillocks, uneven ground, sand bank, often by the roadside. *Feb. - Jul.*

Material: Lined with grass, feathers and leaves.

Parental care: Both actively excavate the nest, incubate and feed the chicks.

Eggs: 4-7, white, round to oval. 19.3 x 17.3 mm.

Call: Call is a pleasant *tit, tit, tit, tree, tree, tee* given from a perch, after a good catch.

Ecological notes: Devour *Hymenopterus* insects on a large scale. Chicks residing in the nest tunnel may rarely survive if the nest bearing earth wall collapses. Nests by the road side are damaged due to the summer road repair activities.

Cultural notes: The vernacular name Veda Raghu, is derived from Sanskrit **Ved-Raghu** (Vedic parrot, a bird known to us since Vedic times). In the epic Mahabharat, prince Abhimanyu's charriot-flag bore the insignia of a golden bee-eater. Inspite of disturbance, the bird returns to the same perch and hence it is called a deaf parrot-Bahira Popat, in Marathi.

Status: Abundant.

Look Alike: The immature ***M**.orientalis may be confused with the young Chestnutheaded -M.leschenaulti, in the overlapping range. Head and back of the latter are brighter chestnut and the neck is more yellow.

Chestnut-headed Bee-eater

Meropidae

Merops leshenaulti (744)

Bramhavaadin

Size: 210 mm.   M / F : Alike   **R, LM**

Distribution : Patchily in the Western Ghats up to 1500 m. Kokan and Malabar.

Nest-site: Self excavated tunnel in an earth bank near a forest lake or stream. Singly or communally. *Feb. - Jun.*
Material: Unlined.
Parental care: Both.
Eggs: 5-6,white. 21.7 x 19 mm.
Call: Sharp chirping notes like *Te, tew, Tu-rit* are uttered in flight. Call can be mistaken with Bluetailed Bee-eater's.
 Ecological notes: Prefers forested areas with moderate rainfall and always keeps near water. Very patchily in our area.
Cultural notes: A bee-eater taking refuge in a rat-hole, is mentioned in the Vedic literature. The bee-eaters were called Prayer-Chanters **(Bramhavaadin)**, in the epic Mahabharat. The perfection with which they catch the bees, gives them the name of **Shaarg-** an arrow.
Status: Rare.

Peetkanthi
Veda-Raghu

Description: Bright yellow chin, throat, chest-nut head, nape and blunt tail. This resplendent bird hawks winged insects over water and returns to the favourite perch. Night roost in reed-beds and trees is communal. Juvenile resembles immature Green Bee-eater, but is larger, has rufous tinge to the throat, and a dark gorget on the neck.

Blue-tailed Bee-eaters.

Blue-cheeked Bee-eater.

Blue-tailed Bee-eater.

Related Species: **Blue-tailed Bee-eater*
(M.philippinus-748) Chestnut throat, ear-coverts, blue rump, tail and green forehead,
supercilium. Hardly any green below the eye mask. WM. Size-300 mm.
****Blue-cheeked Bee-eater** *(M.persicus-747) Rare summer visitor to our northern range, in arid areas near water. Turquoise ear-coverts, supercilium, chestnut throat and green tail.*
*Blue-bearded Bee-eater (Nyctyornis athertoni-753) Patchy. In the Ghats up to 1700m. Beard is seen in silhouette. Heavy, curved beak, large size conspicuous. Eats insects and nectar. Size-360 mm. R.

Chestnut-headed Bee-eater.

Blue-tailed Bee-eater.

Blue-cheeked Bee-eater in flight.

More than sixty Green Bee-eaters
were rescued by these girls after
a hail-storm. They were offered warmth
& protection overnight.
All of them flew to freedom!

European Bee-eater in flight.

Green Bee-eater at the nest hole in an earth bank.　　　　Blue-bearded Bee-eater.　　Green Bee-eater N.E. race.

Indian Roller

Roller, Blue Jay

Coraciidae

Coracias benghalensis (756)

Hemtunda, Swastik

Size: 310 mm.   M / F : Alike   **R, LM**

Distribution : The Deccan, Malabar and Western Ghats up to 1000 m. October to March in Kokan.

Nest-site: Tree-hollow, wall-hole at a moderate height. *Tamarindus indicus, Mangifera indica* trees. Nesting uncommon in this region. *Feb. - Jul.*

Material: Straw, grass, rubbish, feathers.

Parental care: Both.

Eggs: 4-5, glossy round-oval, white. 35.2 x 27.7 mm.

Call: Croaks, harsh chuckles. Loud screams *Kak, kaak*, are given during their aerobatic courtship. Usually silent.

Ecological notes: Insectivorous. Eats pests like field mice, locusts, crickets.

Cultural notes: In NE India it is considered to be an incarnation of lord Vishnu and is worshipped on the auspicious Dasera day, when birds caught earlier are ceremoniously released !

Status: Uncommon.

European Roller.

Related Species:

*European or Kashmir Roller (C.garrulus-754) WM. Migrates from Kashmir to the Middle East and North Africa passing over Rajasthan, Gujarat. Stragglers to Kokan and Malabar. It has a light blue chest, neck and blue-black wings.

*Oriental Broadbilled Roller or Dollarbird (Eurystomus orientalis-758) Patchily and locally Belgaum south, in the Western Ghats. This silent insectivorous bird is rarely sighted.

Neelpankh, Neelkanth, Chassh, Taas, Dhau

Description: Two hues of blue are seen in flight. Streaked brown breast. Inconspicuous while perching, explodes into blue in flight. This experience leaves a lasting memory.

Oriental Broadbilled Roller.

Hoopii, Hudhud

Hoopoe
Upupidae
Upupa epops (763,765)
Putrapriya, Hudhud, Kathaku

Size: 310 mm.   M / F : Alike   **R, LM**

Distribution : Throughout the region up to 2000 m. Uncommon in Ghats. Populations augmented by winter migrants.

Description: Zebra markings on the wings are striking. The long beak is delicate and curved. The crest is flicked open from time to time and when the bird is alarmed. Always remain in pairs and strengthen the bond by feeding each other from time to time!

Hoopoe on the nest. The chicks have a white gape.

Nest-site: Tree-hole, wall-hole, suitable niche in an earth cutting and under thatched roofs in rural areas. Initially clean, becoming very smelly and filthy. Nest-holes near human habitations are often sealed with stones by urchins and the eggs are stolen. *Mar. - Aug.*
Material: Leaves, grass, straw, rags, plastic, paper, hair, foils.
Parental care: Female alone incubates. The incubating female is regularly and meticulously fed by the male from time to time. Chicks are attended by both. The birds are rarely known to squirt liquid excreta at human intruders. Chicks resting in the darkness of the nest hole have a bright white margin to the beaks, probably to enable the parents to locate their mouth, in the poor light.
Eggs: 5-6, white ,round. 25 x 17 mm.
Call: Soft musical run *Hoo-po, hoo-po*, the birds name being onomatopoeic.
Ecological notes: Digs soil for Insects, larvae, pupae, worms. Farmers friend.
Cultural notes: The crest of the bird is supposedly bestowed by king Solomon. Features in Sanskrit and Urdu literature.
Status: Uncommon.

Related Species : None. Hoopoe is wrongly called the woodpecker! This is possibly because of the long bill. Unlike the woodpecker Hoopoe forages on the ground and pecks at cow dung cakes etc. for insects.

Courtship feeding.

Malabar Grey Hornbill

Bucerotidae

Ocyceros griseus(768)

Vardhrinas (A casqued beak, for all hornbills)

Size: 590 mm.   M / F : Dimorphic   **R, LM**

Distribution : The Western Ghats up to1600 m. from Karnala in the north, through Goa up to Kerala in the south.

Nest-site: Secondary cavities in *Artocarpus lakoocha* etc. trees at 10m to 20m. *Mar. - Jun.*
Material: Unlined nests. Nest toilet practiced.
Parental care: Monogamous birds. The pre-hatching phase is 40 days and the post-hatching phase of incarceration is 46 days. The chicks and the female leave the nest together.
(Divya Mudappa-JBNHS 97-1,Apr.2000,15-24)
Eggs: 3-4. 41.8 x 30.3 mm.
Call: Laughter like calls are reminiscent of calls of a hand-held chicken. Also *Kya, kya* ad.lib. Squeaks, cackles announce the arrival of the bird from afar.
Ecological notes: Sugar rich fruit is mainly fed by the male in the prehatching and lipid rich fruit *(Lauraceae, Annonaceae)* and animal matter are fed in the post-hatching phases. Fruits of smaller trees and shrubs also feature in the diet due to the smaller beak size. Effectively helps seed dispersal. Nesting coincides with fruiting figs and the SW monsoon. Endemic.
Cultural notes: The flesh of the hornbills, *Vardhrinas* has been said to be a delicacy in Ravan's Lanka, in the epic Ramayana!
Status: Near threatened.

Female.   **Malbari Dhanesh**

Description: The male has a larger bright orange bill and golden iris; female has a smaller pale bill and a dark brown iris. Contrary to the name it lacks the casque. The male feeds the incarcerated female, berries etc. by regurgitating and ejecting the food from the stomach into the beak, where it is precisely held at the tip of the bill and then delicately offered! This is an exercise to be seen and appreciated!

Male.

Prey.

Common Grey Hornbill.

Look Alike: *The immature Common Grey Hornbill appears confusingly similar to O.griseus. The former however may be hardly encountered alone, separated from the parents.The bill colour is also darker.*

Dhanesh, Rakhi Shingachocha, Dhanchidi

Description: The grey casque on the flat bill (**Vardhrinas**) is smaller in the female. This long tailed, long necked bird has an awkward flight, somewhat reminiscent of an arrow stretched on a bow. Mock nesting with both parents visiting a sealed empty nest has been reported! (Niranjan Sant, Per.Com.)

On the nest with lizard & mantis.

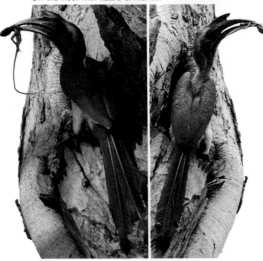

Indian Grey Hornbill

Common Grey Hornbill

Bucerotidae

Ocyceros birostris(767)

Priyatmaja, Matrunindak, Vardhrinas

Size: 610 mm.   M / F : Dimorphic   **R, LM**

Distribution : The Western Ghats Mumbai south excepting the crest line and more to the east. Occasional in Kokan and Malabar.

Nest-site: Natural tree holes in *Mangifera indica, Ficus* spp., *Adansonia digitata, Artocarpus heterophyllus*, etc. *Mar. - Jun.*

Parental care: Female takes self imposed confinement from the time of egg laying till fledging of the chicks. Male feeds her and the chicks during this period. Later the female leaves the nest and both feed the chicks till it is time for them to come out.

Eggs: 2-3, white. 41.9 x 30 mm.

Call: Varied calls like cackles, chuckles, laughter *Ki ,ki, ki* can be heard from afar. Noisy.

Ecological notes: Hornbill numbers are indicative of larger and older trees in the area. Helps in seed dispersal. Does not drink water and this requirement is met through a fruit rich diet.

Cultural notes: As per a misbelief, hidden treasure and ground water are thought to be present under the hornbill's nest trees! The bird generates awe due to the unique nesting habit. It is condemned in the ancient Sanskrit literature due to the habit of imposing captivity to the mother-bird,hence the name ***Matrunindak***!

Status: Occasional to rare.

Sri Lanka Grey Hornbill. Endemic. (Ocyceros gingalensis)

Look Alike: *Malabar Grey Hornbill *(O.griseus-768) may look similar to Grey Hornbill from a distance, but it lacks the casque on the bill, which also differs in colour. The distribution of both may give a clue unless one is in the overlapping range. The Malabar Grey Hornbill has a louder laughter like call.*

Malabar Pied Hornbill

Indian Pied Hornbill

Bucerotidae

Male.

Anthracocerous coronatus (775)

Vardhrinas (For all hornbills)

Size: 920 mm   M / F : Dimorphic **R.**

Distribution : Kokan, Malabar and
the Western Ghats up to 1000 m. Rare in
Raigad, Thane districts and northwards.

Female.   **Kakan, Kakner, Garud**

Nest-site: Existing nest holes in large trees of
the *Ficus* spp.,*Mangifera indica, Adansonia
digitata, Artocarpus heterophyllus,Tamarindus
indica,* etc. are used. Trees in forests, devrai's
and in village groves are used. *Mar. - Jun.*

Material: Unlined, but moulted feathers from
previous years nesting are seen.

Parental care: Incubation by female for three
weeks when the male feeds her. Female
leaves the nest when the chicks are three
weeks old. Both the parents feed the
incarcerated chicks for some time later.

Eggs: 2-3. 55 x 38 mm.

Call: Loud raucous *Kek, kek, kek, kek.*
Squeaks and harsh calls. A very noisy bird.

Ecological notes: Local migration depends
on the availability of fruiting Figs, papayas,
etc. Distribution of the birds overlaps with that
of *Strychnos nux-vomica* tree. Helps seed
dispersal of figs etc. Populations indicative of
old, large fruiting trees in the area.

Cultural notes: Locally well known. In Kokan,
it is called Garud,owing to it's larger size.

Status: Near Threatened. Locally common.

Description: White edge to wings and white
outer tail feathers. The casque of the female
lacks black on the rear. Iris of the male is red,
of the female brown. Orbital skin: male-black,
female-white. Flight is flaps and sail type.

Female.   *Monitor lizard features in their diet.*

*The beak of the Hornbill is used to decorate head gear.
This has significantly contributed to their hunting.*

A flock of the Malabar Pied Hornbill.

Such large & old trees offer nest holes for hornbills & hence should be protected. This Baobab tree has a nest of the hornbills since five decades.

Hornbill House - Eye of the incarcerated female, with chick & shooting out feces.
Male Malabar Pied Hornbill feeding the female in the nest.

Great Pied Hornbill

Great Hornbill

Bucerotidae

Buceros bicornis (776)

Vardhrinas (For all hornbills)

Size: 1300 mm.   M / F : Dimorphic   **R, LM**

Distribution : Patchy. The Western Ghats up to 1500 m.,south of Koyna forest in Maharashtra. Rarely in Kokan and Malabar in SWM.

Nest-site: Natural tree holes in lofty *Canarium strictum, Calophyllum tomentosum, Cullenis excelsa* etc. Same site is used if undisturbed. *Feb. - Jun.*

Material: Unlined. Shed feathers may be seen. Nest entrance is plastered with a cement of females dung, figs, leaves, sticks and wet mud brought by the male, till only a slit is left open.

Parental care: Self imprisoned incubation by the female. The female and later the chicks shoot the excreta out of the nest slit.

Eggs: 1-3. 65.1 x 45.3 mm.

Call: Loud calls like a langur's roar, if heard for the first time, may frighten the solitary listener! Grunts and resonating barks are given. The sound of flapping wings resonates in the valley, which the bird may be crossing.

Ecological notes: Large fruiting trees are a must for their existence, the destruction of which has greatly reduced their numbers. Descend to foothills in monsoon.

Cultural notes: Was extensively hunted for the tasty flesh! It has almost vanished from the north Kokan. Tribals in this area, often tied the lofty hornbill beak to their heads while hunting, as a method of camouflage.

Status: Near threatened.

Nesting in a nest box.

Male.   **Garud**

Description: Male: iris red, yellow casque with red tip and orange mid part. Female: iris white, casque red at the back. Hops and jumps on lofty trees executing the same route. Noisy flight. The white neck feathers are smeared yellow with secretions from the uropygial glands. Does not drink water, a fact well substantiated, since a captive bird 'William', Residing behind the Honorary secretary's chair, was in the BNHS for 26 years.

Female.

Look Alike:
This unique bird does not have any look alike. A new bird watcher may however at times confuse the bird with the Malabar pied hornbill(A.coronatus). The differences are clear cut if observed closely. The black band on the tail and yellow neck of Buceros bicornis are useful markers.

Kuturga, Kartuk

White-cheeked Barbet

Small Green Barbet

Capitonidae

Megalaima viridis(785)

Pippal, Manjuliyak, Vataha

Size: 230 mm.   M / F : Alike   R

Distribution : Throughout the region up to 2000 m. A common bird of our hill-stations.

Nest-site: Self excavated hole in a dry or a rotten tree trunk of *Mangifera indica, Salmalia, Palm* spp.,etc. at a height of 2-10 m. Several holes from previous years may be seen on a single trunk. *Mar. - Jun.*

Material: Unlined. Nest toilet is practiced and yet the nests are smelly!

Parental care: Both. Feeding is initially by regurgitation and later chickoo, figs, berries etc. are fed.

Eggs: 2-4,oval, white. 26.2 x 20.3 mm.

Call: One of the most well known bird calls in the forest. A loud, monotonous, at times tiring *Krrrr-rr* coming from the canopy followed by a succession of *Kutroo, kutroo* incessantly from dawn to dusk. Several birds answer each other and join the cacophony! Often a single *Tuk*.

Ecological notes: Fruiting fig trees are frequented and birds cause seed dispersal of the figs. Sometimes become pests in fruit gardens. Play a role in cross pollination of *Erythrina, Salmalia, Grevillea*, etc.spp.

Cultural notes: Chicks are robbed from the nests to make pets till date! As per an old local belief, an oil thief is reborn as a barbet, the bird's olive-brown color being the colour of oil!

Status: Common.

Brown-headed Barbet (Juv.)

Description: Diagnostic white cheek-stripe from the lores to ear covert. Feeds in the company of mynas, bulbuls, pigeons, babblers, orioles keeping to the canopy, remaining almost invisible. Very vocal in hot weather, the call booming through the forest announcing the presence of the bird to one and all.

Brown-headed Barbet.

*Related Species: *Brown-headed Barbet (M.zeylanica-781,782) has a naked orange-brown patch around the eye extending upto the base of the heavy bill. Undertail coverts are green. The call is similar to M.viridis and evokes a chorus. Birds often visit the preferred water hole, leaking tap on hot noons, to quench their thirst. Cause damage to fruit groves. More heard than seen. Race zeylanica in Kerala, inornata to north.*

White-cheeked Barbet at nest - Chick, Adult with a berry & taking away fecal material.

Feeding fruit pulp, grass-hopper & grub to the chick.

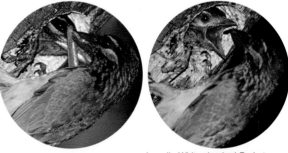

Juvenile White-cheeked Barbet.

Tambat, Pukpukya, Tuktuk

Description: A dull green bird with crimson forehead and yellow neck and streaks on the belly. The birds are seen to nest in trees in the urban areas abuzz with traffic, totally ignoring the human presence. Like in all barbets their toes are zygodactylic.

Coppersmith Barbet.

Yellow-Fronted Barbet (M. flavifrons) Sri Lankan Endemic.

Coppersmith Barbet

Coppersmith, Crimsonbreasted Barbet

Capitonidae

Megalaima haemacephala(792)

Hemak, Manjulitak, Vataha

Size: 170 mm.   M / F : Alike   **R** *Crimson-throated Barbet*

Distribution : Throughout the region.

In urban and forested areas alike.

Nest-site: Self excavated hole in a soft wooded trees. Secondary holes in any other tree. Beak- loads of excavated tree chunks are carried away by the birds during nest boring.
Material: Feathers, cotton, wool, etc.
Parental care: Both.
Eggs: 2-3, pale white. 25.2 x 17.5 mm.
Call: Repetitive, resonant, monotonous call *Puk, puk, puk, puk* commonly at noon is a well known voice in the region. It is one of the most familiar bird calls.
Ecological notes: Prefers dead trees for nesting. An important seed dispersal agent for the trees of F*icus* spp. which otherwise germinate with great difficulty.
Cultural notes: The bird call is likened to the metal hammering of the coppersmith. In the ancient Sanskrit literature it is compared to the tinkering of the goldsmith's hammer, hence **Hemak**! The name **Vataha** links the bird with the fig tree.
Status: Common.

Related Species: *Crimson-throated Barbet (M. rubricapilla-790) South of Sindhudurga in well wooded evergreen localities. Has a crimson forecrown, forehead, chin, throat and upper chest. Large flocks are seen on fruiting Ficus in the company of other barbets. At noon the birds are less active, retiring to trees. Call-rapid 'Poop, poop, poop' in a rising tempo.*
Coppersmith Barbets.

Multiple nest holes on a single tree trunk are often excavated by the Coppersmith Barbets.

Coppersmith Barbets

Orphaned chick can be hand fed or if smaller can be placed in a nest box and fed with pincers.

Singing.

Crimson-throated Barbet - Sri Lankan and Indian spp.

Coppersmith Barbet at nest.

Chitrang

Description: This obliteratively coloured barred bird has a habit of looking for ants, beetles, etc. while hopping on the ground with the tail frequently cocked. It escapes to branches when disturbed and can screw the neck side-ways in the most astounding fashion, hence the names Wryneck and Snakebird in Europe.

Speckled Piculet (Picumnus innominatus-799) A restless bird of the semi-evergreen habitat. Goa south in the Ghats up to 2000 m. Amidst mixed hunting parties of minivets, woodpeckers, barbets, babblers. Call is 'Spit, spit' and also drums on bamboo to which it is partial. It nests in bamboo holes.

Eurasian Wryneck

Wryneck

Picidae

Jynx torquilla (796,797-a)

Chitrang

Size: 190 mm.   M / F : Alike   **WM**

Distribution : Occurs in the Eastern plains of Western Ghats. Rarely in Kokan and Malabar.

Nest-site: In secondary tree holes in willow, mulberry trees in Kashmir, Europe, Russia. *May - Jul.*

Material: Unlined.

Parental care: ? Both.

Eggs: 6-9, glossy white, oval. 21.1 x 15.5 mm.

Call: Silent in winter except for an occasional *chewn, chewn.* Prior to return migration polysyllables *teer, teer* may be heard.

Ecological notes: Though grouped with the woodpeckers it does not drum like them. It picks insects from the crevices and cracks in the tree bark. Prefers forage on the ground to that on the trees.

Cultural notes: Clearly a grossly over-looked bird due to the camouflaging colour and its habit of remaining silent during the Winter, when it comes to our area.

Status: Common to the east of the Ghats and in North Kokan. Vagrant in South Kokan.

Speckled Piculets.

Yellow-fronted Pied Woodpecker

Yellowcrowned Woodpecker

Picidae

Dendrocopus mahrattensis(847)

Kashthakut, Vrukshakukkuta

Size: 180 mm.   M / F : Dimorphic   **R, LM**

Distribution : The Western Ghats up to 2000 m.
Also in Kokan and Malabar.

Nest-site: In holes of *Erythrina* spp.,*Sesbania grandiflora, Moringa oleifera, Phoenix sylvestris, Cocos nucifera, Areca catechu,* etc. Holes are self excavated in tree trunks or in broken branches. The nest hole may be used in successive years. *Jan. - May.*

Parental care: Both.

Eggs: 2-3, white. 22.2 x 16.4 mm.

Call: A prolonged harsh whistle *click, click, click,* hence the sanskrit name **Karkari**. The hammering sound resonates in the vicinity especially if a hollow trunk is being sounded.

Ecological notes: Doctors the trees by excavating and devouring the borers and insects from the tree barks. Dead and rotten trees are often used for nesting. Nests are used by other hole nesting similar sized birds.

Cultural notes: The bird is locally known as the Maratha Sutar since the male bird's red head is reminiscent of the red turban of the Maratha warriors.

Status: Common.

Maratha Sutar, Kawdya Sutar

Description: The scarlet-yellow crown of the male is absent in female. The wedge shaped tripod of the tail is used as a strut while scuttling over the trunks. Strong neck muscles prevent brain injury during repeated rapid drumming with the beak on hard wood.

Yellowcrowned Woodpecker (Juv.)

Brown-capped Pygmy Woodpecker.

Female & Male
with chick.

Related Species:
Brown-capped Pygmy Woodpecker
(D.nanus-852-853) Small pied woodpecker with a white stripe from the eye to the sides of the neck. Female lacks the scarlet hind crown of the male. Usually solitary or in pairs, keeping to the canopy. Flies from one tree top to the other after circumventing the trunks in search of insects. Call- squeaky 'click-r-r' brings this tiny bird to our notice. Up to 1200 m. Race cinereigula Coorg south.

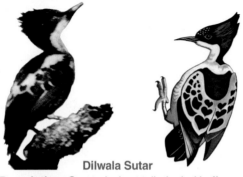

Heart-spotted Woodpecker

Picidae

Hemicircus canente(856)

Darwaghat

Size: 160 mm.   M / F : Dimorphic   **R**

Distribution : Patchy. The Western Ghats up to 1300 m. Hilly areas of Kokan and Malabar.

Nest-site: Small, self-excavated hole in a branch or a tree trunk,2-10 m. up. Dead branches are preferred. *Nov. - Apr.*

Material: Unlined.

Parental care: Both.

Eggs: 2-3, white. 24 x 18 mm.

Call: A sharp multi-syllabic call is given in flight or when looking for grub. *Kwee, kwee, tchlik, tchlik, krit, krit* etc. trilling notes utterred.

Ecological notes: A small woodpecker of the hills and mountains with forest cover.

Cultural notes: The Sanskrit name ***Darwaghat***, of the Vedic literature, means one who produces musical sounds on wood, and is apt for the woodpeckers.

Status: Rare.

Dilwala Sutar

Description: Crested, short tailed, pied buff bird. Heart-shaped spots on the wings. Crown black in male, white in female and streaked in juvenile. Moves along the branches, tapping them for termites, ants, etc. and often perches like a passerine. Partial to bamboo and teak. *Streaked Woodpecker.*

Related Species:

***R**ufous Woodpecker (Celeus brachyurus-804) Rufous brown and barred. Male has scarlet ear-coverts. The nest is excavated in the active carton-nest of the tree ants! Patchily in our area. Size-250 mm. ***L**ittle Scalybellied Green or Streaked Woodpecker(Picus xanthopygaeus -808) Greenish bird with scales on the underside. Male has a scarlet crown. Patchily in our range.Size-290 mm. ***S**mall or Lesser Yellow-naped Woodpecker (Picus chlorolophus -816) Short-crested, yellow naped bird with a barred belly, scarlet and white stripes on the cheeks and dark tail. Patchily in Ghats. Size-270 mm. Rare.

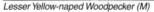

Lesser Yellow-naped Woodpecker (M)   *Laced Woodpecker.*

***L**aced Woodpecker(Picus vittatus-N) New record for India in the NE States. Differs from Streaked Woodpecker by unstreaked chest *Rufous Woodpecker.* and dark submoustachial stripe.

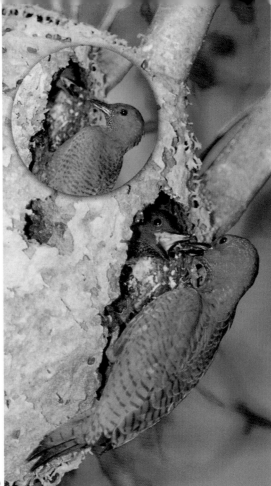

Black-shouldered Woodpecker (M)
Lesser Yellow-naped woodpecker (F)

Rufous Woodpecker on the nest of pyramid ants.
(M & F-inset)

Lesser Golden-backed.

Little Scalybellied Green (F)

Greater Golden-backed Woodpecker
of Sri Lanka-C. lucidus stricklandi.

Male.   **Sonpathi Sutar**

Blackrumped Flamebacked Woodpecker

Picidae

Dinopium benghalense(819,820,821)

Shakun, Darwaghat

Size: 290 mm.   M / F : Dimorphic   **R**

Distribution : The entire region. Race *puncticolle* to the east, *tehminae* Goa south, nominate race in the northern Ghats.

Description: In the male, the crown and the occipital crest and in the female, only the crest, are crimson. The back and the lower rump are black in both sexes (crimson in *D.javanense*).

Related species:*Common Golden-backed Woodpecker (D.javanense-825) South of Goa in the Ghats. *Black shouldered Woodpecker (Chrysocolaptes festivus-858) Deciduous forests and palm groves, is rare in the Deccan. *Greater Flameback (Chrysocolaptes lucidus-862) In the Ghats up to 1800 m. Split sub-moustachial line. Male-crimson head.

D.javanense (M & F)

Greater Flameback (F)

Nest-site: In *Mangifera indica, Albizzia* spp.,*Erythrina* spp.,*Moringa oleifera,* Palm etc. spp. holes on vertical trunks, or on the undersides of branches. If a new nest is to be excavated, boring is begun a few weeks prior to the egg laying. *Feb. - Jun.*

Material: Unlined.

Parental care: Both. Two broods are sometimes raised. A dead chick being removed from the nest by holding it in the beak has been observed! Nest toilet is practiced.

Eggs: 2-3,glossy, white. 28.1 x 20.9 mm.

Call: A harsh laugh *Chi, chi* is given in flight. High frequency drumming *Trr, Drr* is heard as the bird hammers the tree trunk with the beak. The bass of this sound depends on the hollowness of the trunk being struck. A soft call is uttered by the nestlings from within the nest hole, making it seem as if the tree itself is talking!

Ecological notes: In the Kokan, dead tree trunks of Palms are favoured for nest making. Woodpecker nests are used by mynas, tits, magpie robins, Indian robins and parakeets for secondary nesting.

Cultural notes: The names Sutar (carpenter) and **Darwaghat** are in keeping with the birds intimate association with wood !

Status: Common.

Black shouldered Woodpecker (F) V-back typical.

Great Black Woodpecker

White-bellied Black Woodpecker

Picidae

Dryocopus javensis(830)

Shakun

Size: 480 mm.   M / F : Dimorphic   **R, LM**

Distribution : Patchy. The Western Ghats up to 1400 m. Up to Surat Dangs,now rare north of Belgaum.

Nest-site: Self excavated hole in lofty trees, 6-18 m. up. The same hole is often reused. *Jan. - Mar.*

Material: Unlined.

Parental care: Both.

Eggs: 2, white. 35 x 23 mm.

Call: A loud, resonant drumming is heard at the time of breeding. Deep calls and chuckles are given in flight. Monosyllabic notes *Chiank* often repeated.

Ecological notes: Keeps to high rainfall undisturbed forests of the Western Ghats. Partial to teak and bamboo forests. The first bird to abandon the area when forest felling activity begins. Thus an indicator of undisturbed forests. Due to the large scale tree felling, this magnificent bird is likely to be threatened today.

Cultural notes: The woodpecker's calls are said to be auspicious in the epic, Mahabharat! Woodpeckers feature in the Vedic literature, Puranas, Epics and Panchatantra. In Bastar Dist. of Madhya Pradesh, the large black woodpecker is called Buffalo Woodpecker. (Bhainsa Khidree-SA Handbook)

Males on live & dead tree trunks.

Female.   **Kaala Sutar**

Description: A large, crow-sized, black woodpecker with a white rump and belly. Nape, crimson in female. Forehead, crown, crest and cheeks crimson in male. Pairs keep to the higher branches, executing the preferred beat everyday. Loud calls announce their presence in the locality. Often stay for a considerable time on a dead, decaying tree abounding with ants, termites, beetles and pupal galleries. One of the most beautiful forest birds and the largest woodpecker in our range.

Indian Pitta

Navrang

Pittidae

Pitta brachyura (867)

Chitrak, Padmapushpa, Pikaang

Size: 190 mm.   M / F : Alike   **R, LM**

Distribution : Breeds in Kokan during the south-west monsoon. Later on in the Deccan plains, Malabar and Western Ghats up to 1700m.

Nest-site: At the base of trunks and in tree forks up to 5 m. *Mangifera indica, Acacia Catechu, Terminalia elata, T. Bellirica, Garcinia indica, Acacia polycantha,* etc. trees preferred. *May - Aug.*

Material: Twigs, grass, palm leaf shreds, thread, rags, casuarina needles, pods, etc.

Parental care: Both.

Eggs: 4-6, white, thinly lined. Incubation period is 15-16 days. 24.7 x 21.2 mm.

Call: Loud double whistle. Birds answer each other *Wheetoo, Wheetue.* Vocal on cloudy day.

Ecological notes: Insectivorous. Nesting in Kokan coincides with the SWM. In late summer and early winter, when they migrate, birds may be found exhausted in urban areas of the Deccan or Malabar and can be rescued.

Cultural notes: Multicoloured plumage much admired, hence the name Navrang. Features on an Indian postal stamp.

Status: Occasional.

Description: The stub tail is wagged in a clownish manner while rummaging in the undergrowth. White wing spot is seen in flight. The Pitta is more often heard than seen.

Chick after the first flight.

Treating the exhausted Pitta.

Mangrove Pitta of NE India.

Indian Pitta with centipede.

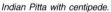

Wing mirrors. This rescued Pitta was released successfully.

Nesting In Kokan: *The nests have a lateral entrance. Nests are often predated by Coucal, Small Greenbilled Malkoha, Shikra. Pitta remains attached to the destroyed nest for a few days. Exhausted migrating Pittas when found in urban areas usually respond well to medical aid, if not fatally injured. (NLBW-41,4,July-Aug. 2001).*

Ashy-crowned Sparrow-Lark

Blackbellied Finch Lark

Alaudidae

Ermopterix grisea (878)

Dhoosarchatak, Bhoomishaya, Kooj

Size: 130 mm M / F : Dimorphic **R, LM**

Distribution : The entire region up to 1000 m.

Nest-site: The neat, delicate cup is placed in a hoofprint or a depression on open ground. Several nests are seen in a secluded area. *Mar. - Sep.*

Material: Grass, fine twigs, rootlets, thread, hair. Lined with cotton and feathers.

Parental care: Both. Incubation 13-14 days. Female does the majority of nest building.

Eggs: 2-3, white, spotted with brown.

Call: Pleasant soft notes are uttered on wings. (Sanskrit name-**Kooj**) Inspiring song, energetic notes *Wheech, wheech.* Shrill calls when diving.

Ecological notes: Insectivorous. Grasshoppers, mantids, weevils, beetles, ants, butterflies, moths are eaten. Rare in Ghats.

Cultural notes: The terrestrial habit of this lark is recognised since long, hence the Sanskrit name **Bhoomishaya**, one who rests on ground. **Dhoosarchatak** means a smoky sparrow, an apt name!

Status: Common.

Dombari

Description: Male has an ashy crown and nape, dirty white ear-coverts and black under-parts. Female lacks the black colour and is pale sandy brown. This obliteratively coloured lark squats on the ground becoming invisible, movement alone betraying its presence. Aerial manoeuvres are superbly executed throughout the day. At noon they rest under some scanty shade, panting with the beaks held open.

*Look Alike: *Blackcrowned Sparrow-Lark E.nigriceps-879) Is seen in NW India and Pakistan and is not recorded in our area. Has a white forehead, pure white ear-coverts and black crown and nape.*

Juvenile Ashy-crowned Sparrow-Lark.

Chicks.

Ashy-crowned Sparrow-Lark (M) with chicks.

Indian Fox-Predator of grassland birds.

Dongari or Malabari-Chandol

Description: Rufous brown and streaked above, black-spotted rufous breast, white belly and vent, crested head and bill longer than other larks. Hops on ground gleaning seeds or actively pecking them from grass. Insects are hawked from air. Confiding. Keeps to hilly areas. Active at dawn and dusk. Cryptically coloured and one is often surprised by the number of larks that flush out while walking.

Syke's Crested Lark.

Related species: **Syke's Crested, Deccan or Tawny Lark (G.deva-902) Our northern limits in plains and Ghat foothills up to 1000 m. Erect crest, finely streaked rufous breast, no white on belly, rump heavily streaked. Call 'Tuee, duii, teoo'. Sings on wings and is a good mimic. A lark of the arid land and scrub country. Flocks. Rarely seen on the grassy coastal areas. *Common Crested Lark (G.cristata-899) Around Pune. Crested, sandy back, streaked chest, white belly, buff underwing coverts and outer tail feathers. Scrub area. Size 180 mm.*

Malabar Lark

Alaudidae

Galerida malabarica(901)

Aranyachatak

Size: 150 mm   M / F : Alike   **R**

Distribution : Western Ghats up to 2000 m.

Nest-site: Under a rock, in a grass clump, a gentle depression on ground in a grassland. *Mar. - Sep.*
Material: Dry fine grass, thin twigs, rootlets, cowdung flakes. Lined with finer, soft grass.
Parental care: Both. Incubation 14-15 days.
Eggs: 2-3, buff, speckled with brown. 21.5x15.5 mm.
Call: *Cheeu, chew, chuii, tuee.* Musical, pleasant calls and song given, though not as fine as the Eastern Skylarks.
Ecological notes: Endemic to Western Ghats. Ground nests are prone to predation by reptiles and mammals and trampling by grazing cattle, goats and herdsmen.
Cultural notes: Larks were popular cage birds due to their singing ability!
Status: Common.

Crested Lark (Adult)

Crested Lark (Imm.) & Greater Hoopoe Lark. (Alaemon alaudipes)

Rufous-tailed Finch-Lark

Rufoustailed Lark

Alaudidae

Ammomanes phoenicurus(882,883)

Kubja Chatak

Size: 160 mm   M / F : Alike   **R**

Distribution : The entire region except Kerala.

Race *testaceus* in Karnataka and Tamilnadu.

Nest-site: The cup is placed under a bush in a depression, cattle-footprint or under an earth clod in a ploughed fallow field. *Mar. - Jun.*

Material: Grass, rootlets, twigs. The nest is supported with pebbles.

Parental care: Both. Incubation 13-14 days.

Eggs: 2-4,white,brown spotted. 21.2x15.7 mm.

Call: Chirping song is uttered in flight. Pleasant multisyllabic musical notes.

Ecological notes: Harmful to agriculture to some extent, as it feeds on grain of standing crop, but insects are also eaten. Heavy unseasonal rains can damage the nests under the earth clods. Preyed upon by shrikes, shikras.

Cultural notes: A well known songster also appreciated for the aerobic displays.

Status: Common.

Murari

Description: Drab, brown, uncrested stout lark. Black terminal band on rufous tail, more rufous in race *testacus* than *phoenicurus*. Large congregations are seen on grasslands in winter. Seeds, insects are taken on ground as the lark hops around. Winged insects are taken in aerobatic sallies. Typical up and down flight executed at dawn and dusk.

Rufous-tailed Finch-Lark.

Redwinged Bush-Lark.

Singing Bush-Lark.

Related Species: *Redwinged or Indian Bush-Lark (Mirafa erythroptera-877) Brown, black-streaked lark. *Endemic* to India. The chestnut wings are best seen in flight. Keeps to arid and scrub areas. Call- 'Chip, chip'. Sings well. The nest cup on ground is domed with grass.

*Singing Bush-Lark(Mirafa cantillans-872) 140 mm. White throat, creamy-buff breast band is faintly spotted. Rufous primaries and white outer tail feathers. Scrub land. Musical song uttered in aerial displays.

Gawai Chandol, Bharat

Indian Small Skylark

Eastern or Oriental Skylark

Alaudidae

Alauda gulgula(907,908)

Swalpsanchar

Size: 160 mm   M / F : Alike   **R, LM**

Distribution : Race *gulgula* in the entire region.

Race *australis* in Nilgiris and southwards.

Nest-site: On ground in a gentle depression under a rock, an earth clod or in a grass clump. *Feb. - Jul.*

Material: Dry grass, tiny twigs are used to weave a cup which is placed on ground.

Parental care: Both.

Eggs: 2-4,white,brown spotted. 20.6x15.3 mm.

Call: Song is typically given on wings, after the bird shoots up in the air. An accomplished mimic. Warbles and melodious notes.

Ecological notes: A grassland bird. Insectivorus and graminivorus.

Cultural notes: Feature in poetry and literature. Wordsworth's poem on the Skylark is well known even today. The song of this drab coloured bird is much appreciated.

Status: Common.

Description: Sparrow-like lark with short crest, fine bill, white outer tail feathers, narrowly streaked rufous breast and unstreaked belly. The aerial display and inspiring song on wing are best heard at dawn and dusk. The vast grassland is never silent at this time, ever resounding with vibrant notes of the skylarks!

Related species:*Greater or Yarkand Short-toed Lark (Calendrella brachydactyla-886) WM, straggler to our area, with one bird recently ringed at Nashik (Satish Ranade). Race dukhunensis in foothills of the region, race longipennis in Gujarat and northern Maharashtra. Keeps to dry areas and gleans seeds. Fat birds on their migration are netted for the dining table, even today!

Yarkand Short-toed Lark.

Indian Small Skylark.

Dusky Crag-Martin

Hirundinidae

Hirundo concolor(914)

Nakuti, Kutidushak

Size: 130 mm   M / F : Alike   **R**

Distribution : The entire region up to 1800 m.

Nest-site: The nest is attached bracket wise to vertical walls, rafters of old caves, rock faces, arches in old fortifications, cliffs and in parking lots in buildings, sheltered by overhangs. The same nest is used subsequently with repairs. *Jun. - Oct.*

Material: Mud pellets, fine grass, feathers.

Parental care: Both build the nest, incubate and feed the chicks.

Eggs: 2-4, white, spotted red-brown. 17.6x12.8 mm.

Call: A soft call *Chit, chit* is uttered in flight, when approaching the nest and from the perch.

Ecological notes: Nesting coincides with the monsoon, when the mud for building the nest is easily available. The bird is inseparably associated with crags, hills hence the name. Nest is parasitized by long-tailed tree mouse (Clement Francis).

Cultural notes: The bird is named *Kutidushak* due to its habit of littering and *Nakuti* for preferring cliffs and ruins. However, the martin is now seen in urban areas, freely nesting under the roofs of basements.

Status: Common.

*Related Species: *Eurasian Crag-Martin (H.rupestris-913) Dusky throat, vent, white belly and black underwing coverts. It is seen with swallows and H.concolor in the Sahyadris in winter.*

Plain Martin (Riparia paludicola-912) is patchily seen in the Northern Western Ghats. The tail has a shallow fork. Throat is pale brown and underparts are white. Flight is weaker than H.concolor.

Northern House Martin (Delichon urbica-930) WM to the Ghats up to 1000 m. Underparts, rump-white. Glossy dark above. Deeply forked tail.

Pale Martin (Riparia diluta-910) White belly, grey breast, faint breast band. Brown above. Vagrant.

Sand Martin (R.riparia-911) White belly, brown breast band. Pale brown above. Vagrant.

Pakoli

Description: In this family, the short tailed birds are the martins and the long tailed ones are swallows. White spots on tail feathers of *H.concolor*, excepting the mid and the outer ones, are seen in flight. Hawk insects in mid air. The flight is swooping, gliding, turning, banking,with intermittent flaps of wings, the dream of any pilot! Rest in communal roosts. The chicks have a peculiar habit of excreting with their rumps jutting out of the nest cup. Droppings are seen littered below the active nests.

Pale Martin.

Nortern House Martin.   Dusky Crag Martin.

Pale Martin.   Plain Martin.   Sand Martin.

Pakoli

Common Swallow

Barn Swallow

Hirundinidae

Hirundo rustica(916,917)

Durbalik, Krusha

Size: 180 mm   M / F : Alike   **WM**

Cliff Swallow nests.

Description: Chestnut forehead, throat, black breast band-complete in race *rustica*, incomplete in *gutturalis*, unstreaked belly. Steel blue above, forked white tipped tail, except on the central pair. On winter mornings flocks are seen on wires, taking to wings from time to time. They spiral up, taking winged insects and at one point speedily (40-50 kmph) shoot down to the perch. Equally at ease amidst busy towns as in fields, near reed beds. Rarely take insects from ground and then clumsily take to wings.

Cliff Swallow.

**Related Species: ***Streak-throated or Cliff Swallow (H.fluvicola-922) Till Nilgiris up to 1000 m. Chestnut crowned small swallow with streaked underparts, shallow forked tail. Patchily near water bodies, dams. The nest colonies are on rock surfaces near water edges, rocky islands, under bridges. Nest-pots have tubular entrance. Dead chicks in the nest tubes may attract birds of prey. The nest pots of Cliff Swallow are parasitized by sparrows sometimes by evicting the rightful owners!*
***H**ouse, Hill or Pacific Swallow (H.tahitica-919) Rufous head, throat. Lacks dark breast band. Tail deeply forked and underwings dark. Greenish glossy back. Coorg south 700m. and above. WM.*

Distribution : The entire region.

Nest-site: In Kashmir, on the sides and hulls of the houseboats, just above the water line. *Apr. - May.*

Material: Mud mixed with grass, twigs is cemented with the salivary glue to a wall or a vertical surface, bracket wise. The nest may be re-used.The nest is very hard when dry.

Parental care: Both. Female alone incubates for 14-16 days. The flying chicks are fed in mid air by both the parents.

Eggs: 4-6, white, red spotted. 19.6x13.7 mm.

Call: Low twitters and chirps are uttered on the wing. Song is uttered by male when breeding.

Ecological notes: Congregations of swallows are seen in winter and their numbers depend on the availability of flying insects.

Cultural notes: These aerial birds have aerodynamically designed wings but weak feet, hence they are called **Durbalik** and **Krush**, in Sanskrit.

Status: Common.

Pacific Swallow.

Barn Swallows on the nest.

Red-rumped Swallow

Indian Striated Swallow

Hirundinidae

Hirundo daurica (925, 927)

Devkulchatak, Bhandik.

Size: 190 mm   M / F : Alike **R, WM**

Gecko in
Red-rumped
Swallow's nest.

Distribution : The entire region up to 1600 m. Race *nipalensis* is a WM.

Nest-site: The retort shaped concrete hard nest is stuck to the roof of temples, mosques, Ruins, natural overhangs, underside of bridges. *Apr. - Aug.*

Material: Mud pellets, fine grass. Egg chamber is lined with down feathers, grass, paper, leaves.

Parental care: Both repair and strengthen the same nest. The family also roosts in the nest.

Eggs: 3-5, white. 21x14.4 mm.

Call: Loud *Cheer*, and chirping soft notes. More vocal and musical during breeding.

Ecological notes: The nest entrance shrinks after drying and instances of the bird getting trapped and dying in the constricted tunnel are reported. The nests are parasitized by sparrow and house swift (if the nests is near a house-swift village).

Cultural notes: The name **Sheetodari** for the swallow means one who survives on air, from their habit of remaining on the wing. People who allowed nesting of the swallows in their premises were respected and the Wiretailed Swallow was **Poojani** (the Worshiped).

Status: Common.

Lalbudi Bhingari

Description: Chestnut rump, deeply forked long tail, steel blue-black back and white chest. Perch on wires, trees in the Ghats and near ship yards in the Kokan. The act of nest building is a treat to the eyes. An instance has been observed, when a nest on a temple roof after being artificially coloured during the renovation of the temple, was subsequently accepted, strengthened and used by the Swallows, a fact important in the viewpoint of conservation.

Related species: *Wiretailed Swallow (H.smithii-921). Prefers proximity of water and roosts in reed beds. Up to 2000 m. The nest built from mud pellets is often close to the surface of water. The swallow was a bird of good omen and was considered auspicious due to its red head, reminiscent of someone who has applied 'Kumkum' on the forehead!*

Wiretailed Swallow on the nest.

Wiretailed Swallow building the nest.

Red-rumped Swallow.

Large Pied Wagtail

Whitebrowed or Indian Pied Wagtail

Motacillidae

Motacilla maderaspatensis(1891)

Kalkanth, Shreekanth, Kakachhadikhanjan

Size: 210 mm.   M / F : Alike   **R**

Distribution : Kokan, Malabar and the Western Ghats up to 2200 m.

Nest-site: On the ground in grass clumps, under rocks and near water beds in dry season. On roofs, tree holes, etc. *Mar. -Sep.*

Material: Twigs, grass, rootlets, paper, thread, Plastic, algae, moss, foils, coir to make the cup. Lined with down feathers, wool, cotton.

Parental care: Both.

Eggs: 3-5, grey-green, brown spotted. 21.9 x 16.2 mm.

Call: Harsh alarm calls are given. The breeding male utters musical whistles and sings perched on a boulder at waters edge or rooftop.

Ecological notes: Breeds as the SW monsoon ceases. Peak activity is noted as the water level recedes. Nesting in urban areas is near water bodies and on roof tops.

Cultural notes: Appears in Sanskrit poetry. In the Vayupuran, it is likened to the eyes of Lord Krishna. The name **Shreekanth** means blackthroated like Lord Shiva. In Manipur, wagtails are considered to be the incarnations of goddess Durga,.

Status: Common.

Thorla Dhobi

Description: Black-headed, white browed pied wagtail. Juvenile is grey-white. The black on the head reaches the tip of the beak. The similar looking Magpie Robin lacks the white eyebrow. Keeps to village tanks, river beds, streams, ponds atop mountains. Also comes near open drains near houses. Often confiding. Feeds at the edge of water by probing mud. Wags the tail as it runs, stops and then moves again.

White Wagtail (M.a.baicalensis)

Forest Wagtails.

Related species: *White Wagtail (M.alba-1885-86) Everywhere in winter; breeds in Kashmir. White forehead in all six races, of which personata and dukhunensis are common to our area.*

Forest Wagtail *(Dendronanthus indicus-1874) Western Ghats and adjoining coasts, Thane south, the Deccan. WM. Olive brown with two black breast bands. Expertly moves on the forest trees along horizontal branches.*

Forest Wagtail.

White Wagtail (M.a.personata)

Large Pied or White-browed Wagtail is a resident
& other Wagtails are winter migrants.

White Wagtail (M.a.leucopsis)

White Wagtail (M.a.dukhunensis)

Yellow Wagtail (M.f. superciliaris) Yellow Wagtail (1st W) Yellow Wagtail (1st W)

Grey-headed Yellow Wagtail (M.f.plexa) M.f.beema 1st Winter.

Yellow Wagtail in moult. Esturine habitat of Wagtails.

Grey Wagtail

Motacillidae

Motacilla cinerea(1884)

Khanjanak, Gopit

Size: 170 mm. M / F : Alike in winter   **WM**

Distribution : Kokan, Malabar and the Western Ghats up to about 2000m.

Nest-site: In the Himalayas above 2100 m. The nest cup is on the ground. *May - Jul.*
Material: Grass, rootlets, hair, wool, weeds.
Parental care: Both.
Eggs: 4-6, yellowish green, brown freckled. 19 x 14.2 mm.
Call: Chirping notes are continually uttered as the bird hops on the ground or flies in a flurry.
Ecological notes: Winter visitor to our water bodies in forested areas and open country alike. Brood of *P. citreola* is parasitized by *Cuculus canorus*. Birds fall prey to harriers.
Cultural notes: The wagtail running with the head thrust forward is likened to a horse and is the Diwali Ghoda, as it arrives in winter, the time of Diwali. Since it appears as if suddenly from the sky, it is the ***Khanjanak***. Kings considered sighting of the wagtail as a good omen before proceeding on expeditions.
Status: Common.

Karda Dhobi, Parit

Description: The white wing margins form a 'V' on the back. Hind claw is shorter than the hind toe. Runs while wagging the tail and guards its territory, avoids mingling with other wagtail species. Also hawks insects in the air and has been seen to steal prey from the beaks of bee-eaters. Roosts communally. Likes to take bath at noon. Insects are taken from the ground near the water's edge.

Yellow-headed Wagtail.

Grey Wagtail.

Yellow Wagtail.

Related Species:
*Yellow Wagtail
(M.flava-1875, -76, -78) Grey, blue and black-headed species concern us. This group is under taxonomic reconsideration. Hind claw is longer than the hind toe. Male has yellow underparts and throat.
*Citrine or Yellow-headed Wagtail (M.citreola-1881-82) Lemon yellow head and belly, grey back and rump, brown, white-edged wings and black flanks. Flocks throughout the region. WM. Uncommon.

Panthali Charchari

Description: Brown above, buff streaked below. White edges of the tail are best seen in flight. Like wagtail, run and hop on the ground while wagging the tail and shoot up in the air like a lark. Do not crouch when disturbed but perch in an upright stance. Injury is feigned by some females when nest is approached, while others give a threat display. Frequents water holes on hot noons. They are seen with larks, buntings, warblers. The nominate race is smaller and paler brown than *richardii* (1857).

Eurasian Tree Pipit.

Related species: *Eurasian Tree Pipit (A.triviallis- 1854, 55) Streaked underparts. Seen on trees and wires in winter.*
***Bl**yth's Pipit (A.godlewskii-1863) Adults have square black centres on median wing coverts. Winter migrant to the area.*
***O**riental Tree or Olivebacked Pipit (A.hodgsonii-1852) The head pattern is more prominent than in A.trivialis. In our southern range in winter.*
***R**ichard's Pipit(A.richardii-1857) Long hind-claws, pale lores. WM.*
***B**rown Rock or Longbilled Pipit(A.similis-1868, 69)A small resident population near Pune and Satara.*

Paddyfield Pipit

Motacillidae

Anthus rufulus (1859,60)

Sphotika, Tulika, Dulika

Size: 150 mm.   M / F : Alike   **R, LM**

Distribution : Kokan, Malabar and the Western Ghats up to 1800 m.

Nest-site: The nest cup with a lateral approach tunnel is placed on an embankment in agricultural area, grass clump, dyke, shallow depression such as a hoof print or an impression of an upturned stone. *Mar. - Jun.*
Material: Grass, rootlets, hair, twigs. A partially curved incomplete roof is sometimes seen.
Parental care: Both.
Eggs: 3-4, grey, brown spotted.
20.2 x 15.4 mm.
Call: *Tsip, tseep, pipit, pipit* is uttered in flight. Clicks while hopping on the ground.
Ecological notes: Brood parasitized by *Cuculus canorus.* Bird of arid areas.
Cultural notes: Sanskrit names are descriptive of the tail wagging habit of the pipits and of suddenly shooting up into the air and uttering a song on the wing !
Status: Common.

*Nilgiri Pipit (A.nilgiriensis-1870) Nilgiri south. 1000 m-2300 m. Streaked back, belly,flanks. Buff below. Size-170 mm. Endemic.
*Tawny Pipit (A.campestris-1861) WM up to south Karnataka. Unstreaked below. Size-150 mm.

Red-Throated Pipit (A.cervinus) vagrant to our region.

Buff-bellied Pipit (A.rubescens-1872) recorded at Satara.

Richard's Pipit.

Longbilled Pipit.

Blyth's Pipit.

Paddyfield Pipit.

Nilgiri Pipit Endemic.

Oriental Tree Pipit (A.hodgsoni-1852)

Richard's Pipit.

Tawny Pipit.

Rosy Pipit (Vagrant)
Anthus roseatus.

Female.   **Motha Kahua**

Description: A stubby bird partial to the canopy, usually noticed by the call. The dark eye streak of the male is paler in the female. Her underparts are also barred. The bird is more often heard than seen.

Black-headed Cuckoo-Shrike.

Related Species: **Black-headed Cuckoo-Shrike (C.melanoptera-1079)* Throughout the region up to 2100 m. Partial to mango and tamarind groves. Encountered with mixed hunting parties. Feeds on caterpillars, insects, fruits etc. The nest is a cup. Both parents share domestic duties.
*Black-winged Cuckoo-shrike (Coracina melaschistos- 1077) Winter vagrant. Upright stance. White banded undertail pattern is typical. Female is paler. Can be confused with Greybellied Cuckoo.

Large Cuckoo-Shrike

Campephagidae

Coracina macei (1072)

Pikaang

Size: 280 mm   M / F : Dimorphic   **R**

Distribution : The entire region up to 1200 m.

Nest-site: On the outer branches of tall trees. *May. - Oct.*
Material: Lichen, bark, cobweb, twigs.
Parental care: Both.
Eggs: 3, pale green, blotched. 31x22.4 mm.
Call: Loud, pleasant call coming from the canopy-*Tii-ee,tii-ee,*is heard over a long range.
Ecological notes: Helps in seed dispersal, especially of the figs. Eats large insects.
Cultural notes: The Sanskrit name ***Pikaang*** means a bird with a double call like that of a cuckoo, and it may also be indicative of either the Cuckoo-Shrike or the Pitta.
Status: Uncommon.

Large Cuckoo-shrike on the nest (M)

Black-winged Cuckoo-shrike (M)

Large Cuckoo-shrike (M)
(C.m.macei)

Small Minivet

Campephagidae

Pericrocotus cinnamomeus (1093,1094)

Saptaswasaraha

Size: 150 mm   M / F : Dimorphic   **R**

Distribution : The entire region up to 1000 m.
Race *malabaricus* occurs south of Goa,
nominate race northwards.

Nest-site: An open cup adherent to the upper
surface of a fork of *Mangifera indica,*
Euphorbia tirucali tree, very well matching in
colour with the branch. It is often difficult to
locate the nest again even after the first
sighting ! *May - Sep.*

Material: Fibres, rootlets, twigs, coated and
cemented with cobwebs and lichen. The cup is
tightly woven, thin,yet strong.

Parental care: Both construct the nest and
feed the chicks. Female alone incubates, when
the male feeds her on the nest.

Eggs: 2-3, pale, greenish white or buff cream
with brown spots. 16.6x13.4 mm. Nestlings fit
in so snugly, that they are not noticed unless
they move.

Call: The musical *Swee, swee* is a familiar call
in semideciduous forests.

Ecological notes: Insectivorous.

Cultural notes: It is called Nikhar, meaning an
ember! Sanskrit name **Saptaswasaraha**,
Hindi-Saheli, and Bengali Saat Sayali, all
indicate the bird's habit of flocking.

Status: Common.

Female.   Chota Nikhar

Description: Head, throat and back are grey-
black in male. Breast is brilliant orange red,
which is more appreciable in sunlight. In
female and young male-orange is replaced by
yellow and throat is white. During courtship,
the male chases the female on the wing,
singing all the while. Flocks of this arboreal
bird search the canopy for insects and nectar,
often hovering like sunbirds. The chicks are fed
even after they abandon the nest. Cooperative
feeding of the chicks has been reported.

Related species:
*White-bellied Minivet (P.erythropygius-1096)
Occasional on the eastern fringes of the
Western Ghats, in scrub country and grass-
lands. Not on coasts. Call is 'Treep, treep' and
an alarm call 'Chit, chit' is also uttered.*

Small Minivet male on the nest.

White-bellied Minivet (M)

Scarlet Minivet - females.

Small Minivet (M & F) feeding the chick.

The flying Minivets appear like embers. Unlike the true embers which cause destructive forest fires they please the eye through soothing beauty!

Scarlet Minivet

Campephagidae

Pericrocotus flammeus(1081)

Vishpulingaka

Size: 200 mm   M / F : Dimorphic   **R**

Distribution : Throughout the region up to 2000 m.

Long-tailed Minivet.

Nest-site: The cup is plastered 6-18m. up on the surface of a branch, blending well with its colour, almost invisible. Trees on forest fringes or near cliffs are preferred. *May - Aug.*

Material: Rootlets and fibres are plastered with cobwebs to make the tight cup. It is concealed with lichen, moss and spider's egg-cases.

Parental care: Both. Female alone incubates.

Eggs: 2-4, greenish, brown-spotted. 23x17 mm.

Call: Repetitive, musical, pleasant whistle *Whee-wheeriri, whee, twit, twee, twee.* Song *Tweety-wee, tweety-wee.*

Ecological notes: Invariably a forest canopy bird, not seen in the urban habitat.

Cultural notes: Finds mention in the Rgveda, as a charm against the poison of insects, whom the minivets devour! When a forest fire is ablaze, embers (**Vishpulingaka**-also the birds name) scatter hither and thither. So is the impression one gets, while watching the minivets fly over the forest canopy! In the British times, it was known as the Fiery-red Bird.

Status: Uncommon.

Ashy Minivet.   *Rosy Minivet.*

Scarlet Minivet (M)-Imm.

Female.   **Thingi, Nikhar**

Description: Male has fiery red underparts and a scarlet wing patch. Female has lemon yellow underparts. This bird of the canopy flits in search of spiders, cicadas, winged insects, which they catch in flight. Display of the breeding male is a superb spiralling flight, coming down in a swift sally, ending with a neat landing on a tree top, repeated. This is an exciting and memorable experience.

Related Species: * *Rosy Minivet (P. roseus-1089) WM, vagrant. 180 mm. Pink bands on dark wings, rosy belly. Female olive yellow.* ***Long-tailed Minivet (P. ethologus-1085) is a bird of the N and NE Indian region, vagrant in our area in winter. Red wing patch extends down the secondaries as a line, in males.* ***A**shy Minivet (P. Divaricatus-1089a) Vagrant to Karnala. 180 mm. White forehead. Pied.*

Scarlet Minivet (F & M)

Male.   **Kabra Khatik**

Description: Typical hunchback posture. Black parts of the male are sooty-brown in the female. Often encountered in the deciduous and evergreen jungle interface, chasing one-another from tree to tree. Perch on branches jutting over the forest streams, from where they make frequent sallies after winged insects.

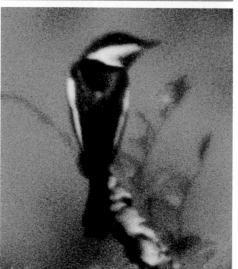

Pied Flycatcher-Shrike

Barwinged Flycatcher-Shrike

Campephagidae

Hemipus picatus(1065,1066)

Size: 140 mm   M / F : Dimorphic   **R**

Distribution : The entire region up to and above 2000 m. Race *leggei* in south Kerala.

Nest-site: A small cup matching well with the leafless branch, is placed in a fork 3-10 m. up. It is extremely difficult to locate. *Mar. - Jun.*
Material: Roots, grass, twigs, moss, lichen, cobwebs, etc. Lined with down feathers.
Parental care: Both.
Eggs: 2-3, greenish-white, blotched with black. 15x12.5 mm.
Call: A high pitched Whi-rir-ri.
Ecological notes: The birds are often a part of mixed hunting parties of smaller forest birds. Insectivorous.
Cultural notes: A less noticed bird !
Status: Occasional.

Note: *This bird has features common to both, flycatchers and wood-shrikes.Their flycatcher-like habit of catching insects, and wood-shrike -like behaviour, warrant a separate genus for these small, elusive forest birds.*

Common Woodshrike

Lesser Wood Shrike

Campephagidae

Tephrodornis pondicerianus (1070)

No sanskrit name recorded

Size: 160 mm   M / F : Unlike   **R**

Distribution : The entire region up to 1000 m.

Nest-site: A cup plastered to the fork of a bare branch of *Frangipani* spp. etc. trees,2-8 m. up. Colours of the nest and branch match hence nest escapes detection, inspite of being exposed. *Feb. - Apr. & Jun. - Oct.*

Material: Bark, lichen, moss, grass, twigs.

Parental care: Both. Chicks are of the same colour as that of the nest. The chicks are fed every 15 mins. or so.

Eggs: 2-4, greyish-green, blotched. 19x15.1 mm.

Call: A whistle *Weet, weet*,and a song *Whi-whi-whi-whee*, are pleasant to the ear. Utters many notes during breeding. A special warning call to the chicks to remain immobile.

Ecological notes: Insectivorous.

Cultural notes: Largely overlooked due to the drab and cryptic colour.

Status: Common.

Raan-Khatik

Description: Eye stripe paler in female. The nest is often mistaken for a branch knot. Seen in mixed hunting parties. The stout bill is hooked. Winged insects are caught in air, in a flycatcher-like manner. On hot noons, chicks exposed to the direct sunlight, are taken under the wings by the parents. They probably lose heat by panting.

Common Woodshrike (F) with chick.

Common Woodshrike (M)

Large Woodshrike.

Related Species: **Large Woodshrike** *(T.gularis-1068) Larger in size than Common Woodshrike, has rufous tips to outer tail feathers. Tail square, beak stout, hooked. Call is a loud resounding 'Witoo, witoo, chack, chack, chirr'. Encountered in small flocks in well wooded country up to 1800 m.*

Pycnonotidae

Pycnonotus cafer(1127,1128)

Angarchudak, Fenchak, Krushnachud

Size: 200 mm   M / F : Alike   **R**

Distribution : The entire region up to 1500 m.

Race *humayuni* occurs Mumbai south.

Nest-site: Forks of bushes, creepers, trees up to 10 m. Houses, lampshades, lofts, wire-bundles, electric housings and similar places. *Feb. -Jul.*

Material: Grass, twigs, rootlets, paper, plastic, cobwebs, foils, etc..

Parental care: Both.

Eggs: 2-3, pink-white, laid over 2-3 days. 21.1x15.5 mm.

Call: Pleasant and musical. Well known song- *Pit, wit, wit, peep, peep.* Alarm calls.

Ecological notes: Adapted well to urbanization. Pollinator, insect controller. Held responsible for the spread of lantana. Flocks seen on flowering *Salmalia, Woodfordia, Capparis, Erythrina, Bombax, Loranthus* and fruiting *Ficus, Zizyphus, Santalum, Salvadora, Lantana.* Destructive to fruit gardens.

Cultural notes: Bulbul is a Persian name for Nightingale, which featured extensively in their poetry. It was given to the Red-vented Bulbul of Bengal and the actual bird was forgotten! It now features extensively in our poetry. Bulbul fights with betting of high stakes were conducted in the past.

Status: Abundant.

Lalbudya Bulbul

Description: White rump more prominent than crimson vent. Seen in forests and urban areas alike. The display of the breeding male and the broken wing display of the parents, to divert the intruder when their nest is approached, are spectacular! Noted to bully smaller passerines and rob food from robins, tits, larks. Whether this is a mere sport or true robbery is uncertain!   *Yellow Throated Bulbul.*

Bulbul fight conducted with stakes in spite of being banned.

***Related Specie:** *Yellow Throated Bulbul (P.xantholaemus-1135) On boulder studded slopes, deciduous forests of South India with a few records from southern Western Ghats, BR Hills. Eat berries, winged insects. Breed from March till July. Female builds the nest and incubates. Both feed the chicks.*

(Clement Francis, Chaitra, Rajesh) ***Vulnerable.***

Red-vented Bulbul nesting on a tube-light.
These bulbuls commonly nest in houses, balconies,
terraces, on hanging wires, bulb-holders etc.
Care should be taken to avoid electrical fires. Nests can
be successfully relocated if found near live electric wires.

Yellow-browed Bulbuls.

Red-vented Bulbul with various food items -
grasshopper, cockroach and gecko.

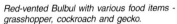

Red-vented Bulbul feeding the chicks in a garden nest.

Shipai Bulbul, Narad Bulbul, Bulandi

Description: Red and white whiskers, black crest, breast band, and red vent. Territorial, keeping to the nest locality, confiding and freely entering houses. Seen with Red-vented Bulbul, but not with others of the species. Displaying male fans the tail and drops the vibrant wings while singing. *Paper-to make a nest.*

Red-Whiskered Bulbul nesting on a wire.

Red-whiskered Bulbul

Pycnonotidae

Pycnonotus jocosus (1120)

Karnikar,Pushpavasantak

Size: 200 mm M / F : Alike   **R**

Distribution : The entire region up to 2100 m.

Nest-site: In the fork of garden trees and bushes within 3 m. At any suitable place in houses. *Feb. - Aug.*

Material: Leaves, grass, twigs, hair, paper, plastic, bark, rootlets, thread.

Parental care: Both. A pair was once seen to strengthen an active nest hastily and with extreme urgency, till late evening; amazingly, it rained heavily that night. Nest intrusion is vocally resented and diversion tactics are used.

Eggs: 3-4, pink, oval, brown and red mottled. 24.4x16.1 mm. Two broods are common.

Call: Accomplished songster. Sings an array of melodious notes when breeding. A soliloquy is often uttered at noon.

Ecological notes: Prefers humid areas, avoids dense forests. Nesting near electric appliances, meter-boxes carries the risks of fire, electrocution. Such nesting is best prevented and if seen, nest translocation to safer adjacent places is accepted, if the nest has chicks.

Cultural notes: Narad, the devotee saint, kept a vertical lock of hair on the head, hence the name Narad Bulbul. The red ears of this bird are as if adorned with ornaments (***Karnikar***), like the apsara Karnika, or one wearing floral ornaments (***Pushpavasantak***).

Status: Common.

Immature Red-Whiskered Bulbul.

Look Alike: The immature Red-Whiskered Bulbul lacks crimson-red ear patch and may be mistaken for *White-eared Bulbul in the overlap range. Yellow vent of the latter is different.

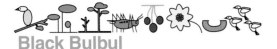

Black Bulbul

Pycnonotidae

Hypsipetes leucocephalus(1149)

Paryantika

Size: 230 mm   M / F : Alike   **R, LM**

Distribution : Western Ghats 400 m up to 2100 m. Mumbai south.

Nest-site: Fork of a tree, creeper or thorny bush, about 6 m. up and above. *May - Aug.*
Material: Grass, leaves, moss, lichen, cobwebs.
Parental care: Both. Two broods are raised.
Eggs: 2-3,whitish, brown spotted. 26.6x19.6 mm.
Call: A vocal bird. Chirping notes *Che, che, Chirp, chirr,c hrip.* Also whistles, alarm calls.
Ecological notes: Frugivorous and nectar seeking forest bird always found above an altitude of 700 m. throughout its range. Keeps to evergreen and humid forests. Causes seed dispersal of berries and cross pollination.
Cultural notes: The Sanskrit name **Paryantika** means a wanderer and it is descriptive of the restless activity of this bulbul!
Status: Locally common.

*Look Alike: *Himalayan Bulbul (P.leucogenys-1127) has a pointed prominent crest, often dropping forwards. Not in our range. Black Bulbul can be mistaken for *Black Drongo.*
*Related species: *White-eared Bulbul (P.leucotis-1123) Has uncrested black head and throat, white cheeks, yellow vent, brown back and white underparts. White tipped tail. Resident population around Mumbai for the past 50 years. A confiding bird.*

Kala Bulbul

Description: Restless large bulbul with black crested head, gently forked tail, orange bill and legs. This bird of the forest canopy is seen with flycatchers, ioras, nuthatches, white eyes chasing each other.

White-eared Bulbul

Himalayan Bulbul.

Kajal, Pivlya Bhivayicha Bulbul

Description: Upper parts of the northern race *ictericus* are grey-green and belly dull, as compared to bright yellow of the southern race *indicus*. Black eye is prominent. Looks for berries and nectar in the canopy. Also forage on ground in the undergrowth and in groves.

Ruby-throated Yellow Bulbul.

Other Bulbuls: **White-browed Bulbul (P.luteolus-1138) Uncommon, patchily in Ghats, in scrub and drier areas. Dull olive back, whitish underparts, prominent eyebrow,forehead and a yellow vent.*
*Grey-headed Bulbul (P.priocephalus-1114) Light iris, black chin,yellow forehead, grey head, barred rump and banded tail. **Endemic** *to Western Ghats, Belgaum southwards. Keeps to the canopy in wet forests.*
*Variable or Ruby-throated Yellow Bulbul (P.atriceps-1116) Black headed olive bulbul with a ruby-red throat. Western Ghats, Goa, Belgaum southwards up to about 1200 m.

Yellow-browed Bulbul

Pycnonotidae

Iole indica(1143,1144)

Peetatanu

Size: 200 mm   M / F : Alike   **R**

Distribution : Western Ghats up to 2000 m.

Race *ictericus* till Goa, *indicus* southwards.

Nest-site: Unlike the usual cup, the nest is a hammock, slung on branches of a low bush in the forest within 4 m. *Feb. - Jun.*

Material: Leaves, grass, thin twigs, cobweb, Moss, rootlets. Lined with dark roots.

Parental care: Both. Incubation period is 15-16 days. Though frugivorous, the parents feed larvae and insects to their chicks.

Eggs: 2-3, pale pink, red spotted. 23.1x16.6 mm.

Call: Soft bisyllabic whistle. Breeding male sings. Alarm calls distinct.

Ecological notes: Frugivorous and nectar eater.

Cultural notes: The vernacular name Kajal, indicates the black eyes of the bird. A less noticed bird, often mistaken for a female Iora, when glimpsed in dense foliage. Ruby-throated Bulbul is the State-bird of Goa.

Status: Locally common.

Grey-headed Bulbul.

White-browed Bulbul.

White-browed Bulbul on the nest.
Yellow-throated Bulbul on the nest.

Partial albino Red-vented Bulbul.

Yellow-eared Bulbul-
Sri Lankan *Endemic*.

Black-headed Bulbul (P. atriceps) is seen in the NE States & Andaman Is.
It may be mistaken for the Black-crested Bulbul (P.melanicterus) seen in our region.

Female.   Subhaga

Common Iora

Juvenile.

Iora

Irenidae

Aegithina tiphia (1100,1101)

Shreevad, Madhuka, Sukarika

Size: 140 mm   M / F : Dimorphic   **R**

Distribution : The entire region up to 1500 m.
Race *deignani* upto Palghat, *multicolor*
southwards.

Nest-site: The cup is placed in the fork of
*Mangifera indica, Terminalia crenulata, Acacia
catechu, Acacia nilotica, Tamarindus indica,
Ziziphus mauritiana,* etc. spp. *Jun. - Sep.*
Material: Grass, rootlets, threads are bound
with abundant cobweb, hence nest looks white.
Parental care: Both.
Eggs: 2-4, pale pink-white. Brown blotched.
18x13 mm.
Call: Sweet musical whistles *Pheeou, whee,
whee* and chirruping notes are uttered more so
on cloudy days. During courtship, varied notes
like *Chee, chee, Tsee-uu,s oo, sooo, soo*
(Hence the name **Sukarika**). Songster. The
very name Iora comes from Latin: Io-a joyous
note.
Ecological notes: Breeding coincides with the
SW monsoon. Insects, mainly beetles and ants
are devoured in large amount, hence
beneficial to fruit gardens. Nests are
sometimes parasitized by the Bay-banded
Cuckoo.
Cultural notes: Names **Shreevad** and
Madhuka are in appreciation of the melodious
song of this bird.
Status: Common.
Common Iora (M)

Description: Black and yellow with white
wing bars. In female, black is replaced by
greyish green. During breeding, male displays
by showing the yellow wings. Seen looking for
insects in dense foliage, assuming acrobatic
positions. The yellow Iora nesting on a Babul
tree with its yellow blossoms, is an exquisite
sight! The nests of the Iora are often predated
and their success rate is one of the lowest. The
birds however do not give up, and rebuild the
nests, often without a fruitful outcome!

*Related species: *Marshall's or Whitetailed
Iora (A.nigrolutea-1102) Occasional in the
northernmost limits of our range in scrub
country and groves. Black tail has a prominent
white edge and breeding male has a broken
yellow collar.*

Whitetailed Iora.

Gold-fronted Chloropsis

Gold-fronted Leafbird, Green Bulbul

Irenidae

Chloropsis aurifrons (1104,1105)

Pakshagupta, Patragupta

Size: 190 mm.   M / F : Dimorphic   **R**

Distribution : Race *frontalis* in the entire region, up to 1800 m; *insularis* Palghat south.

Nest-site: On a branch, high up in a dense tree, well concealed. *May - Aug.*

Material: Roots, bark-fibres, soft grass.

Parental care: Both.

Eggs: 2, creamy, red spotted. 23.4 x 15.5 mm.

Call: Continual chattering and loud shrill whistles *Chich-wee, chich-chich.* An excellent mimic of calls of iora, tailor-bird, magpie-robin, white-breasted kingfisher, drongo, sunbird, sparrow-hawk. All calls are uttered in succession, making one believe that a mixed bird party is present on the fruiting tree.

Ecological notes: Frugivorous, insectivorous and nectarivorous. Visits flowering *Butea, Salmalia,Erythrina,* spp.,and fruiting *Achras sapota, Mangifera indica* trees.

Helps pollination and seed dispersal.

Cultural notes: Was a popular cage bird, owing to its skilful ability for mimicry. The name Chloropsis is derived from Greek: Chloros-light green, Opis-appearance.

Status: Occasional.

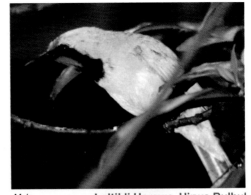

Male.   **Laltikli Harewa, Hirwa Bulbul**

Description: Male is leaf-green, has purple black throat, golden orange patch on forehead and absence of blue on flight, tail feathers. Female and young birds lack the forehead patch and black throat, are dull-coloured. Frequent fruiting and flowering trees, in the company of bulbul, iora, minivet. The leafbird is invisible amongst the green leaves, hence the Sanskrit name **Patragupta**!

Goldmantled Chloropsis Female & Imm.

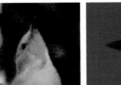

Gold-mantled or Blue-Winged Chloropsis on the nest.

Related Species:

Jerdon's or Goldmantled Chloropsis (C.cochinchinensis-1107) Lightly wooded areas. No golden orange colour on the fore-head. Has purple blue moustachial stripes. Female has a blue green chin. In parties in groves around villages. Occasional.

Asian Fairy Bluebird

Fairy-Bluebird

Irenidae

Irena puella(1109)

Neelachhawi, Neelachatak

Size: 270 mm.   M / F : Dimorphic   **R, LM**

Distribution : The entire region up to 1800m. Rare in Northern Ghats.

Nest-site: In dense jungle upto 5 m. up in a tree fork. *Feb. - Apr.*

Material: Rootlets, twigs, green moss are used to build a thick nest platform.

Parental care: Female builds the nest and incubates, during which time the call of the male can be heard. Both feed the chicks.

Eggs: 2, olive grey, blotched in brown at the broad end. Chicks have a grey-brown down. 28.2 x 20.2 mm.

Call: A fluid whistling *Whit-tu, peepit, weet, Weet, what's it, be-quick* given from a fruiting tree. *Chi,chi,chi* in flight.

Ecological notes: Pollinators of *Erythrina, Grevillea*, spp. Aid seed dispersal of *Ficus* spp. Also eat insects. Prefer heavy rainfall evergreen, semi-evergreen forests. They are now rarely seen in the northern Sahyadri range, due to deforestation and habitat loss.

Cultural notes: The Sanskrit name **Neelachhawi**, mentioned in the Kalpadrukosh, means the Blue Beauty !

Status: Uncommon.

Malabar Whistling-Thrush

Description: Male-shining ultramarine blue above, has blue undertail-coverts, black bill. Female-dull blue-green,has black lores, brown-black bill. One of our most beautiful birds. Flocks are seen moving from one fruiting or flowering tree to the other, in the company of green pigeons, barbets and hornbills. One of the first birds to arrive on a fruiting tree, just after the break of dawn.

Lalita

Look Alike: **Malabar Whistling-Thrush (Myiophonus horsfieldii-1728) may be mistaken for Fairy Bluebird, if seen in shady foliage. The former is terrestrial and is cobalt blue in colour. Latter is arboreal and rarely descends to fruiting bushes or to the ground.*

Fairy Blue Birds - Male, Immature & Female.

Rufous-backed Shrike

Longtailed Shrike

Laniidae

Lanius schach (946,947)

Latushak

Imm.

Size: 250 mm.   M / F : Alike   **R, LM**

Distribution : Race *caniceps* in the entire range up to 2000 m. and *erythronotus* WM Belgaum south.

Nest-site: In thorny trees in scrub country. Old babblers' nests are sometimes used. The same nest is used if available. *Mar. - Jun.*

Material: Thorny twigs, rootlets, wool, hair, rags, paper, rubbish, to build the loose cup.

Parental care: Both. Brood-parasitized by *Cuculus canorus, Clamator jacobinus.*

Eggs: 3-6, grey-green, spotted. 23 x 17.9 mm.

Call: Harsh calls *Krr, chrr, grekk.* Sings when breeding. Mimics calls of 30 or more birds and mammals! Calls of migratory birds are remembered correctly and reproduced, long after they have left, thereby confusing birdwatchers.

Ecological notes: Some specialization in predation is seen amongst the shrikes of the same species in a given area. One pair going after small birds, the other taking lizards, yet another preferring mice. This reduces the food conflict amongst the shrikes of the same locality. (Satish Pande, Amit Pawshe)

Cultural notes: In Mogul times, shrikes were trained to hunt. The Brown Shrike is the first Indian bird to feature in the classification by Carl Linnaeus! (SA Handbook)

Status: Common.

Naklya Khatik

Description: Rufous flanks, long tail. Surplus prey, fecal sacs, vomitus balls are impaled on thorns of *Acacia* spp., *Gymnosperma montana, Flacourtia indica,* trees. Birds, reptiles and mammals are decapitated, brain eaten and the heads impaled, as if trophies! Hence the name Butcher Bird. The larder is guarded, its position altered from time to time, and yet prey is stolen by crows, thereby arousing much commotion! The hooked bill is adapted for predation and tearing the prey apart. Prey is impaled with a single lateral sweep of the head. How the shrike avoids injury from the thorns is a matter of awe! Parents teach chicks to search for prey by dropping food on the ground.

Phillippine Shrike.

Related species:*B*ay-backed Shrike(L.vitta-tus-940) 180mm. Black ear to ear band across the forehead, chestnut back, fulvous breast, white rump and wing mirrors. Affects dry areas near fields. *Brown Shrike(L.cristatus-949) WM. White forehead, black lores and ear coverts. Brown above and pale rufous below. No white in wings. Brown tail; not rusty-red as in Rufous tailed Shrike-L.isabellinus-943.

*P*hillippine Shrike is a new record for the Western Ghats (Bird Banders Training Programme, 1998-02, Final report, BNHS).

Bay-backed Shrike.

Brown Shrike.

Southern Grey Shrike

Grey Shrike

Laniidae

Lanius meridionalis (933).

Latushak, Latwa

Size: 250 mm.   M / F : Alike   **R, LM**

Distribution : North Kokan, Deccan and eastern foothills of Western Ghats up to Belgaum. Unrecorded in Central, South Kokan and Malabar.

Nest-site: Deep cup in a thorny tree in scrub country. *Acacia* and *Zizyphus* spp., *Prosopis cineraria, Parkinsonia aculeata*, preferred. *Mar. - Jun.*

Material: Hair (human, hare, goat), rags, Feathers, paper, wool, thorny twigs.

Parental care: Both. Small insects, caterpillars are fed to the chicks in the first week, mantids, grasshoppers, in the second and bird, reptile, mammal flesh from the third week onwards. To avoid the attention of predators, prey is not impaled on the nest tree but on other thorn trees. Larder is guarded by one parent as the other tears the flesh piece-meal and takes it to the chicks in the nest(Satish Pande).

Eggs: 3-6 whitish, blotched. 25.9 x 19.7 mm.

Call: A mimic. Harsh chuckles are aggressively uttered when an active nest is approached.

Ecological notes: Predator. The same nest may be used in successive years. A dead blind chick was noted in a nest, next to a healthy chick. It was not removed by the parents. The other chick grew in the shadow of death!

Cultural notes: *Gandhari*, the blind- folded queen mother of the Kauravas in the epic Mahabharat, is an apt name for this masked bird.

Status: Uncommon.

Juveniles.

Khatik, Gandhari

Description: Black forehead and wing mirror conspicuous. Takes a low parallel flight after the initial dive. Eats insects, larks, sunbirds, small green bee-eaters, warblers, buntings, brahminy mynas, sandgrouses, mammals, reptiles and bird-eggs(dove,bunting). Was seen to eat a dry shrivelled tomato probably mistaking the red colour for flesh(Amit Pawshe).

Juv. near a decapitated Bee-Eater.

**Other Shrike: *****Rufoustailed or Pale-brown Shrike (L.isabellinus-943) Uniform brown head, back, rufous rump and tail, white belly, wing speculum and black ear-coverts, which are brown in female. WM to Goa, east of Ghats, not in Kokan.*

Look Alike:
**Great Grey Shrike(L.excubitor-936)* *White wing patch is larger and there is no black on the forehead. Not in our area.*

Southern Grey Shrike with impaled rat on a thorn.

Grey Shrike impales food on thorns for future use.
A dried tomato was brought by the Shrike,
probably the red colour is linked with flesh.

Prey of the Grey Shrike.
Adult feeding the chick outside the nest.

Nesting habit of the Grey Shrike -
Eggs, and chicks in two phases.

Parents intentionally drop food on the ground
so that the chicks can develop the ability of finding it.

Rufous-tailed Shrike.

Hypocolius

Grey Hypocolius, Shrike-Bulbul

Bombycillidae

Hypocolius ampelinus(1063)

No Sanskrit name recorded

Size:250mm   M/F:Dimorphic   **WM**

Distribution: Recorded in Maharashtra at Kihim,dist.Raigad in Kokan and nearPune.

Nest-site: Extralimital-Iraq, Iran. 1-3 m up. *Jun. - Jul.*

Material: Twigs, grass .

Parental care: Both.

Eggs: 4-5, white, blotched. 26x19 mm.

Call: Squeaky, fluid, pleasant notes. Silent in winter.

Ecological notes: Mainly frugivorous. *Lantana aculeata, Salvadora persica, Zizyphus* berries noted in stomach (Handbook,Salim Ali). Rarely eats insects in native land-S.Iran, Arabia, Iraq, Afghanistan and N.Africa.

Cultural notes: Unknown bird in our region.

Status: Vagrant.

Male.

Khatik-Bulbul

Description: Head feathers are erected when excited and then the eye band appears larger giving the bird an angry look. Keeps to bushes like the bulbul and rarely takes an insect from the ground swooping in a low shrike like flight. Both with blue grey back and long tail with terminal black band which is white tipped in female. Female lacks black eye to nape band.

Male.   *Female.*

Hypocolius may be encountered in such habitat in our region.

Blue Rock-Thrush

Muscicapidae

Monticola solitarius pandoo(1726)

Durga, Pandavika, Pandushama, Uma

Size: 230 mm   M / F : Dimorphic   **WM**

Distribution : A winter migrant to the Kokan and the Western Ghats keeping to the drier and rocky areas. Also on the coasts.

Nest-site: In Kashmir and in the Himalayas. The nest pad is hidden in crevices in earth banks, walls and embankments. *Apr. - Jun.*
Material: Moss, grass, hair, leaves, twigs, rootlets.
Parental care: Both. Incubation by female.
Eggs: 3-5, pale blue with brown spots. 26.7x19.7 mm.
Call: Silent in winter. Whistles of the male can be heard in March, at the time of departure.
Ecological notes: A winter migrant (October-March) to hilly, rocky scrub country.
Cultural notes: Name *Durga* is after the goddess who has a blue complexion! The name *Pandushama* indicates its light blue colour and it is also the name of the species!
Status: Common.

Related species: *Blue-headed Rock-Thrush (M.cinclorhynchus-1723) WM to the Western Ghats, keeping to the wetter areas. The white wing bar of this secretive bird is conspicuous in the shady areas, where it likes to rummage.*

Blue-headed Rock-Thrush.

Female.   **Shailkastur, Parvatkastur, Pandu**

Description: The blue gloss of the male is best seen in direct sunlight. Female is barred brown. Perches erect and hops on ground, vibrating the tail, while searching for insects under boulders. Chosen water hole is visited daily at noontime, when all the rock-thrushes in the arid locality gather with buntings, pipits, Larks, etc. to quench their thirst.

Blue-headed Rock-Thrush.

Blue Rock-Thrush (M)

Parwat or Sheelkari Kastur

Description: Cobalt blue colour on forehead and wings glistens in sunlight, when the thrush, which usually keeps to the shades and undergrowth, comes in the open. Small snakes, snails, crabs, molluscs are battered and then eaten. Shy. Often hops on ground in search of food. Chicks have a yellow gape.

Pied Thrush.

Spot-Winged Thrush (Z. spiloptera) Sri Lankan Endemic.

Tickell's Thrush.

Siberian Thrush.

Malabar Whistling-Thrush

Muscicapidae

Myiophonus horsfieldii (1728)

Shreevad, Seekaram

Size: 250 mm   M / F : Alike   **R, LM**

Distribution : In Western Ghats up to 2200 m.

Nest-site: The bulky nest is placed under an overhang on the ledge of a vertical cliff near a waterfall. Though protected from direct rain it gets wet in the spray. Sometimes built in the ruins of hill forts or in tunnels. *Jun. - Oct.*

Material: Grass, roots, twigs, stems, plastered with mud. Moss and lichen grow on the outer mud wall making the nest invisible.

Parental care: Both. The thrush does not approach the nest in the presence of observers.

Eggs: 3-4, buff, speckled. 33.1x23.9 mm.

Call: The sonorous whistle has a hypnotic effect, and it is a familiar sound near mountain streams. The melodious song, with much tonal variation, once heard in the picturesque settings in which the thrush dwells, cannot be forgotten. Call- *Tzeet, tzee* (the Sanskrit name **Seekaram**) is heard from much before dawn.

Ecological notes: The eggs are eaten by jungle cats and snakes. A bird ringed at Mahabaleshwar was taken at N. Coorg, 650 Km. Away (SA Handbook).

Cultural notes: The whistle of the thrush has a human quality, hence the bird is called the 'Whistling or Idle Schoolboy'.

Status: Uncommon.

Related Species: **Pied Thrush (Zoothera wardii-1731) Rare WM Karnataka south. 220 mm. Pied with yellow beak, black throat, spotted plumage, black tail. Female brown, lacks black throat.*

*Tickell's Thrush (Turdus unicolor-1748) Rare WM to Maharashtra. 210 mm. Male-grey back, white belly, yellow beak. Female-brown,white throated, spotted breast, dark malar stripe. Open forests.

*Siberian Thrush (Z.siberica-1732) Winter vagrant. Male-slaty, female brown with scaly breast, white brow. White outer tips to tail.

Malabar Whistking Thrush eating a rather large snake & finding it difficult to swallow!

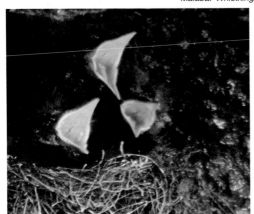

Eurasian Blackbird shares the same habitat & also breeds in the monsoon.

Chicks of Malabar Whistling Thrush, Their golden yellow gape is typical.

The typical nesting site of the Malabar Whistling Thrush.

Kadukhaoo,Shubhrakanthi Kastur

Description: White throat, ear-coverts and two brown cheek stripes are typical. Keeps to the shade, hopping on the ground, looking for berries and insects under the leaves. Seen in the company of the pitta and blackbird. Male displays and sings during courtship.

Orange-headed Thrush on the nest (Z.c.cyanonotus)

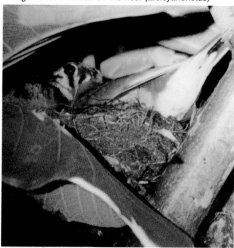

Nilgiri Thrush.

Z.c.citrina

Whitethroated Ground Thrush

Muscicapidae

Zoothera citrina cyanotus(1733,1734)

Size: 210 mm   M / F : Alike   **R**

Distribution : The entire region up to 2000 m. Race citrina is vagrant to Maharashtra till about Ratnagiri in winter.

Nest-site: The bulky shallow cup is placed in the fork of a low bush or as high as 6 m. up in a tree near a forest road or in the garden. *Jun. - Sep.*

Material: Roots, twigs, moss, fibres, coir and plastered with mud.

Parental care: Both.

Eggs: 3-4, greenish, red-mottled. 25x18.5 mm.

Call: Mimics calls of bulbul, tailor bird, babbler. Alarm calls, songs and whistles are given.

Ecological notes: Centipedes, millipedes, spiders, worms, small snakes, lizards are fed to the chicks. Nests in sacred groves on mango and *ficus* spp. in Kokan.

Cultural notes: The vernacular name Kadukhaoo means one who eats earthworms, an apt description of the thrush's food habit.

Status: Common.

Orange-headed Thrush (Z.c.citrina)

Related species: *Scaly or Nilgiri Thrush (*Z. dauma-1742*) Occasionally in the Western Ghats, Goa and Belgaum south. Muddy brown with prominent scales all over. Belly whitish.
*Zoothera citrina citrina (1733) is a winter vagrant to our area. It lacks the black cap of the Western Ghat species.

Eurasian Blackbird

Blackbird

Muscicapidae

Turdus merula (1753,1755,1756)

Kalchatak

Size: 250 mm   M / F : Dimorphic   **R, LM**

Distribution : The entire region up to 2000 m.

Race *nigropileus* Jog Falls north, *sinillimus* Jog

Falls to Palni Hills, *bourdilloni* southwards.

Nest-site: Large deep cup in the fork of a bush, tree 1 to 8 m. Up, on hill slopes or groves. *Jun. - Sep.*
Material: Grass, twigs, roots, moss, bark, palm leaf shreds. Lined with vegetable down.
Parental care: Both. Incubation by female. Parents leave the nest when approached.
Eggs: 3-5, bluish, red spotted. Chicks are lighter in colour. 27.4x20.9 mm.
Call: Mimics calls of cuckoo, kite, partridge, babblers, etc. Specific alarm notes.
Ecological notes: Eggs and chicks are often marauded by snakes, even in the presence of vocal helpless parents. Help offered by alert observers, by driving away snakes, when the nest is built near a house, is not infrequent. The empty nest may be used by magpie robin(Vishwas Joshi).
Cultural notes: This songster features in poetry.
Status: Common.

Male.   **Kaloo**

Description: The darker head as compared to the body, gives the male birds head, a cap like effect. Female is paler. Both have yellow bill, legs and eye-rings. This shy bird flies at the slightest intrusion. Only the male sings. More terrestrial, with thrush like habits. Keeps to hills. Flocks are seen on flowering trees in the plains, in winter.

Magpie Robin (F) using a Blackbird's nest.

Blackbird's eggs & chicks, lower nest is made of moss.

Eurasian Blackbird (F)

Male.   **Shankar**

Description: Arrives in October and departs by March. Come to the same feeding ground on identical dates. These drab coloured, silent and shy birds are often overlooked. They guard the feeding territory and when encroached upon by wagtails, sandpipers the rufous tail is fanned and cocked aggressively, revealing the angry chestnut colour! If disturbed, they fly into dense foliage.

White-bellied Shortwing (Race-major)

Bluethroat (F)

Bluethroat

Muscicapidae

Luscinia svecica(1644,1645)

Neelagreeva

Size: 150 mm   M / F : Dimorphic   **WM**

Distribution : To the region up to North Karnataka and Bangalore.

Nest-site: Breeds in North Eurasia, on wet ground, with the nest cup hidden in dense vegetation. One race, Ladakh Bluethroat (Syn. No.1646) breeds in Ladakh. *May - Jul.*
Material: Leaves, sticks, fibres, rootlets.
Parental care: Both.
Eggs: 3-4, greenish brown, red spotted.
Call: Silent in winter. *Churr-r-r, chuck, chuck* uttered from a perch when disturbed. A song may be heard in April, if bird over-winters.
Ecological notes: Affect reed beds, irrigated fields and marshes which are diminishing.
Cultural notes: As per the Puranas, during the Samudramanthan (churning of the oceans by the gods and demons, in search of nectar), along with other things, a flask of poison was obtained. This was swallowed by Lord Shiva or Shankar, which imparted a blue hue to his throat. The vernacular name Shankar, of the Bluethroat, is in keeping with this popular tale!
Status: Common.

White-tailed Rubythroat.

Related Species:
Indian Blue Robin or Blue Chat (L.brunnea-1650), winters in Western Ghats, Deccan. Keeps to shady areas. White eyebrow, black lores and ear-coverts, blue back and orange breast.
White-bellied Shortwing (Brachypteryx major-1637,1638) 150 mm. Endemic to Ghats Karnataka south,900 m-2000 m. Race albiventris-short tail, white belly, slaty blue forehead, long legs. Terrestrial. Palghat south. Race major-rufous flanks and belly. Vulnerable.
Siberian Rubythroat (Luscinia calliope-1643) Vagrant. White brow, belly, brown back, black lores, pink chin throat. Tail cocked.
Himalayan Rubythroat (L.pectoralis-1647) One record from Londa (Koelz,JBNHS 43:14).

White-bellied Shortwing (Race-major)

White-bellied Shortwing (Race major) on the nest with chicks.

White-bellied Shortwing (Race-albiventris)

Bluethroat - threat display.

Siberian Rubythroat (F)

Siberian Rubythroat (M)

*Indian Blue Robin
(Male & Female)*

Magpie Robin is the national bird of Bangladesh & features on the currency.

A Magpie Robin chick being eaten by the crow.

Magpie Robins attending the chick.

Magpie Robin (F)

Magpie Robin (Juv.)

Magpie Robin (M)

Oriental Magpie-Robin

Robin Dayal

Muscicapidae

Copsychus saularis(1661,1662)

Ashwak, Dadhika, Shreevadpakshi

Size: 200 mm   M / F : Dimorphic   **R**

Distribution : The entire region up to 2000 m.
Race *ceylonsis* south Karnataka southwards.

Nest-site: Tree hole, wall hole, broken lamp shades, signal boxes, open switch boxes. Also in the crotch of an epiphytic fig. *Apr. - Jul.*

Material: Rootlets, hair, threads, grass, paper, plastic, cloth and feathers.

Parental care: Both feed the chicks and build the nest. Incubation is by the female alone.

Eggs: 3-5, pale blue green, brown spotted. 21.9x17.1 mm. Incubation-13 days.

Call: A monotonous *Swee, swee-ee, churr*, in the non-breeding season. Pleasant song uttered during breeding. Mimics bird songs. Soothing notes of the happy Dayal heard in the tranquil afternoon is a pleasant experience.

Ecological notes: Visit flowering *Erythrina, Salmalia, Butea*, trees for nectar. Cross pollination follows. Insects are also eaten.

Cultural notes: Name **Ashwak** indicates the birds habit of flicking the tail like a horse (**Ashwa**). The name Dayal has roots in the Sanskrit word **Dadhika** (One who sells curd), signifying a black bird, as if splashed with curds.

Status: Common.

Dayal, Dominga, Chitko

Description: Pied, cocked long tail; male black and female grey. White wing bar. Feeds on ground, remaining in the vicinity of human dwellings in gardens. When alighting, the pied tail is fanned, producing a dazzling, shimmering visual effect! The breeding male displays with wings and tail fanned, performing aerial sallies, singing all the time.

Oriental Magpie Robin Female.

Oriental Magpie Robin Juvenile.

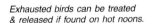

Exhausted birds can be treated & released if found on hot noons.

White-rumped Shama

Shama

Muscicapidae

Copsychus malabaricus (1665)

Krushnika, Kumarishama

Size: 250 mm.   M / F : Dimorphic   **R**

Distribution : Patchily throughout the region up to 700 m.

Nest-site: Bamboo clump or tree hollow, within 2m. in dense thorny clump, difficult to approach. *Mar. - Jul.*

Material: Leaves, grass, roots, rubbish.

Parental care: Nest built by the female, both feed.

Eggs: 3-5, pale, blue-green, brown-speckled. 22x17mm.

Call: Vocal in summer. Melodious notes Oio-lee-nou. Call-*Chir-churr, chi-churr*. Also gives harsh alarm calls when the nest is approached by others. More heard than seen. A peculiar wing flapping is heard during the courtship.

Ecological notes: Associated with bamboo forests.

Cultural notes: It was a popular cage bird due to its pleasant song, long life and adaptation to captivity. The name **Kumari** is that of a goddess, created with the confluent energy radiating from the eyes of Bramha, Vishnu and Mahesh. The colour of the bird Shama compares well with the description of the goddess.

Status: Uncommon.

Sham, Shama

Description: White rump of the Shama as it rummages in dark undergrowth, quickly meets the eye. Cocks tail like the Magpie Robin. Loud chattering on the sighting of a cobra has been noticed. Usually keeps to shady areas. Male has a glossy-blue back. Female has a grey-brown head, dull orange chest, short square tail. During the breeding period much chasing and singing is noticed.

Indian Robin

Muscicapidae

Saxicoloides fulicata (1719,1720)

Krushnapakshi, Shakuntabaalak, Podaki

Size: 160 mm   M / F : Dimorphic   **R**

Distribution : The entire region up to 1800 m.

Race *fulicata* is seen Belgaum south.

Nest-site: Holes in compound and well wall, earth cutting, in pipes, discarded cans, etc. The same site is often used subsequently. *Apr. - Jun.*

Material: Grass, hair, twigs, feathers, plastic. Snake-skin is seen in many nests!

Parental care: Both build and incubate. The chicks are fed even after they leave the nest.

Eggs: 2-3, creamy white, blotched with red. 21.1x14.9 mm.

Call: *Chreep, chreep, churr, churr*, giving them the name-Chirak. Male sings when breeding. A whistle *Wich, wheech, swee, sweet* is uttered. Very vocal when the nest is approached.

Ecological notes: Chicks from the ground nest are eaten by mongoose, snake, monitor lizard, etc. Birds frequent termite mounds, rotten trunks, sewage tanks, looking for insects, cockroaches. A pair in an urban area was seen to feed cockroaches to the chicks.

Cultural notes: The robin features in nursery rhymes and poetry!

Status: Common.

Race *intermedia* (Brown crown)   **Chirak, Lalbudya**

Description: Male-black brown with a blue gloss and a white wing patch. Female-brown and lacks the wing patch. Both sexes-wings rounded, thin bill curved and the vent is chestnut, best seen when tail is cocked. The breeding male executes a dance, flies and puffs the feathers. He offers food to the female, which if she accepts, mating follows. Bulbuls often bully the robins while they attend to the chicks, by stealing the grub.

A rare albino Indian Robin.

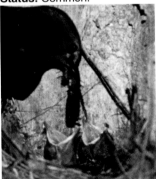

Female Robin feeding a cockroach to the chicks.

Nest in a niche.

Courtship feeding. Race *fulicata* has the male with entirely dark crown.

Female Robin feeding a chick.

Male.   **Thirthira, Kapra**

Black Redstart

Muscicapidae

Phoenicurus ochruros (1672)

Kapekshuk, Khanjari, Chalpichhak

Size: 150 mm   M / F : Dimorphic   **WM**

Distribution : In the region till Palni Hills, up to 600 m.

Description: The vibrating, shivering tail is diagnostic. Arrives at the same spot at the same time every winter, male coming earlier than the female. Prefers shady area. Rummage on ground and launch aerial sallies after winged insects. Confiding, freely entering gardens, verandas, parking spaces under buildings. On cold winter mornings, it is fun to see shivering people squatting by the blazing fire the Redtstart, hopping in the vicinity with its trembling tail!

Black Redstart (F)
White-capped Water Redstart (M)

Nest-site: Western Himalayas, Kashmir, Ladakh, etc. Nest is placed in a wall hole or rock crevice. *May - Aug.*

Material: Grass, rootlets, fibres, moss, hair, lined with feathers.

Parental care: Both. Incubation more by the female.

Eggs: 3-6,white to blue.

Call: Silent in winter. Alarm calls and multisyllabic warbles are sometimes uttered. Call- *Tic, tic, whit, whit*. Sings when breeding.

Ecological notes: Brood parasitized by *Cuculus canorus*.

Cultural notes: Was considered to be a bird of augury. Sanskrit name **Kapekshuk** means one trembling like a kind of Grass (*Ikshu*). The vernacular names are also descriptive of the ceaseless vibrations of the tail of this bird.

Status: Uncommon.

Look alikes:

Rusty-tailed Flycatcher (Muscicapa ruficauda-1409) Looks similar to female Black Redstart. The tail-shivering habit is absent. Vagrant to our area.

White-capped Water Redstart (Chaimarrornis leucocephalus-1675) Himalayan, not seen in our region.

Rusty-tailed Flycatcher

Pied Bushchat

Muscicapidae

Saxicola caprata(1700,1701,1702,1703)

Chatak

Size: 130 mm   M / F : Dimorphic   **R, WM**

Distribution : The entire region up to 2300 m.
Race *burmanica* till Nilgiris, race *nilgiriensis*
Nilgiri south, race *bicolor* WM up to north
Karnataka and race *atrata* Palghat south.

Nest-site: Niche in a compound wall, on
ground under a jutting rock or under a bush.
Feb. - May.

Material: Hair, goat hair, threads, twigs,
feathers.

Parental care: Both build the nest and attend
to the chicks. Female alone incubates.

Eggs: 3-5, pale blue, brown spotted, blotched.
17.6x13.9 mm.

Call: A song *Whee-te-too* and intermittent
harsh calls *Chek, chek, chek,*with wings
shaking.

Ecological notes: Insectivorous. The nest is
rarely brood parasitized by *Cuculus canorus*.

Cultural notes: It is called Gappidas-a great
chatterer, due to its hypervocal habit!

Status: Common.

Races-caprata
& bicolor.

Male.   **Gappidas, Kawdya Vatavatya**

Description: Male-small black with white
rump, abdomen and lower breast. White wing
bar and upper tail coverts are pointers.
Juvenile is scaly brown overall. Female is dull
brown. Perch on a wire, bush top, high mound,
boulder, with the pair calling each other. Tail
and wings are fluttered while singing. Seen in
the same area, guarding the chosen, scrub
and arid grounds, from other chats.

Pied Bushchats juvenile, male and female.

Pied Bushchats with gecko, grasshopper, mantice and cricket.

Male (Br)   **Rangit Vatvatya**

Description: The white collared male has a black head, throat, rufous belly, white wing patch and rump. Female has a streaked head, back and rufous rump. Perches upright on bushes, launching sallies after winged insects rarely coming to the ground. Beautiful bird of scrub country when in fresh plumage.

Related Species:
Indian Brown Rock Chat (Cercomela fusca-1692) Brown with rufous underparts and black tail. Sexes alike. Confusable with female Indian Robin and Blue Rock-Thrush. Recorded as a vagrant, at Saswad near Pune (authors). Usual southern limit is till the Narmada river.

Indian Brown Rock Chat.

Common Stonechat. (Non-Br. male-S. maura)

Common Stonechat

Collared Bushchat

Muscicapidae

Saxicola torquata(1695,1697)

Kharpidwa

Size: 130 mm   M / F : Dimorphic   **WM**

Distribution : The entire region. Race *maura* in North Maharashtra, *indica* up to south Karnataka.Up to 2000 m.

Nest-site: On ground in a grass clump, niche in a stone wall. In the Himalayas. *May - Jul.*
Material: Grass, fibres, goat-hair, rootlets, moss. Lined with feathers.
Parental care: Both, but incubation only by the female, for 13 days.
Eggs: 4-5,blue-grey, red spotted. 17.3x13.5 mm.
Call: Silent in winter. Call *Chek, chek, ke*.
Ecological notes: Insectivorous. Brood-parasitized by *Cuculus canorus*. A bird of drier terrain, but is also seen near the seashore and tidal creeks on dry sand banks.
Cultural notes: The name Kharpidda is interesting.(Khar-a thorny plant, Pidwa-Goral goat or Pija, in Kashmiri dialect). It may indicate the birds preference for goat-hair entangled in thorny bushes, which it uses for making the nest.
Status: Occasional.

S. maura is now the Asian spp.
Common Stonechat. (1st Year male-S. maura indica)

Desert Wheatear

Muscicapidae

Oenanthe deserti(1710)

Kharchatak

Size: 150 mm   M / F : Dimorphic   **WM**

Distribution : Rare winter visitor to the region.

Nest-site: The cup is placed under a stone or a boulder. Breeding is extralimital in Central Asia, Europe, N.Africa and also in Kashmir. *Apr. - Jun.*

Material: Rootlets, fibres, moss, goat hair.

Parental care: Both. Female guards the nest and does not approach it in the presence of intruders. She has been observed to drive away the male, lest he should go to the nest and reveal its location! (SA,Handbook.)

Eggs: 4-5,in shades of blue, blotched. 19.7x15.4 mm.

Call: No song on winter grounds. A sharp call *Chi, chee-ti-ti, chee-ti-ti,* is heard. This familiar call heard on cold mornings, in the scrub country, helps one find this drab coloured bird.

Ecological notes: Commonly seen on winter crops like Jowar, near scrub and desert country, eating crop pests.

Cultural notes: A little known bird in the wintering grounds.The name Rangoja means a bird of the desert.

Status: Uncommon.

Male.   **Rangoja**

Description: Cryptically coloured sandy brown above, whitish below. Dark wings and tail. Ear coverts and throat of the male are black during breeding. On winter grounds, the otherwise black throat of male is spotted grey. Female has a dull colour and brownish ear coverts. Perches silently on a thorny shrub and launches aerial sallies after winged insects.

Desert Wheatear (M)

Related species:
*Isabelline Wheatear, Chat (O.isabellina-1706) The plump build, white eyebrow, slightly shorter tail, more white at base and buff throat help differentiate from female Desert Wheatear. Occasionally recorded along the eastern fringes of Western Ghats, with a few records from South Kokan. Prefers scrub and arid terrain. Perches erect.

Isabelline Wheatear.

Wynaad Laughingthrush

Rufousvented Laughing Thrush

Muscicapidae

Garrulax delesserti delesserti(1287)

Wakaar, Kuruwaak, Kurubahu   Malabar Nymph.

Size: 230 mm   M / F : Alike   **R**

Distribution : Western Ghats up to 2000 m. southwards from Goa and Belgaum. Not in Kokan.

Nest-site: Bamboo brakes, *Strobilanthes* etc. dense bushes, thorny shrubs, within 2 m. *Jul. - Aug.*

Material: Rootlets, twigs, creepers, dry leaves to make the half-domed cup.

Parental care: Both.

Eggs: 3-4, white. 27.5x21.3 mm. Juveniles are less chestnut coloured and duller.

Call: A garrulous bird. Screeches, shrill whistles, chatters and laughter are uttered by the flock in unison. Call *Tree, kree, truu, truu.*

Ecological notes: The nominate race is Endemic to Western Ghats, keeping to rain forests with dense undergrowth.

Cultural notes: Sanskrit names are aptly descriptive of the cacaphonous nature of these birds, who take pleasure in uttering harsh notes. A bird known since Buddhist times.

Status: Uncommon.

Pandhrya-galacha Gappu

Description: Black masked with grey head, chestnut brown back, white throat, grey breast and chestnut belly. Hops on ground, upturning rotten leaves and rubbish on forest floor with its beak and legs in search of insect, grub and berries. Flocks take to wings and retreat to the safety of dense undergrowth at the slightest intrusion, uttering harsh squeaks. At noon flocks perch huddled togeather uttering laughter like calls.

Grey-breasted Laughingthrush (G.j.fairbanki)

Related Species: *****N**ilgiri Laughingthrush (G.cachinnans-1307-08) 200 mm. Endemic to Nilgiri Hills 1200 m to over 2000 m. Rufous below, brown above with a white brow, dark chin. Bushes and thickets. Call-'peo,peo,tioo'. *****G**rey-breasted or Jerdon's Laughingthrush (G.jerdoni-1309-11) 200 mm. Endemic to Western Ghats Coorg south, from 1000 m to 2000 m. White brow, greyish chin, throat and rufous breast. Flocks in bushes. Call-'pu, pu, pee, pee, tui'. Race jerdoni in Coorg, fairbanki in Cardmom, Palni Hills upto Anchakovil gap, meridionale southwards in Ashambu Hills. Both Endangered.*

Nilgiri Laughingthrushes.

Ashy-Headed Laughingthrush
Sri Lankan Endemic.

Grey-breasted Laughingthrush
(G.j.meridionale)

Indian Rufous Babbler

Rufous Babbler

Muscicapidae

Turdoides subrufus (1259,1260)

Gokirati, Kurubahu (For all babblers)

Size: 250 mm   M / F : Alike   **R**

Distribution : Western Ghats southwards up to 1400 m. Race *hyperythrus* Palghat south.

Nest-site: In dense, often thorny bushes, vines and trees within 5 m. The nest-cup is placed in a fork of *Zizyphus,Lantana*, etc. trees. *Feb. - Jul.*

Material: Grass, twigs, leaves, bark, lined with rootlets and soft grass.

Parental care: Both.

Eggs: 3-4, deep blue, glossy. 25x18.5 mm.

Call: Loud high pitched call *Tree, tree, tree* and harsh squeaks in a low volume. Alarm calls.

Ecological notes: Fond of *Lantana* berries. Take nectar of *Capparis, Erythrina, Salmalia, Butea,* spp. aiding cross pollination. **Endemic** to Western Ghats.

Cultural notes: Sanskrit name ***Gokirati*** features in the Atharvaved (2500 BC). Girls of forest dwelling Kirat tribe search the forest floor for medicinal roots. Their name is given to these ground rummaging babblers!

Status: Common.

Lal Satbhai

Description: Forehead grey,blackish yellow bill, dark lores. Race *hyperythrus* is more rufous than the northern race. Moves swiftly in bamboo brakes, dense scrub and thorny bushes, skulking, not easily seen, only the call betraying their presence. Secretive vanishing in the undergrowth at the slightest intrusion. Does not mingle with other babblers.

Related Species:**C**ommon Babbler (*T.caudatus-1254) Scrub country, coasts and Western Ghats up to 1500 m. Terrestrial, but ascends trees for nectar and berries. Brown streaked back and finely barred tail. Call-'Which, whiri, whiri-ri'. Unlike the other babblers, the nest cups are well made, and may be brood parasitized by the cuckoos.* ***S**potted or Puff-throated Babbler (Pellorneum ruficeps-1152) common in wooded areas of Ghats and Kokan, not on coasts. Skulker, with reddish head, white eyebrow and spotted breast.*

Common Babbler.

Common Babbler.

Spotted or Puff-throated Babbler.

Shwetkanthi Satbhai

Description: The northern race *abuensis* has reddish head, race *albogularis* has red only on the forehead. Race *hyperythra* without the white throat, is encountered along eastern fringes of Ghats around Pune. Flocks move in follow the leader fashion, calling each other, especially when crossing forest clearings. Rummage in the undergrowth with warblers, robins and tailor birds.

Rufous-bellied Babbler (D.h.abuensis)

Slatyheaded Scimitar Babbler's nest.

Related Species:
*Indian or Slatyheaded Scimitar Babbler (Pomatorhinus horsfieldii-1173, 74)
Elusive. White belly, eyebrow, olive brown back and yellow curved bill. In Ghats and Kokan till 2000 m. Race horsefieldii up to Goa, travancorensis southwards. Call 'Kuli, kiwi, kwa, kluwi, kqui, qua, quak'. The nest ball is seen on ground, bush or niche in an earth bank.*

Rufous-bellied Babbler

Tawny-bellied or White-throated Babbler
Muscicapidae
Dumetia hyperythra (1219-21)
Bhoosarika
Size: 130 mm   M / F : Alike **R**

Distribution : The entire region up to 1800 m.
Race *albogularis* Pune southwards, *abuensis* in the northern range.

Nest-site: The nest-ball is hidden in a dense or thorny bush within 1 m. of the ground. *May -Sep.*
Material: Bamboo leaves, grass, twigs, paper woven into a ball. Lined with fine grass, rootlets.
Parental care: Both.
Eggs: 3-4,pink white, brown blotched. 17.4x14.1 mm.
Call: Twittering, squeaking calls *Chik, chik, Sweech, sweech*, uttered incessantly. Alarm calls.
Ecological notes: Cross pollinator. Brood is parasitized by Banded-Bay Cuckoo.
Cultural notes: The Sanskrit name **Bhoosarika** (Ground-bird) indicates their terrestrial habit.
Status: Uncommon.

Slatyheaded Scimitar Babblers.

Large Grey Babbler

Muscicapidae

Turdoides malcolmi(1258)

Haholika, Kutsitaangi

Size: 280 mm   M / F : Alike   **R**

Distribution : The entire region up to 1200 m.

Uncommon on coasts.

Nest-site: On a horizontal branch of thorny trees 3-5 m. up. *Mar. - Aug.*
Material: Twigs of *Acacia nilotica, Azadirachta indica, Delonix regia*, threads, grass to build the shallow untidy cup lined with rootlets.
Parental care: Both and the sisterhood.
Eggs: 3-5,shiny blue, unmarked.
24.2x19.5 mm.
Call: A loud *Ke, ke, kay, kay*, initiated by one and repeated by the entire group, becomes hysterically noisy. (Hence the name-*Haholika*).
Ecological notes: Play a role in cross pollination of flowering plants.
Cultural notes: Their habit of following each other noisily from tree to tree, is similar to that of the Gosavis (people who move from house to house, vociferously asking for alms), hence the name Gosavi. Satbhai are seven brothers. The Sanskrit name **Kutsitaangi** refers to their untidy and dishevelled appearance.
Status: Common.

Gosavi, Satbhai, Kekatya

Description: Grey forehead, black lores, bright lemon-yellow iris and white bordered long tail. Move clumsily on ground looking for insects, seeds, berries. Flower nectar is taken. Warning calls uttered when intruders are sighted. The attention of one author (Satish Pande), while sitting in a hide, was directed towards a cobra, by the alarm calls of these birds!

Communal feeding.

Brood Parasitism: *Possibly more than one pair lay eggs in one nest. Communal feeding is noted with some birds performing sentry duty. Yet, nests are brood parasitized by the cuckoo species. A Common Hawk-Cuckoo chick was seen to be fed by the sisterhood both on trees and ground. An anxious vibrant wing activity of chick precedes feeding. A fight breaks out when foster parents and cuckoo chick enter the territory of other babblers.*

Large Grey Babbler feeding the Brainfever Bird's juvenile.

Raanbhai, Bhoosarin

Description: Western Ghat or Bombay species (Syn.No.1263) is less striated and more brown, with blacker flight feathers. Sisterhoods roost communally close to each other. Predators are warded off by puffing the feathers and by vocal and physical aggression. They are poor fliers. Flowering *Salmalia, Butea,Erythrina* spp. are habitually visited.

Jungle Babbler's nest.

Related Species: *Yellow-billed or White-headed Babbler (T.affinis-1267) Head, neck creamy-white, ear coverts dark, neck mottled and beak yellow. In Ghats up to 1000 m. Belgaum south in scrub, secondary forests, in gardens near villages. Call 'Tri-ri-ri' and chuckles uttered in a chorus.*

Yellow-breasted or Striped-Tit Babbler (Macronous gularis-1228) 110 mm. Rufous cap, brown tail, pale yellow eyebrow, streaked chin and chest. Small population near Mysore, other in NE India.

Jungle Babbler (T.s.malabaricus)

Jungle Babbler

Muscicapidae

Turdoides striatus(1262,1263,1264)

Wakkar, Kuruwak

Size: 250 mm   M / F : Alike   R

Distribution : Race *somervillei* till northern Karnataka, *malabaricus* southwards, *orientalis* widespread up to 1200 m.

Nest-site: Fork of a horizontal branch of a thorny tree 2-5 m. up. *Mar. - Oct.*

Material: Leaves, grass, twigs,paper, thread and lined with rootlets.

Parental care: Both parents share the duties, with the sisterhood assisting them. More than one pair may lay eggs simultaneously in the same nest, but usually in succession.

Eggs: 3-6, turquoise blue. 25.2xz19.6 mm. Identical eggs are laid by parasitic cuckoos.

Call: Loud calls *Ke, ke, ke, ke* uttered by the flock while rummaging or when alarmed. Squeaks, chatters, squabbles uttered in a commotion.

Ecological notes: Nests are brood parasitized by Pied Crested Cuckoo and Common Hawk-Cuckoo. Cause seed dispersal of *Lantana, Zizyphus* and Fig spp. Cross pollination of flowering forest trees.

Cultural notes: Sanskrit names **Vakkar** and **Kuruvak** indicate their noisy nature and this habit is recorded from before the Buddhist times (600BC).

Status: Common.

Jungle Babbler Dark morph.

Orange-Billed Babbler Sri Lankan Endemic.

Yellow-eyed Babbler

Muscicapidae

Chrysoma sinense(1231)

Peetanetra

Size: 180 mm   M / F : Alike   **R**

Distribution : The entire region up to 1500 m.

Nest-site: Well hidden in *Carissa congesta*, *Lantana*, etc. bushes and in grass clumps. *May - Sep.*

Material: Leaves, grass, twigs are plastered with cobweb to make a deep cup within 1 m.

Parental care: Both. Nest is reluctantly approached in the presence of observers and displeasure is loudly expressed. A broken wing display to divert the intruders is common.

Eggs: 4-5, pink, with chestnut blotches. 17.9x14.9 mm.

Call: Harsh calls Cirr, chirr are uttered when alarmed. Breeding male sings *Cheep, cheep* from an exposed branch-hence its name Chipka.

Ecological notes: Help pollination and insect control.

Cultural notes: Often heard than seen, hence the onomatopoeic name!

Status: Common.

Chipka

Description: Orange eye ring is more conspicuous than the yellow iris! Secretive, keeps to bushes and disappears at the slightest disturbance. Flight is broken, jerky and clumsy. Clinging to twigs in an acrobatic manner in search of insects, berries and nectar.

Dark-fronted Babbler.

Brown-cheeked Fulvetta.

Brown-cheeked Fulvetta's hammock nest.

Related Species: *Black-headed or Dark-fronted Babbler(Rhopocichla atriceps-1224, 25) Yellow eyes and black head.
Western Ghats Goa south till 1800 m. Race bourdilloni Palghat south. Evergreen forests. Loose parties, close to streams and ravines. Call-'Churr, churr'. Roost in old nests.
*Quaker Tit-Babbler, Brown-cheeked Fulvetta (Alcippe poioicephala-1389,1390) Vocal and keeps in flocks, mixed parties, in moist forests up to 2100 m. Nominate race Goa south. Constant 'Chirr, chirr' is uttered to keep in touch with one another. Breeding coincides with SW monsoon,and nests are often near houses.

Jungle Babblers taking a mud bath.

Striped Tit Babbler (Macronous gularis-1228)
Rufous crown, thinly streaked breast & yellow eye brow.
Rarely in the southern Ghats near Mysore. Size-110 mm.
Keeps to undergrowth & bamboo clumps.

Yellow-billed or White-headed Babbler-note the pale rump.   Yellow-billed or White-headed Babbler.

Jungle Babbler.

Rufous-fronted Prinia

Rufousfronted Wren-Warbler

Muscicapidae

Prinia buchanani(1506)

Chatakika

Size: 120 mm   M / F : Alike   **R**

Distribution : The entire region up to North Maharashtra till about Pune.

Nest-site: Woven under 1 m. in a grass tussock or thorny hedge, suspended by twigs. Bungalow gardens, neglected corner in a plant nursery. *Jun. - Sep.*
Parental care: Both.
Eggs: 3-5, white, red blotched. 15.9x12 mm.
Call: Inspiring warbles and calls *Chirrup, chirrup* are continually uttered in a low volume, while moving in the undergrowth. Call *Chirrit, chirrit* is uttered with tail fanned and cocked.
Ecological notes: Frequents dry areas and low hilly terrain. Moves with *P.hodgsonii, Orthotomus sutorius, Dumetia hyperythra.*
Cultural notes: None in particular.
Status: Uncommon.

Lal-top Phutki

Description: Usually seen in pairs, hopping on ground, climbing twigs or a low branch. The graduated tail is flicked and fanned from time to time. Rufous head and white tail-tips help field identification. Pairs often roost at night in a bungalow garden, near edges of cultivations or scrub country.

Rufous-fronted Prinia.

Related Species: **Streaked Fantail-Warbler or Zitting Cisticola (Cisticola juncidis-1498, 1499) Restless, tiny, streaked brown bird with white belly and short white tipped tail, often fanned. They shoot up in a zigzag manner, calling 'Chip, chip' proclaiming their presence. Nests in monsoon. Seen up to 2100 m. Race cursitans replaced in Kerala by race salimalii.*
*Golden-headed Fantail-warbler or Brightheaded Cisticola (C.exilis-1496) Unstreaked brown head. Recently recorded near Pune. Resident in Western Ghats Coorg south and BR Hills above 900 m.

Brightheaded Cisticola.

Zitting Cisticola on the nest.

Zitting Cisticola.

Franklin's Prinia removing the fecal pouch from the nest. **Kantheri Watwatya**

Description: Long tail with black and white tips. Grey breast band of the breeding birds is inconspicuous in non-breeding adults and in juveniles. Race *albogularis* has the band only in summer. Jerky flight. They have an obliterative colour and are not as common as Plain Prinia. Flocks are seen in winter on flowering trees.

Franklin's Prinia (Juv.)

Franklin's Prinia

Greybreasted Prinia

Muscicapidae

Prinia hodgsonii (1503,1504)

Patangika, Putrika

Size: 110 mm   M / F : Alike   **R**

Distribution : Race *hodgsonii* up to 1500 m. in the entire region. Race *albogularis* Coorg south.

Nest-site: Tailorbird-like nest on short trees, dense bushes, dense undergrowth or tussocks. *Jun. - Oct.*

Material: Leaves are stitched to make the outer oblong sheath within which a cup made from grass, fibres and rootlets is placed.

Parental care: Both. Parents continue to feed the chicks for some time after they leave the nest. At this time they are prone to predation.

Eggs: 3-5, white, oval, red spotted. 14.7x11.7 mm. The immature birds have yellow underparts.

Call: Song is uttered from a perch or when nosediving during the breeding display. Low incessant chatter *Chee, chee, pree, pree, wich*, and warbles are uttered, as the birds stealthily move in the grass clumps.

Ecological notes: Cause cross pollination. Breeding depends on SW monsoon.

Cultural notes: The bird is not generally known as a separate species and is mistaken for any other warbler or a tailorbird.

Status: Uncommon.

Birds often feed fecal sacs to the chicks in first few days.

Plain Wren-Warbler

Muscicapidae

Prinia inornata (1511,1513)

Chatika, Phutkari

Size: 130 mm   M / F : Alike   **R**

Distribution : The entire region up to 1800 m.

Race *franklinii* is seen Coorg South.

Nest-site: In grasslands, grassy field bunds, and *Lantana, Zizyphus, Croton* spp. bushes. *May - Sep.*

Material: Unlined. Grass blades, fibres, threads, are woven around approximated grass stems to form the longish pouch, within 1m. from ground. A tailor bird like nest with leaves stitched together is also seen.

Parental care: Both. Wing flapping sound as the bird leaves the nest hurriedly when it is approached, gives away the site of the well concealed nest! Multiple broods are reared.

Eggs: 3-5, bluish green, blotched, lined. 15.6x11.5 mm.

Call: Prolonged calls *Tee, tee, cree, cree, click, Click, phut, phut* with vibrating wings and fanned tail are uttered from a bush top.

Ecological notes: Nests are predated by coucal, shikra, shrike, snake, mongoose. Nests in tall grass clumps suffer damage in heavy rains. Nests are sometimes trampled by grazing cattle. An incidence was noted, where the nest bearing grass was cut for fodder and chicks smothered in the stack!(Satish Pande)

Cultural notes: This bird is called **Phutkari** due to it's call *Phut, phut*.

Status: Common.

Vatvatya

Description: Back is grey brown, lores and eyebrow white. The graduated erect tail is loosely flicked, as it scuttles in grass and reed beds. Tail has a terminal white band and subterminal dark band. Iris is orange red. Seasonal plumage variation is significant. in fresh plumage, the otherwise cream coloured breast is yellow.

Plain Prinia in Giant Wood Spider's web of death !

Juvenile.   *Jungle Prinia.*

Related species:
**Jungle Prinia
(P.sylvatica-1521) Iris orange red, longer drooping tail with white outer-tail feathers. Kokan, Western Ghats up to 1500 m. Prefers stony and scrub habitat. Call- 'Pretty,Pretty, pit, pit'. The unlined nest, woven with grass, has a side hole and is close to the ground. The male displays by nose diving and repeatedly shooting up in to the air.*

Rakhi-vatvatya

Ashy Prinia

Ashy Wren-Warbler

Muscicapidae

Prinia socialis (1517)

Purullika

Size: 130 mm   M / F : Alike   **R**

Distribution : The entire region up to 1500 m.

Nest-site: Nest pouch is seen in grass tussocks, the tailor bird like nest is seen in *Croton* spp; *Annona squmosa, Exora* spp; *Arisaema murraye* etc. Within 1 m. from the ground. *May. - Sep.*

Material: Grass blades, thin twigs, cobwebs, thread-often green in colour, cotton, hair. Three types of nest are built, often by the same pair! **a)**Pouch of grass **b)**Leaf cup-tailor bird like and **c)** Grass ball-munia like, with lateral entrance.

Parental care: Both. Nest toilet is practised.

Eggs: 3-5, reddish, shiny, beautiful. 16.2x12 mm.

Call: A sharp repetitive *Tee, tee*, hence the name *Tuntuk*. Male sings when breeding. Wing and beak clapping is also heard.

Ecological notes: Nests are brood parasitized by Grey-bellied cuckoo. Nest predation is common. Nesting is often in gardens. Birds rid plants of pests. Repair of inadvertently damaged nests is accepted by birds.

Cultural notes: The name *Purullika* means one who inhabits gardens. (Sanskrit: Paurak-garden)

Status: Common.

Description: Drooping tail flicked often, ashy grey-brown back, yellowish white belly and red iris. Birds undergo spring and autumn moults. Active, restless, confiding, vocal. Rummaging in dense foliage appearing now and then on an upper branch. More than one nest ibuilt simultaneously.

Eggs & chick.

Juvenile.

Munia like nest.

Accepting a repaired nest.

Indian Great Reed-Warbler

Clamorous Reed Warbler

Muscicapidae

Acrocephalus stentoreus(1550)

Trunchatika

Size: 190 mm   M / F : Alike   WM, **R**

Distribution : The entire region up to 1600 m.

Also resident patchily in Maharashtra and Kerala.

Boru Vatvatya

Nest-site: Breeds locally in North India and also outside India. *May - Aug.*

Material: Leaves, reed and bark are used to make the large cup. It is attached hammock-wise to reeds, barely 50 cm above the water surface.

Parental care: Both feed the chicks.

Eggs: 3-6, ashy, green, brown spotted. 22.7x15.9 mm.

Call: Loud enthusiastic calls are uttered-*Churr, churr, ke, ke, karra, karra, kreet.* Mimic.

Ecological notes: This winter visitor is partial to wet lands and reed beds. Ringing data of extralimital migration, Calcutta to Uzbek, near Samarkand is available (SA,Handbook). Brood parasitized by *Cuculus canorus.* The nests are prone to drowning in floods.

Cultural notes: A less known bird. The call is however linked to the arrival of winter in Kokan, where reed-rich wet lands and swamps are still encountered close to houses.

Status: Uncommon.

Description: Olive brown with a noticeable white eyebrow, white throat and buff belly. Tail lacks white tips. Similar looking *Thick-billed Warbler lacks white supercilium, orange mouth and gape. Essentially a bird of marshy habitat. Keeps to inner reed beds and often climbs to the top. Shares the habitat with ruddy crake, bittern, waterhen. More heard than seen. Its Hindi name Tiktiki is onomatopoeic.

Related Species: *Pale Grasshopper Warbler (Locustella naevia-1545) WM to drier areas, fields. White tips to tail and stripes on forehead. Call 'Chek, churr'.*
* *Blyth's Reed-Warbler (A.dumetorum-1556) Common WM to bushy areas, bamboo clumps, away from water and up to 2100 m. Call-'Chuk, chuk, chuk'.*
Paddyfield Warbler (A.agricola-1557) To wet land, covered with reeds, elephant grass or paddy. Rufous rump is best seen in flight. A secretive WM.
Thick-billed Warbler (A.aedon-1549) 200 mm. WM to marshes, gardens. White below, olive brown above. No eye brow.

Pale Grasshopper Warbler.

Paddyfield Warbler.

Thick-billed Warbler.

Jungle Prinia in the nest.
Plain Prinia.

Jungle Prinia on a blossom.

Paddyfield Warbler.

Blyth's Reed-Warbler during ringing. Bird ringing is done to study migration.,

Common Tailor Bird

Tailorbird

Muscicapidae

Orthotomus sutorius(1535)

Tuntuk, Putini (A leaf envelope)

Size: 130 mm   M / F : Alike   **R**

Distribution : The entire region up to 2000 m.

Nest-site: On broad leaved trees like *Canna indica, Bixa orellana, Pothos, Monstera, Annona* spp.,*Tectona grandis, Mangifera indica*, also *Salvadora oleoides* and other small leaf trees, in gardens, forest fringes. *Apr. - Sep.*

Material: A single large or many small leaves are skillfully stitched with thread, wool, twigs, bark shavings, hair, without placing knots, to make the outer funnel after several attempts. Nest cup is placed in it, lined with down, wool, moss.

Parental care: Female alone incubates, both build the nest and feed the chicks.

Eggs: 3-4, pink or blue-white,red spotted. 16.4x11.6 mm.

Call: A monotonous *Towit, towit* becomes high pitched if the nest is approached. Also *Pretty, pretty* and a soft call *Cheep, cheep*.

Ecological notes: Cross pollinator and insectivorous, beneficial to garden plants. Nest predation and brood parasitizm by Plaintive Cuckoo common. If cotton is offered while nest making, it is used for the nest! It was seen to use cotton from an old nest of an Ashy Wren-Warbler.

Cultural notes: Offered repair of broken active nests, with cloth and threads, is accepted! Being a garden bird, many calamities can be noticed when help can be successfully offered.

Status: Common.

Shimpi

Description: Green above. White belly and undertail-coverts with rufous head. Restless and confiding. Tail is cocked up and is a bit longer in male. The pointed beak is useful for nest stitching. Small black skin patches are seen on the throat of the breeding birds. Chicks emit a hissing sound when the nest is approached.

Look Alike: *Rufous-fronted Prinia (Prinia buchanani-1506) can be mistaken for Tailorbird, if seen in dense foliage. Both often move together in sparsely wooded areas. The rusty head is similar, but olive back of the tailorbird and its long tail streamers are distinct.*

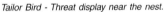

Tailor Bird - Threat display near the nest.

Dusky Warbler.

The leaf nest of this Tailor Bird suffered
premature collapse. The leaf stalk snapped from the stem
because of excess weight of the nest.
This nest was repaired with cloth and twine.
The Tailor Bird accepted the help
& chicks flew to freedom.

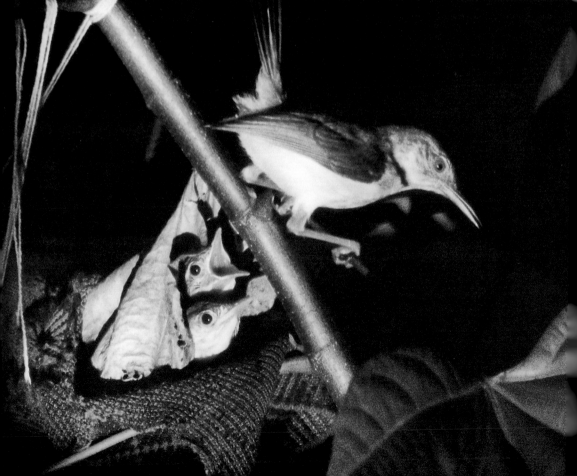

Common Chiffchaff

Brown Chiffchaff

Muscicapidae

Phylloscopus collybita (1575)

Jalaphutki

Size: 100 mm   M / F : Alike   **WM, PM**

Distribution : The entire region up to North Karnataka.

Nest-site: The domed pouch is placed in reed-beds or low thorny bushes. *Jun. - Jul.*
Material: Grass, fibres, twigs.
Call: The onomatopoeic *Chiff-chaff, chip, chalp, chivet, wichoo-weet-weet.* Sings in summer.
Ecological notes: Beneficial to winter crops since it devours insects and larvae, while perching on cobs.
Cultural notes: The influx of these drab coloured winter migrants is well noticed and it is called the winter sparrow.
Status: Common.

Panphutki

Description: Leaf warbler with an olive brown back, buff breast, whitish belly, eyebrow, unmarked and unbarred wings, black bill and legs. Tail is flicked as it hops on ground or clambers on standing crop, bushes and reeds in swamps. Hawks insects in air or from water surface, by steeply bending over, at times practically wetting itself and losing its balance, when it lets go the foothold and clumsily finds another hold! Restless bird with a jerky flight and horizontal perch.

Tytler's Leaf-Warbler(P.tytleri) Winter migrant.

Typical habitat.   Greenish Leaf-Warbler.

Hume's Warbler.

Related Species: *Hume's Warbler (P.humei-1590)* White wingbars and eyebrow. Seen up to Belgaum and Goa. *Greenish Leaf-Warbler (P.trochiloides-1602-5) Fine wingbar and thick eyebrow. The entire range. *Western Crowned Warbler (P.occipitalis-1606) Barred wings and bold eyebrow. Olive above, grey white below. In Western Ghats, not Kokan. *Tickell's Warbler (P.affinis-1579) Y ellow brow, belly. Lacks wing bars. Green above. Patchily in our area till 2100 m. *Dusky Warbler (P.fuscatus-1584-86) Rare. Recorded near Pune. Brown above, buff flanks, white belly, brow. Call-'chak'. Scrub areas.* Above Phylloscopus warblers are winter visitors.

Shwetkanth

Description: Restless brown-backed and grey-capped warbler with white underparts and tail with white edges and tips. When catching insects, it has a habit of over-reaching with the beak, to the point of toppling over! Keeps to the undergrowth and shrubs. Also takes nectar. Relatively confiding.

Orphean Warbler.

Booted Warbler.

Large-billed Leaf-Warbler.

Common Lesser Whitethroat

Lesser Whitethroat

Muscicapidae

Sylvia curruca(1567)

Shwetkanthi Chatakika

Size: 120 mm   M / F : Alike   **WM**

Distribution : The entire region, common in the Deccan.

Nest-site: Nesting is extralimital in Siberia.
Material: Grass, rootlets, hair, wool. *Apr. - Aug.*
Parental care: Probably both.
Eggs: 3-4.
Call: Silent on wintering grounds. Clicking notes *Tik, tik* are repeated in a low tone.
Ecological notes: Arrive on the same spot at the same time every year. A bird ringed in Kathiawar on 24-9-61 was retrieved at the same place on 26-9-62 (Shivrajkumar,JBNHS 59:963). Numbers apparently reduced when rains are less.
Cultural notes: None in particular.
Status: Uncommon.

*Related species: *Booted Warbler (Hippolais caligata-1562,63) A widespread WM to arid country. Pale brown back, white belly and supercilium. Call 'Chuck, chuck, churr.'*
Orphean Warbler (Sylvia hortensis-1565) WM. Keeps to fields and bushes. Secretive. Alarm calls 'Trr, trr' . A dark -headed grey warbler (150 mm),with white underparts and white edged dark tail. Female has a brown back.
*Large-billed Leaf-Warbler (Phylloscopus magnirostris-1601) WM Pune south. Call-Dirr, tee, wee, wee. Olive brown back, yellow brow, dark eye streak Buff belly, fleshy bill.
*Olivaceous or Sulphur-bellied Leaf-Warbler (P.griseolus-1581) Yellow brow, belly. Up to Uttara Kannada.
Yellow-browed Warbler (P.inornatus-1592) 100 mm. Wing bar, crown band. Yellow brow. Green back, yellow belly. WM till Belgaum.

Yellow-browed Warbler.   Sulphur-bellied Leaf-Warbler.

Broad-tailed Grass-warbler

Broad-Tailed Grassbird

Muscicapidae

Schoenicola platyura *(1546)*

Trunchatika

Size: 180 mm   M / F : Alike   **R**

Distribution : Very patchily, Pune and southwards in Western Ghats 900 m. and above.

Nest-site: 1-3 m. Up in grass tussocks.
Mar. - Aug.

Material: Grass blades, thin twigs.

Parental care: Probably female alone builds the nest.

Eggs: 2-3, pink-white, red spotted. 19.4x14.7 mm.

Call: A songster amongst warblers. Trills and warbling notes.

Ecological notes: Insectivorous.

Cultural notes: None in particular. Less noticed bird..

Status: Rare. Likely to be Endemic to the Western ghats. One old record from Sri Lanka. **Vulnerable**.

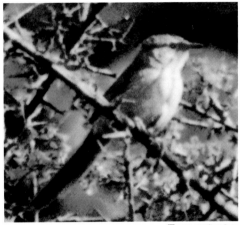

Trun-vatvatya

Description: Brown above, white below with broad faintly barred white tipped graduated tail. Secretively moves in grass and shrubs occasionally coming to the top. *Andropogon* grass preferred (Salim Ali, Handbook) Breeding birds perform spectacular displays as they soar in air..

Bristled Grass-warbler.

Bristled Grass-warbler.

***Related Species: *S**triated Marsh-warbler or Grassbird (Megalurus palustris-1548) Size-250 mm. Prominent white brow, boldly streaked above. Fine streaks on breast. Tail and bill longer that Bristled Grass-warbler. Rarely in our Northern range in marshy areas.
B**ristled Grass-warbler or Grassbird (Chaetornis striatus-1547) Size-200 mm. Bristles in front of eyes are movable and protective when the bird moves through coarse grass(Whistler, JBNHS 33:783) Streaks on upper part and throat. Buff tipped broad tail. Patchily and locally in the region. **Vulnerable

Striated Grassbird.

Male (F.p.parva)   **Tambula**

Description: Rufous chin and throat, dark tail with white proximal outer tail feathers, brown back and white belly. The tail is cocked rather jerkily, as it hawks insects in the canopy or on the ground. Female and young male lack red on the throat. Usually confiding, freely visiting gardens near bungalows. Arrives in September and leaves by May. Restless.

Brown-breasted Flycatcher.

Related species: *Asian Brown Flycatcher (Muscicapa dauurica-1407) White throat, eye-ring and dark legs are prominent. Keeps to deciduous forests. Breeds in South Ghats.*
Grey-headed Flycatcher (Culicicapa ceylonensis-1448,1449) Race calochrysea is WM till Maharashtra, race ceylonensis is R Karnataka south. Head and neck grey, belly yellow and back olive green.
Brown-breasted Flycatcher (M.muttui-1408) Rare WM to Ghats Goa south. Lower mandible is pale, breast dusky white and legs yellow.

Red-throated Flycatcher

Redbreasted Flycatcher

Muscicapidae

Ficedula parva (1411,1412)

Raktakantha Chatak

Size: 130 mm   M / F : Dimorphic   **WM, PM**

Distribution : Race *parva* up to Karnataka, race *albicilla* in the entire region up to 2100 m.

Nest-site: Extralimital, in eastern Europe and Russia, in a tree hole at medium height. *May - Jun.*
Material: Moss, dry leaves, bark, hair.
Parental care: Both ?
Eggs: 3-5, pale green, spotted.
Call: Song is not heard during the winter. Calls *Click, click, chick, chic* are uttered on the wing.
Ecological notes: An extralimitally breeding winter and passage migrant to our area. Preys upon winged insects, mosquitoes, etc.
Cultural notes: The name Tambula highlights the red coloured throat of the breeding bird, which is more conspicuous by May, when the birds commence their return migration.
Status: Uncommon.

Asian Brown Flycatcher.

Black-and-Orange Flycatcher

Black-and-Rufous Flycatcher

Muscicapidae

Ficedula nigrorufa (1427)

No sanskrit name recorded

Size: 130 mm   M / F : Dimorphic   **LM, R**

Distribution : Western Ghats Wynaad and BR Hills southwards till Ashambu Hills. 700 m. up to and above 2000 m.

Nest-site: Bushes, ferns up to 1 m. *Mar. - Jul.*
Material: Dry leaves, grass to make a nest ball.
Parental care: Both. Female builds the nest.
Eggs: 2, grey white, freckled with pink. 18.4x13.1 mm.
Call: High pitched *Chiki, ri, ri.*
Ecological notes: Occurs Nilgiri south and is altitude specific, common above 1500 m. Insectivorous. Seen in *Strobilanthes, eeta* bamboo undergrowth, cardamom and coffee plantations.
Cultural notes: None in particular.
Status: Endemic to Southern Western Ghats. Near Threatened.

Little Pied Flycatcher (M)

Related Species: *Kashmir Flycatcher *(Ficedula subrubra-1413) Male-rufous throat, belly with black lateral edge. Female is less rufous. Tail black with white edge. Rare WM. Vulnerable.*
*Rusty-tailed Flycatcher (Muscicapa ruficauda-1409) 140 mm. Rare WM. Brown above, white below. Lower mandible orange.(See page 250)
*Little Pied Flycatcher (F.westermanii-1419) 110 mm. Vagrant. Male -white belly, brow, wing band. Female-brown back, grey white below, rufous rump, tail, pale wing bar. Recently near Pune (Satish Pande).

Male.   **Narangi Nartak**

Description: Rufous orange and black-male, female is duller. Keeps to undergrowth. Confiding when not disturbed. Often forages on ground like a babbler. Sallies after insects. Very beautiful when seen in sunlight.

Rusty-tailed Flycatcher.   *Little Pied Flycatcher (F)*

Kashmir Flycatcher (M)

Grey-headed Canary Flycatcher.
Tickell's Blue Flycatcher (F) with gecko & centipede.

Grey-headed Bulbul.

Kashmir Flycatcher & Red-throated Flycatcher (F.parva albicilla)
Tickell's Blue Flycatcher in the nest during the Monsoon.

Tickell's Blue-Flycatcher

Muscicapidae

Cyornis tickelliae (1442)

Neelchatak

Size: 140 mm   M / F : Dimorphic   **R**

Distribution : The entire region up to 1500 m.

Nest-site: Tree holes, bamboo clumps, palm leaf crotch and unused arches in walls. *May - Aug.*

Material: Leaves, grass, twigs, moss, rootlets, to make a bulky cup.

Parental care: Both. The chicks move with the male after they leave the nest.

Eggs: 3-5,brownish,red spotted.18.4x14.2mm. Juvenile birds have buff streaks on the throat, head and neck and the back is blue.

Call: A musical multisyllabic song, crescendo-decrescendo. Vocal when breeding.

Ecological notes: Chiefly eat dipterous insects. Risk of predation from cats.

Cultural notes: This well known songster features in poetry owing to its blue colour. It freely nests within the premises of houses and the nests are not disturbed by people.

Status: Common.

Neelaang

Tickell's Blue-Flycatcher Juv.

Description: A familiar garden bird in the hilly country. Male blue above, rufous breast and white belly. Female greyish above. Perch upright on branches overhanging streams, from where they launch aerial sallies.

Other Flycatchers:**N**ilgiri Flycatcher (Eumyias albicaudata-1446) Indigo blue, black lores and a white patch on the base of the tail. An occasional winter visitor to our area. Resident in southern Western Ghats. **Threatened.** *Endemic.*

Nilgiri Flycatcher (F & M)

Dull-Blue Flycatcher (E.sordida) Sri Lankan Endemic.

Male.   **Pandharpotya Nartak**

Description: Male indigo-blue except for white belly and black lores. Female brown, chestnut tail, orange throat, grey head (which female Blue-throated Flycatcher lacks). Sallies launched after flying insects from low bushes or from ground. Secretive nature. Fans tail and screws it from side to side when perching.

Verditer Flycatcher (M)

Related species: *(Winter Migrants)*
*Verditer Flycatcher (Eumyias thalassina-1445) Blue-green with black lores. No white on the tail base. Rare WM till 1000 m.
*Ultramarine or Whitebrowed Blue Flycatcher (Ficedula superciliaris-1421) Male-white eye-brow, deep blue above, white below. Female-brown with a brown shoulder patch. Rare.
*Blue-throated Flycatcher (Cyornis rubecu-loides-1440) Winter migrant Goa south. Male like Tickell's Flycatcher but with blue throat, female has a rufous tail.

Whitebrowed Blue Flycatcher (M)

White-bellied Blue-Flycatcher

Muscicapidae

Cyornis pallipes(1435)

Neel-latwa

Size: 150 mm   M / F : Dimorphic **R**

Distribution : Western Ghats
Mumbai south up to 1700 m.

Blue-throated Flycatcher.

Nest-site: The cup is placed on ground, in a tree hole or on a rocky moss covered ledge. *Jun. - Oct.*
Material: Fibres, rootlets, moss, lichen, bark. Lined with soft moss.
Parental care: Both. Chick is brown with streaked head and spotted back.
Eggs: Usually 4, greenish, spotted with brown. 20.2x15.5 mm.
Call: Notes *Tsk, tsk, tsk.* Song of varied notes.
Ecological notes: Endemic to the Western Ghats. Keeps to evergreen patches, in the low bushes, trees and undergrowth.
Cultural notes: None recorded.
Status: Uncommon.

Verditer Flycatcher (F)
Whitebrowed Blue Flycatcher (F)

Black & Orange Flycatcher (F)

White-bellied Blue Flycatchers (F)

White-bellied Blue Flycatcher (Juvenile)

White-bellied Blue Flycatcher (M)

(F)   Banpakhroo, Surangi, Swargiya Nartak

Description: The adult male is white, black crested with two long tail streamers. Young male, female have dull white breast, chestnut back, the male having chestnut tail streamers. Intermediate phases of male are seen. Aerial sallies of the male ,with tail ribbons undulating gracefully is a sight not easily forgotten! This heavenly aerial dance gives it the apt name of a fairy-Swargiya Apsara! The birds also fly from tree to tree, with the long tail trailing behind, like an arrow shot from a bow, hence the name Banpakhroo-Arrowbird!

Male.

Paradise Flycatcher

Muscicapidae

Terpsiphone paradisi (1461)

Rajjuwal, Arjunak, Shwetvanwasin, Nandan

Size: 500 mm. with tail M / F :Alike   **WM**

Distribution : The entire region up to 2000 m. Race *paradisi* is resident and *leucogaster* is winter migrant.

Nest-site: 2-4 m. Up, in the crotch in a bamboo clump or thorny tree. *Mar. - Jul.*
Material: Fibres, lichen, fine grass bound with cobweb to make the tight nest cone.
Parental care: Both. The incubating male with the long tail feathers dangling clumsily, is a funny sight. Silent when nesting.
Eggs: 3-5, pale creamy pink, brown blotched. 20.2x15.1 mm.
Call: Harsh *Chuik, cheew, chui-ssk, chui-sssk.* A pleasant song is given during breeding.
Ecological notes: Catch winged insects. Partial to bamboo clumps, shady well watered palm, mango groves and forest streams.
Cultural notes: *Shwetvanwasin* means white forest dweller. In appreciation of its beauty and grace, the ancient law makers, Manu and Yadnyavalkya specifically prohibited eating the flesh of these small birds!
Status: Uncommon.

Female building the nest with nesting material brought by the male.

Indian Birds Of Paradise :

Greater Racket-tailed
Drongos.
(Lok Wan Tho-1952,
BNHS Collection)

Asian Paradise
Flycatcher (M) on nest.

Yellow-billed
Blue Magpie.

Female.   Neelmani

Description: This azure blue Monarch is not a true flycatcher. The belly is white, a black thin foreneck gorget and nape patch are present. Female lacks the nape patch and has brownish wings. They catch insects in the flycatcher fashion with the tail fanned out. The black nape patch is erected when excited. Azure blue colour becomes bright during breeding.

Male.

Black-naped Monarch- Flycatcher

Blacknaped Blue Flycatcher

Muscicapidae

Hypothymis azurea(1465)

Neelachatak (Blue flycatchers)

Size: 160 mm   M / F : Dimorphic   **R, LM**

Distribution : The Western Ghats up to 1500 m. Fond of bamboo clumps.

Nest-site: Within 5 m. in the fork of *Mangifera indica, Carissa congesta, Garcinia indica. Jun. - Aug.*

Material: Bark, twigs, soft grass, green moss to build the nest cone. Covered with lichen, cobwebs and spider's egg-cases.

Parental care: Both. Incubation-12 days.

Eggs: 2-3,creamy, with red brown blotches. 17.4x13.3 mm.

Call: Lively calls *Sweech-which, chwee,chwee.* Usually silent. Vocal during breeding.

Ecological notes: Partial to mango groves and bamboo. Insects are caught in flight, flitting over forest streams and clearings. The open nest is often predated.

Cultural notes: Nesting is with rains when the sky is overcast with sable clouds. Hence, a glimpse of fresh blue of the Monarch's wings at this time, arouses much interest! The name Neelmani is after a precious blue stone.

Status: Uncommon.

Tickell's Blue Flycatcher can be mistaken for Monarch Flycatcher if seen from behind, in dense foliage.

Look Alike: **White-bellied Blue Flycatcher (Cyornis pallipes-1435) Male is indigo blue(not azure blue like H.azurea), and lacks the black nape patch. Female is brown, with an orange throat and chestnut tail vs. the pied tail of the *Redbreasted Flycatcher (Ficedula parva-1411) which is a winter migrant to our region.*

Paradise Flycatcher (M)
in brown plumage on the nest
with four chicks.

Eggs of White-throated
Fantail Flycatcher.

Skull of the White-throated Fantail Flycatcher
showing sensory bristles attached to the bone.
White-browed Fantail Flycatcher.

Nachra, Nachan, Nhavi

Description: Tail is fanned and moved from side to side as the joyous bird flits energetically with the grace of an accomplished dancer, performing one waltz after the other. Much time is spent in the practice of nest building, outside the breeding season. The white spots on the throat and the tail have given it the Sanskrit name- **Bindurekhak**.

White-browed Fantail-Flycatcher.

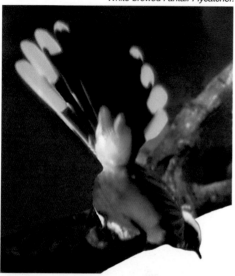

Whitespotted Fantail

Muscicapidae

Rhipidura albicollis (1458)

Latwa, Vihar, Bindurekhak

Size: 170 mm   M / F : Alike   **R**

Distribution : The entire region up to 2000m.

Nest-site: From 1-4m. up, in a fork of *Lantana* bush, *Acacia nilotica*, etc. tree, standing close to a stream or in a garden near a house. *Apr. - Aug.*

Material: Tidy, compact, neat cup built from grass blades, twigs, threads, fibres, cobwebs.

Parental care: Both.

Eggs: 2-3, pink-white, spotted at the broad end. 16.2x12.7 mm. High risk of nest predation. Eggs from as many as three successive clutches being devoured by coucal, shikra, koel are recorded. Decoy nests are built!

Call: Loud continual *Chuk, chuk, chuk*, with harsh notes in between when the bird is disturbed. Multisyllabic song.

Ecological notes: Insectivorous. Like the iora, has an overall poor outcome from nesting, with significant egg loss.

Cultural notes: The local name Nhavi, meaning a barber, comes from the birds call-*Chuk, chuk*, reminiscent of a barber at work with his scissors! The name **Latwa** means a dancing boy, an apt epithet for this lively bird.

Status: Common.

White-browed Fantail-Flycatcher.

Related Species: *White-browed Fantail-Flycatcher* (R.aureola-1451-52) White forehead and eyebrows aid in field identification. Chest spotting absent. Uncommon. The nest is a fine cup, which is rounded at the bottom (The nest of R.albicollis has an inferior tail-like extension.) Commoner in our southern range than in the north.

Great Tit

Grey Tit

Paridae

Parus major (1794,1795)

Shamavalguli, Kshudravalguli

Size: 130 mm M / F : Alike **R,LM**

Distribution : The entire region up to and above 2000 m. Race *mahrattarum* in Kerala.

Eating the worm after extracting it from its housing!

Tit, Topiwala

Nest-site: Tree-hole, wall-hole, pipes, poles. In *Erythrina variegata & suberosa, Sesbania grandiflora, Butea monosperma,Mangifera indica,Adansonia digitata, Moringa oleifera* etc. trees. *Feb. - Nov.*
Material: Hair, rags, moss, feathers, paper, wool.
Parental care: Both.
Eggs: 4-6, pink-white, spotted, with red-brown specks. 17.5x13.6 mm.
Call: *Wit, wit, whee-chi-chi*, song. Pleasant notes uttered during the breeding season.
Ecological notes: Pollinator, insect controller. Nesting in old nests of barbets, woodpeckers, magpie robins, blossom headed parakeets.
Cultural notes: A little known bird.
Status: Occasional.

Description: A pied, grey acrobatic, energetic bird. Moves with white-eyes, ioras, sunbirds, leaf warblers. Clings to sprigs upside down.

Related Species: *Black-Lored or Yellow-cheeked Tit (P.xanthogenys-1810) is common in Kokan and Western Ghat. Black crested head, black lores and yellow underparts. Black chest band in male which is olive green in female. Call usually-'Chee, chee', and in breeding - 'Cheewit-pretty-cheewit'. Sanskrit name of the yellow cheeked tit is* **Patti**.
***W**hite-naped or Whitewinged Black Tit (Parus nuchalis-1798) Black back, yellowish belly, white nape and dark linear throat stripe. In BR Hills; (E.J.Lott,C.Lott,Forktail 15,1999) keep to the same locality. R in scrub area. Rare. Vulnerable.*

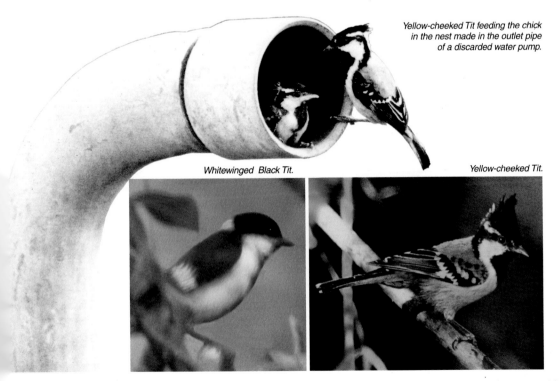

Yellow-cheeked Tit feeding the chick in the nest made in the outlet pipe of a discarded water pump.

Whitewinged Black Tit.

Yellow-cheeked Tit.

Great Tit.

For more information on the status of the occurrence of White-naped Tit in South India, please refer to an article by Eric J. Lott & Christine Lott in Forktail 15 (1999), 93-94.

This nest site in a pipe hole of a discarded water pump was later used by the Magpie Robin for nesting. It was seen to exhibit keen interest in this site even when active nesting of the Blacklored Yellow Tit was going on. The latter is seen to show resentment at this intrusion.

Velvet-fronted Nuthatch

Sittidae

Sitta frontalis(1838)

Kapoti, Shilindhri, Kandali

Size: 100 mm   M / F : Dimorphic   **R**

Distribution : The entire region. Western

Ghats up to and above 2000 m.

Nest-site: Secondary tree hole, 5m. and above, abandoned by woodpeckers, barbets, etc. Also excavates holes in soft wooded trees. *Feb. - Apr.*

Material: Moss, wool, leaves, etc. The entrance is partially sealed with mud-plaster.

Parental care: Both.

Eggs: 2-6, white, red-spotted. 17.2x13.2 mm.

Call: *Chwit, chwit, tsit, tsit, chilp, cheep* in a low volume. High pitched whistles are uttered.

Ecological notes: Eat wood boring insects. Seen in mixed hunting parties of tits, minivets, flycatchers, woodpeckers. In drier regions, they tuck nuts in trunk crevices, and open them by hammering with the bill, hence the name nuthatch!

Cultural notes: As per a tale in the 'Kandagalaka Jataka', a nuthatch befriended a woodpecker and came to a catechu forest, abandoning its home of soft wooded trees. While attempting to bore a new nest-hole in the hard wooded catechu tree, it broke its neck and died!

Status: Uncommon.

Related species:
*Chestnut-bellied Nuthatch (S.castanea-1830) Uncommon. Up to 1000m. Pointed bill is heavy. Has a slaty blue back and deep chestnut belly. Usually a part of mixed hunting parties,moving hither and thither on tree trunks and under side of the branches, with astonishing agility. Eat insects. Call-soft 'Chilp, chilp'.

Chestnut-bellied Nuthatch.

Shilindhri

Description: Yellow eyed, short tailed, coral red-billed small bird with strong feet. Purplish blue above and lilac below with a jet black velvety forehead. Female lacks dark eye stripe. It continually creeps on the tree trunk moving to and fro keeping to shadowy portions hence difficult to photograph. Guards nest site by an aggressive wing display.

Velvet-fronted Nuthatch.

Spotted Creeper

Spotted Grey Creeper

Certhiidae

Salpornis spilonotus (1841)

Chikurchatak, Godha

Size: 130 mm   M / F : Alike   **R**

Distribution : Very patchily and locally, only in Northern Western Ghats and on the adjacent plains. Not in Kokan, Malabar.

Khod-Chimni

Description: White eyebrow, long, slender, curved bill, longer than the head. Tarsus scaly. The barred rounded tail is not used as a strut. Dull brown, white spotted creeper, slithers up and down the trunk, like a mouse, in search of insects. It often works on the underside of a branch or a bole, when the curved claws come handy. Flies from one tree top to the base of another tree, zigzags upwards and does not spiral like the woodpecker. Does not perch.

Nest-site: Exposed fork of a bare branch or the crotch of branch-trunk junction, in a thorny tree. The deep, soft nest cup is well hidden. *Mar. - May.*

Material: Leaf bits, tree-bark ,twigs, lichen, spider egg-cases.

Parental care: Both.

Eggs: 2-3, greenish, brown spotted. 18x13 mm.

Call: A low volume sunbird like tinkle *Chi-chi, chiu, chiu.* Short whistles, chirps Kek, kek.

Ecological notes: Partial to *Acacia* spp.,mango groves and trees with fissured trunks. Insectivorous.

Cultural notes: Yajurved mentions the name of treecreeper, along with the drongo and the woodpecker, as a befitting sacrifice to the Deity of the Trees. This is due to the inseparable relationship of these birds with lofty trees! The Sanskrit names **Chikurchatak** and **Godha**, respectively mean a Mousebird and a Lizardbird, both indicative of similar agile movement of the bird on the vertical tree trunk!

Status: Rare in our area.

Himalayan Creeper -Lok Wan Tho, 1951.
It is not seen in our region.

Habitat of the Spotted Creeper.

Tickell's Flowerpecker

Palebilled Flowerpecker

Dicaeidae

Dicaeum erythrorynchos(1899)

Chatika

Size: 80 mm   M / F : Alike   **R**

Distribution : The entire region up to 2000 m.

Nest-site: 2-15 m. up, draped on a twig, often amidst red ants, hidden in dry leaves of a creeper or a tree. *Feb. - Jun.*
Material: Grass, twigs, fibres, moss, down, lined with soft silk-cotton floss. Has a lateral entrance hole without the projecting portico. Nest pouch is draped on a branch and not suspended.
Parental care: Both. Chicks are fed pulp of berries of *loranthus* spp.
Eggs: 2-5, white. 14.4x10.5 mm.
Call: *Chit, chit, chip, chip* with tail screwed from side to side. Twittering song on the wing.
Ecological notes: Birds cause damage to forest plantations and fruit groves due to the spread of parasitic *Loranthaceae*. The sticky seeds are excreted and they readily adhere to teak, mango, guava trees where they sprout, causing damage.
Cultural notes: Unnoticed due to its small size and obliterative colour. Often mistaken for a female sunbird.
Status: Common.

Legge's Flowerpecker. Sri Lankan *Endemic*. (D.vincens)

Related Species: *Thick-billed Flowerpecker (D.agile-1892) Faintly streaked breast, red eyes, white terminal tail band and dark bill. Spits seeds of berries on the same tree, after eating the jelly. Call 'Chik, chik'. Nest typically of red-brown down, suspended from a dry branch. Up to 1000 m. in our area.*
*Plain or Nilgiri Flowerpecker
(D.concolor-1902) White supercilium, blue-grey bill, yellow -white underparts. Stouter than Palebilled. Call is 'Chek, chek'. Central Kokan southwards and Western Ghats Mahabaleshwar south, up to 2000 m.*

Phooltochya

Description: This smallest member of Indian avian world is olive-brown above, white below, with a curved flesh coloured pale bill. Visits orchards for berries of parasitic *loranthus* family, fruit pulp, flowers and insects, keeping to the canopy. The sticky jelly of ripe berries gives the birds a difficult time while cleaning their bill! They move with sunbirds, white-eyes, along a chosen beat, flitting from tree to tree.

Thick-billed Flowerpecker.

Plain or Nilgiri Flowerpecker.

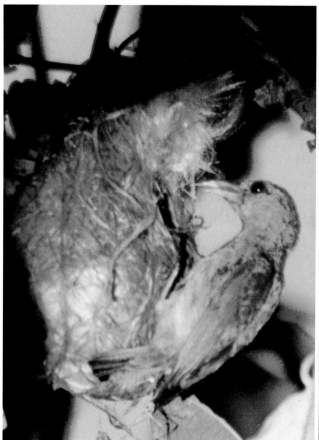

The wrap-around nest of the
Thick-billed Flowerpecker.

Tickell's Flowerpecker on the nest.
Floss can be seen in the nest chamber.

A Tickell's Flowerpecker chick was found under a destroyed nest.
Two of the three chicks were predated. The rescued chick was placed in
an open nest & it was accepted by the parents.

Scarlet-backed Flowerpecker (M) Vagrant.

Purple-rumped Sunbird

Nectariniidae

Nectarinia zeylonica(1907)

Madhupkhag

Size: 100 mm   M / F : Dimorphic   **R**

Distribution : The entire region up to 2000 m. Not in Gujarat.

Nest-site: In and around houses, branches of thorny trees, creepers and near nest masses of spiders,2 to 5 m. *Mar. - Aug.*

Material: Fine grass, twigs, to make the pear shaped pouch with a portico and lateral entrance hole. Cobwebs and rubbish are liberally coated on the outer surface, to make it appear like a ball of rubbish hanging on a tree.

Parental care: Both build the nest in a week, female incubates for 14 days, chicks leave the nest in 16-17 days. Both feed the chicks.

Eggs: 2-3,grey green, brown spotted. 16.4x11.8 mm.

Call: *Sisiwee, sisiwee* with fluttering wings. Call *Tit-tue,tit-tui* is uttered as it feeds and flits from flower to flower.

Ecological notes: Old nests are reused if available. Hence a nest if found should be left alone. Help cross pollination especially of *loranthus* spp.

Cultural notes: It is mentioned in the epic Ramayana as the **Madhupkhag** (Nectar-bird).

Status: Common.

Female.   **Shinjir**

Description: Male does not have an eclipse moult. Sides of head and throat band deep chestnut, purple rump, metallic green head and yellow breast. This beautiful bird has a resplendent plumage, best seen in sunshine. Also eats spiders as it flits around for nectar.

Accepting a repaired nest (F)

On a natural nest (F)

Male non-br. purple & male purple-rumped sunbirds.

Male.   **Shakkarkhora, Jambhala Surypakshi**

Description: Black in shade, metallic purple in sunlight. Faint red mark on throat and yellow pectoral tufts in breeding male. Eclipse male is brown with black wings, yellow under parts and has a central black chestband, which the female lacks. Visits every coloured flower which has nectar to offer! Rarely small insects are caught in midair. Long tubular flowers are punctured at the base to suck the nectar.

Sunbirds are rarely seen on the ground. This bird was collecting nest material & grub.

Sunbirds suck nectar from long tubular flowers by puncturing them at the base.

Nectariniidae

Nectarinia asiatica (1917)

Madhukar,Pushpandhay

Size: 100 mm   M / F : Dimorphic   **R**

Distribution : The entire region up to 2400 m.

Nest-site: The nest pouch is suspended at the most unexpected places. On roadside hedges, arched trellis, gardens of occupied houses. *Mar. - Jul.*

Material: Leaves, grass, twigs, plastic, paper, aluminium foils, cobwebs, spider egg cases, rubbish. Lined with cotton, feathers, wool, soft grass.

Parental care: Female incubates and builds the nest silently, approaching it stealthily. The male rarely assists her. Both feed the chicks.

Eggs: 2-3, pale green, brown blotched. 16.3x11.6 mm.

Call: A *Wich, wich, chip, chip* is uttered monotonously. The breeding male's song *Chee-wit, chee-wit* is heard the year round.

Ecological notes: Aids cross pollination. The often damaged nests are accepted if repaired, but only when chicks are present. One damaged nest was replaced by another recently vacated sunbird's nest, the chicks were relocated in the new nest, which was supended at the site of the damaged nest. This surrogate nest was also accepted(Satish Pande).

Cultural notes: Flowering plants are specially potted by people on their terrace flats to successfully attract these charming birds!

Status: Common.

Non-breeding male purple sunbird.

An injured, rescued Purple Sunbird sucking honey from a syringe and nectar from an offered flower!

Purple Rumped Sunbird
(M & F)

Crimson Sunbird

Nectariniidae

Aethopyga siparaja vigorsii(1929)

Suvarnapushpa

Size: 100 mm.   M / F : Dimorphic   **R**

Distribution : The Western Ghats upto about 1400 m. and Kokan. Common upto North Karnataka, rare southwards. Also Nilgiris.

Nest-site: Suspended from trees and bushes. Also in gardens near occupied houses from roof rafters, dangling cords, wires with an empty bulb-holder (Sachin Palkar,per.com.). *Jun. - Oct.*

Material: Palm leaf shreds, fibres, twigs, cobweb, rubbish, dry grass, paper. Lined with wool, cotton, hair, feathers. The nest is an oblong, untidy looking pouch with a long characteristic tail.

Parental care: Female builds the nest and incubates for 14-15 days. Both feed, female does more work, male doing the noon shift. Chicks leave the nest on day 16 or 17. A second brood is raised if a nest is unsuccessful.

Eggs: 2-4, salmon, oval, tiny, blotched. 15.1x11.4 mm.

Call: *Chii, chii, chwee, chwee.* Utter sharp calls while leaving the nest after feeding the chicks, as if to tell them to keep still in their absence. The breeding male is a songster.

Ecological notes: Partial to crimson and scarlet flowers (Hibiscus, Canna, Ipomoea, Bottlebrush, Powder-puff). One nest suspended from a cord in the backyard of a house, was mistakenly fumigated by a muncipal sanitation worker and the chicks died (Rohan Lowlekar). So well are the nests camouflaged.

Cultural notes: Beauty of this bird has been justly praised since the post-vedic period (1200BC). This bird with bejewelled plumage is called the Golden Flower-***Suvarnapushpa***, and is said to dwell in the heavenly garden of Indra, the king of Gods!

Status: Rare.

Male.   **Pivlya Pathicha Suryapakshi**

Description: In male, central feathers of the green tail are elongated and scarlet breast is streaked yellow. Yellow back is seen in flight. Female is brown, with yellow belly and dark tail. Hover in front of blossoms in an aerobatic manner, sipping nectar with the suctorial tongue. Execute a daily beat. Its mere glimpse is inspiring.

Female.

Nest on a bulb holder.

Male.

*Look Alike: *Small Sunbird may be mistaken for Yellowbacked, since yellow rump of the latter is seen only in flight. Latter is larger.*

Small Sunbird

Crimsonbacked Sunbird

Nectariniidae

Nectarinia minima(1909)

Madhup, Shinjirika

Size: 80 mm.   M / F : Dimorphic   **R**   *Male.*

Distribution : The entire region up to and over 2000 m. Gujarat southwards.

Nest-site: Suspended on a roadside *Strobilanthes* etc. spp. within 2 m. at forest edge. *Jan. - Apr. Sep. - Nov.*

Material: Fibres, twigs, moss, cobwebs, to knit the well made globular pouch.

Parental care: Both.Female incubates.

Eggs: 2,white,ovoid,red spotted. 14x10.2 mm.

Call: *Chik, chik* calls as it moves in the foliage. Sings from a perch *Si-si-swee, si-si-si.* Mousy squeaks are uttered.

Ecological notes: Endemic to Western Ghats. Aphids hitch-hike on sunbird's beaks, from flower to flower. Their beaks have a finely serrated margin to catch tiny insects.

Cultural notes: This Indian equivalent of the Hummingbird has an onomatopoeic name in Sanskrit-**Shinjirika** (**Shija-Shinja** means to jingle like a tinybell).

Status: Uncommon.

Male.   **Chota Suryapakshi**

Description: Exquisite with crimson chest-band and mantle, metallic green head, violet throat and yellow belly. Full glory and charm of this tiny gem are evident only in sunlight. Eclipse male is like the female, but with a crimson back. Female has only a crimson rump. On flowering *Erythrina* and *Loranthaceae* spp. in evergreen forests and near streams. This restless bird is sighted as a passing glimpse of shimmering colour.

Related Species:*Loten's or Maroon-backed Sunbird (N.lotenia-1911) Sickle shaped long bill. Male has a brown belly, crimson breast-band, yellow armpits and is metallic purple above. Female is yellow below and muddy above. Western Ghats, Thane south, keeping to moist deciduous areas up to 1600 m. Avoids evergreen forests. Breeds in summer.*

Loten's Sunbird (Eclipse M)

Loten's Sunbird (M & F)

Loten's Sunbird (M), yellow armpits.

Kolikhaoo

Description: White-tipped dark brown tail, long curved bill, olive back, white throat, orange tinged flanks and yellow underparts. Flits like a sunbird from tree to tree. Hovers in front of spider webs and gently hawks the unsuspecting spider with the long beak! Keeps to dense, wet, forested areas. Fond of banana flowers. Meticulously looks for insects by probing flowers and leaves.

Loten's Sunbird (F) can be confused with Spiderhunter.

Look Alike: *Female sunbird, especially *Loten's sunbird, looks very similar to the Little Spiderhunter. But the latter has an unmistakable long curved bill and a white throat. Female Sunbird has yellow throat and belly. The Little Spiderhunter has a restricted range and it is an uncommon bird.*

Little Spiderhunter

Nectariniidae

Arachnothera longirostra(1931)

Kshudravalguli

Size: 140 mm   M / F : Alike   **R**

Distribution : Western Ghats up to 2100 m. and in South Kokan and southwards. Prefers banana plantations near rivers.

Nest-site: The edges of the inverted dome-nest are stitched to the underside of a broad leaf like banana. Oval side entrance. *May -Aug.*
Material: Dry leaves, grass, cobwebs, fibres, to make the tiny nest lined with down feathers.
Parental care: Both.
Eggs: 2-3, pinkish, streaked. 18.4x13.1 mm.
Call: *Wich, wich, cheep, cheep.* Pleasant song and high pitched notes are uttered in flight. Monotonous notes are uttered from a bare branch.
Ecological notes: Banana and *loranthus* flowers are chiefly cross pollinated. Brood parasitized by *Hierococcyx fugax* in NE India, which is the disjoint range of the Spiderhunter. It is seen only in the Western Ghats in peninsular India.
Cultural notes: *Kshudravalguli* is a small bird with a pleasant song and residing in holes. This is a befitting description of the Little Spiderhunter, whose peculiar nest indeed has a tiny hole-like semicircular side entrance!
Status: Rare.

Loten's Sunbird (F) may be mistaken for Spiderhunter.

Spiders feature largely in the diet of Spiderhunters.

Oriental White-eye

White-eye

Zosteropidae

Zosterops palpebrosus (1933,1935)

Sarang, Chatakika, Putrika

Size: 100 mm   M / F : Alike   **R**

Distribution : Race *palpebrosus* northwards, race *nilgiriensis* BR Hills south.

Nest-site: In *Lantana* shrubs, *Citrus limon*, *Santalum album,Pongamia pinnata, Annona* spp. Trees. Nest is hidden in leaves. *May - Sep.*

Material: Grass, fibres, threads, twigs, cobwebs used for making the hammock cup.

Parental care: Both.

Eggs: 2-3,pale blue. 15.2x11.5 mm. All the chicks leave the nest simultaneously.

Call: A soft musical, nasal call with pleasant notes is given. *Chirr, prrr-u, tzip, tzip, sisifsisi,* twittering is uttered. Flocks are noisy.

Ecological notes: Seen on flowering *Erythrina, Madhuca, Bombax, Grevillea robusta, Moringa oleifera* trees. Helps in pollination. Can cause but a little damage to fruit gardens. A single ripe custard-apple opened by a koel, has been seen to meet the demands of flowerpeckers, white-eyes and bulbuls for three days!

Cultural notes: Name Chashmewala means bespectacled, owing to its white eye-ring!

Status: Common.

Bathing & eating fruit pulp.

Chashmewala

Description: Small bird with white eye ring, greenish-yellow back, bright yellow throat and white belly. Square tail, strong legs, short curved bill and protractile tongue with two stiff brushes aid them in seeking nectar, fruit pulp and insects in acrobatic manner. Often with sunbirds, tailor-birds, minivets, ioras.

Sri Lanka White-Eye (Z.ceylonensis)

Oriental White-Eye feeding the chicks.

Look Alikes: The chicks of White-eye can be mistaken for *Tickell's Flowerpecker. On close observation however, the chicks show a faint green tinge to the head,and have shorter bill. The eye-ring is indistinct, developing later.*

Male.   Yuvraj, Turewala Bharit

Description: The black male has chestnut wings and a graceful crest, like a pygmy crested coucal. The female is short-crested and brown streaked. They are seen along the winding mountain roads, in scrub and secondary forests near farms and in the hills. Males have a courtship display. Mixed flocks with pipits and larks are sometimes seen on the ripening winter crop.

Female.

Male.

Crested Bunting

Emberizidae

Melophus lathami(2060)

Kiriti, Vanchatak

Size: 150 mm.   M / F : Dimorphic   **R**

Distribution : The Western Ghats till about Kolhapur, up to 1400 m. and adjacent Kokan foothills. Not on the coast.

Nest-site: On a sloping hill in Ghat country, often on the roadside, under a rock, grass clump, bush, in a shallow depression on the ground. *May - Aug.*

Material: Grass, rootlets, twigs, hair; lined with feathers, soft fibres, wool, to make a tight cup.

Parental care: Female incubates; both build the nest and feed the chicks. The female is bolder than the male when feeding the chicks.

Eggs: 2-4, white, brown spotted. 20.1 x 15.6 mm.

Call: A melodious song is given by the male from a bushtop, *Chwee-chi-chi-chi, chwee-chi-chi-chi-chwee, chwee.* A *Chip, tip* call is also uttered.

Ecological notes: Prefers to stay on burnt, black ground and grey rocky terrain, where it is not easily seen. Fond of lantana berries. Occasionally, brood parasitized by *Cuculus canorus.* This berry and seed eating bird, feeds insects to the chicks. Ground nest is predated by reptiles, mammals and birds.

Cultural notes: The Sanskrit name ***Kiriti*** means a crown. The vernacular name Yuvraj is descriptive of the princely bearing of this bird.

Status: Common.

Male feeding a grasshopper to chicks & Immature.

Grey-necked Bunting

Emberizidae

Emberiza buchanani(2050)

Shaishir Chatak, Atak Chatak, Bharit

Size: 150 mm.   M / F : Dimorphic   **WM, PM**

Distribution : Eastern fringes of the Western Ghats and adjoining Deccan up to about Kolhapur. Not in Kokan and Malabar.

Nest-site: Extralimital (Baluchistan). The ground-cup is placed under a boulder or in a grass clump. *Mar. - May.*
Material: Grass, twigs, fibres, rootlets. Lined with soft fine grass.
Parental care: ? Both.
Eggs: 4,white, red blotched. 20 x 15 mm.
Call: A multisyllabic song is given when breeding. Silent in winter. An ocassional *Chirp, click* is uttered when feeding on the ground.
Ecological notes: Arrive in October, leave in May .Prefer dry, stony and scrub country. Avoid forests. Destructive to the winter crop, more so in north India, where larger flocks are seen.
Cultural notes: These buntings have been regarded as the wandering sparrows-**Atak Chatak**. Since they are seen in winter-**Shaishir Chatak** (Shishir-winter, Chatak-a small bird).
Status: Uncommon in our area.

Related species:*Ortolon Bunting (Emberiza hortulana- 2049) WM. Has a pale yellow throat, olive grey head and breast, white eye-ring, yellow stripes on the throat and submoustachial area. Vagrant to the eastern fringes of the Western Ghats, seen in the scrub, arid country. Recently recorded at Saswad near Pune and at Nashik. Larger and bolder than the Grey-necked Bunting. Rests under thorny bushes on hot noons.

Ortolon Bunting.

Karda Bharit

Description: An orangish bill, grey head, buff submoustachial stripe, rusty breast and lightly streaked back in the male. The female is duller. Seen in flocks on the ground in stony terrain with pipits, larks and other buntings. At noon, they rest in the shade of *Acacia nilotica, Ziziphus* etc. Spp. remaining immobile and very reluctantly abandoning the cool shelter.

Grey-necked Bunting.

Grey-necked Bunting.

Patteri Bharit

Description: Striated head and throat, brown back, black eye and submoustachial stripes and rufous wings. Female duller, merging well in the arid, hilly terrain. Birds cool-off under the scanty shade of boulders or bushes on hot noons, after drinking water from a favourite puddle, with pipits, larks, blue rock-thrushes, mynas, bulbuls and robins. Seeds are gleaned by hopping on the ground.

Striolated Bunting

House Bunting

Emberizidae

Emberiza striolata(2057)

Red-headed Bunting.

Annadushak (Black & Redheaded Buntings)

Size: 140 mm.   M / F : Dimorphic   **R, LM**

Distribution : Patchy. Eastern fringes of the northern Western Ghats up to about Sangli. Not in Kokan and Malabar.

Nest-site: On ground, under a jutting rock, niche in an earth bank or on a ledge in an earth cutting. Several nests are often seen in the vicinity. *Feb. - Oct.*

Material: Grass, threads, rootlets, goat-hair are woven to make the cup. Lined with feathers, soft flowering grass.

Parental care: Both share all the duties.

Eggs: 2-3, blue, brown spotted. 20 x 15 mm.

Call: The multisyllabic, enthusiastic song is very refreshing, especially when heard in the arid landscape, where the bird dwells! Song *swich, wich, wich, sweech, whee-oui, whee-chii*, is given from a bush top or a high perch.

Ecological notes: Nests of Zitting Cisticola, pipits, larks, buntings are often found near each other. A misplaced nest of the Striolated Bunting, on a developing termite mound, was eaten by the termites and the eggs were eaten by red ants! (Amit Pawshe, Satish Pande)

Cultural notes: The gregarious Red and Blackheaded Buntings cause damage to the winter crop and hence were called **Annadushak** (Destroyers of food), by Sushruta(400 BC).

Status: Uncommon.

Striolated Bunting (Juv.)

Related Species: **Black-headed Bunting (E.melanocephala-2043) and *Red-headed Bunting (E.bruniceps-2044) Buntings. WM. The northern Ghats up to 1400 m. in passage. Rare in Kokan, former not in Malabar. Habits similar to E.striolata; gregarious. Flocks on a babool tree appear as if the tree is covered with yellow flowers! More time is spent on the wintering than on breeding grounds.*

Black-headed Buntings.

Male.

Female.

Common Rosefinch

Rosefinch

Fringillidae

Carpodacus erythrinus(2010,2011)

Raktashirsha

Size: 150 mm.   M / F : Dimorphic   **PM, WM**

Distribution : Kokan, Malabar and the Western Ghats, up to 2000 m. Race *ferghanensis* up to Belgaum.

Nest-site: In Baluchistan and in the Himalayas from Kashmir up to Tibet. The nest-cup is placed up to 2 m. up in a rose bush, Willow tree etc. *May - Aug.*
Material: Grass,hair, rootlets, juniper fibres.
Parental care: Both. Feeding is by regurgitation of food.
Eggs: 3-4, blue, red spotted. 20.8 x 14.5 mm.
Call: Somewhat similar to the call of a blossom-headed parakeet, but low in volume *To-ii,to-ii*. A murmur is heard, when one passes by a field of sugarcane etc. where a flock of Rosefinches has settled. Sing just before return migration.
Ecological notes: Help pollination of flowering plants, being fond of flower nectar. Can be harmful and destructive to the standing winter crop like jowar, bajra, cereals, pulses. A bird ringed near Bharatpur in India was found in Ulyanovsk region of the former USSR (53,50'N,46,21'E). SA Handbook.
Cultural notes: To the farmers, the arrival of these winter migrants, means an extra vigilance duty for the protection of ripening crop!
Status: Uncommon.

Male (Non-Br; Br.) **Gulabi Chatak, Gulabi Chimni**

Female.

Description: The beautiful male has a crimson head and rump, chestnut brown back, pink chin and pale belly. The female is streaked brown with two white wing bars and has a whitish belly. Both have a stout black bill and a forked tail. In late September, they start arriving with the buntings, share their habitat, and leave by May. Flocks seen on winter crop, fruit gardens, flowering trees around scrub country,strewn with *Acacia, Ziziphus, Euphorbia, etc.* They are more active in the late morning and in the evening.

Female.

Red Munia

Red Avavdavat

Estrildidae

Amandava amandava(1964)

Sunal, Sevya, Latukika

Size: 100 mm.   M / F : Dimorphic   **R, LM**

Distribution : Patchy. Throughout the Kokan and the Western Ghats in wetter areas, up to 2100 m. Probably not in Malabar.

Nest-site: Swamps, marshes, grass and reed clumps. Also lantana bushes, close to the ground within 1 m. *May - Sep.*

Material: Grass blades, palm leaf, paddy, *Eleusine coracana*-shreds. Lined with feathers and soft grass. Globular nest with side entrance.

Parental care: Both. Incubation-14 days. The male was seen accompanying the female, when she laid the clutch. Very alert when brooding; a pair was seen to hurriedly leave the nest much before a Shikra soared over their nest!

Eggs: 4-8, white. 14.4 x 11.2 mm.

Call: Vocal when breeding. A low pitched call *Chee, chee, chee* is also given when feeding.

Ecological notes: Prefers marshes. Rarely nests away from water bodies. Nest predation by mammals and birds. **Sunal** and **Sevya**, in Sanskrit, are grasses growing near water and the Red Munias nest in these grass beds, hence the names!

Cultural notes: A popular cage bird. The name Avadavat is two centuries old and it is a corrupt version of Ahmedabad (JBNHS36:837). A story of the Red Munia's active nest being trampled by a rogue elephant, is cited in the Latukika Jatak (600 BC).

Status: Uncommon.

Lal Munia

Description: The female and the non-br. male are brown, with dark, white-spotted wings, buff belly and red beak. Male gets the magnificent red breeding plumage gradually, which is fully acquired in May. Intermediate phases are seen. Flocks are seen near the grassy shores of ponds, rivers, mangroves, well watered paddy, sugarcane fields, etc.

Female Red Munia & intermediate phase of breeding male. The male gradually acquires the resplendent red eye catching plumage.

Male Red Munia.

Green Munia. Its popularity as a cage bird has made it vulnerable.

Related Species: *Green Munia (A.formosa-1965) Rare. Western Ghats, up to about Mahabaleshwar. Not on the coasts. A few birds released from captivity may be encountered in urban areas. More commonly seen caged than free, even today! Voice similar to the call of Rufous Babbler or Small Minivet. Red bill on the green coloured bird is its attraction as a cage bird! **Vulnerable.**

White-throated Munia

Plain Munia, Whitethroated Silverbill

Estrildidae

Lonchura malabarica(1966)

Pirili

Size: 100 mm.   M / F : Alike   **R**

Distribution : Kokan, Malabar and the deciduous and lightly wooded patches of the Western Ghats, up to about 1000 m.

Nest-site: The self made nest is placed in a thorny bush or under thatched roofs. Nests are sometimes seen within the fabric of eagle and vulture nests! (SA Handbook) Old Baya nests are parasitized through the natural tube entrance, or through an extra side hole to the egg-chamber, bored from inside. *May - Jan.*

Material: Grass, paper, twigs make the globular nest lined with down feathers, cotton, grass inflorescence.

Parental care: Both.

Eggs: 4-8, white. Up to 25 eggs in a single nest, laid by more than one female are known. 15.7 x 11.7 mm.

Call: A low chirping sound. An occasional whistle after taking nectar or a ripe berry. Fly away noisily when an active nest is approached.

Ecological notes: Old Baya nests with pre-existing unhatched eggs are parasitized at times. 12 eggs have been seen in one such nest. Nests are also used for roosting.

Cultural notes: The Sanskrit name *Pirili* means a bird of the hedge, an apt description of the birds habitat!

Status: Common.

Related Species:

*White-rumped Munia (L.striata-1968) Pied munia with white rump and abdomen. Occurs in Kokan and Malabar, less so in the Ghats. Semideciduous, scrub areas, fields and gardens are their chosen habitat. Relatively confiding. Flocks move from bush to bush, occasionally landing on the ground, in search of seeds and grain.

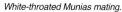
Shubhrakanthi, Phiki Munia

Description: A pale brown munia with pointed black tail, white rump and underparts. Flocks are seen perched in rows on wires, fences, branches, in the early morning, taking to wing when approached. They prefer a dry habitat. Courtship display of the male is a relatively short ritual. Mating takes place after the nest is partly begun, indicating that pairs are formed much earlier.

White-throated Munias mating.

White-rumped Munia.

White-rumped Munia.

Thipkedar Munia

Description: This chocolate brown munia has chestnut throat, conical bill and a white scaly breast, as if adorned with tiny flowers (Sanskrit name **Pushpabhushi**). Scales are less prominent in non-breeding birds. The Thick tongue helps in dehusking seeds, gleaned from glebes. Gregarious grassland bird. Grass blades bend with the weight of the birds when they settle, making the birds recover their balance frantically!

Related Species: *Black-headed Munia (L. malacca -1978) In the entire region. *Blackthroated, Hill Munia (L.kelaarti-1971) In grasslands, Goa south. Uncommon. Dark brown above, pinkish below; lacks white rump.*

Black-headed Munia.

Spotted Munia (Juv.)

Spotted Munia

Scaly-breasted Munia

Estrildidae

Lonchura punctulata (1974)

Pushpabhushi,Paroshni

Hill Munia.

Size: 100 mm.   M / F : Alike   **R**

Distribution : Throughout the region up to 2100 m.

Nest-site: Several munia nests of more than one species, may be seen on a single thorny bush like *Phyllanthus distichus, Terminalia, Acacia, Ziziphus* spp.1-5 m up. *Jun. - Sep.*

Material: Blades of standing green grass are cut, held in the beak and carried to the nest site. The flying bird with the grass blade streaming behind is a funny sight. The globular nest is lined with feathers and soft grass.

Parental care: Both. Pairs roost in the nest at night. More than one pair occasionally use one nest simultaneously.

Eggs: 4-8, white. 16.4 x 11.6 mm.

Call: A low volume *Chweep, cheep, cheewp, Chee, chee*. Chirping calls are common.

Ecological notes: The birds have adapted well to urban areas, freely nesting in gardens, creepers on trellis in verandahs, etc. A grass nest having caught fire due to Diwali crackers, and eventually burning the tree, has been observed. A nest placed on a Christmas tree was predated by a cat, due to the easy access. Old munia nests are used by field mice.

Cultural notes: A popular cage bird even today. The name Munia is derived from the Hindi name Munni, meaning a girl.

Status: Common.

Hill Munia of Sri Lanka (L.k.kelaarti)

Red Munia pair in the nest. Male is brilliant red in colour.

Spotted Munia.

Red Munia males in breeding plumage. This Munia is also known as Strawberry Finch.

Indian Silverbill or White-throated Munias.

Female. **Ghar-chimni, Chimni, Chioo**

Description: This bold bird is closely linked to human beings. One of the first birds to visit a bird-feed. Often perch with muddy feet on washed linen causing annoyance. Features in the diet of predators. Sparrows freely mix with bulbuls, white-eyes, munias, sunbirds. Alarm calls are given. Favourite nesting sites are not easily given up even if the nesting is discouraged. Nesting in their homes is often permitted by people, after observing the persistence of the sparrow!

House Sparrow (F)

Yellow-throated Sparrows.

House Sparrow

Passeridae

Passer domesticus(1938)

Chatak, Vartika

Size: 150 mm. M / F : Dimorphic **R**

Male.

Distribution : Throughout the region; on hills and coasts alike, but always around human settlements.

Nest-site: Wall-hole, under roofs and any place in the house where nesting material can be placed and eggs can be laid. In nests of cliff swallow. *Jan. - Dec.*

Material: Grass, twigs, fibres, plastic-shreds, feathers, wool, paper, etc.

Parental care: Both.

Eggs: 3-5, greenish-white, brown spotted. 20.6 x 14.9 mm.

Call: Familiar chirping call when feeding or roosting. Breeding male sings *Tsi-tsi, chip, chip, chew, cheer* when displaying with flapping wings.

Ecological notes: Flocks cause damage to standing crop. Freely enter and litter houses, especially when nesting. Chicks often fall from their nests. Birds are prone to injury from ceiling fans. Grains and spices kept for sundrying are soiled with droppings of the feeding sparrows.

Cultural notes: Feature extensively in nursery rhymes. In the Rigveda, a reference is made to a sparrow injured by a wolf, which was treated by Ashwinikumar twins, the physicians of Gods.

Status: Common. Numbers supposedly declining currently.

Java Sparrow.

Related Species:

Yellow-throated Sparrow or Chestnut-shouldered Petronia (Petronia xanthocollis-1949) In lightly forested areas of the region. Nests in tree-holes. Plain grey head, neck. Chestnut coverts and white wing-bars in the male, brown coverts and buff wing-bars in the female. Males and few females have a yellow throat patch. This is the famous sparrow which is mentioned by Salim Ali in his autobiography.
Java Sparrows may be encountered occasionally but these are released birds.

Streaked Weaver

Passeridae

Ploceus manyar(1962)

Kalvink, Kaulik, Chanchusuchi

Size: 150 mms.   M / F : Dimorphic   **R, LM**

Distribution : Patchy. Kokan, Malabar and the Western Ghats up to 1000 m.

Nest-site: *Typha, Phragmites*, Bulrush clumps in marshes or even on the branches of trees. *Jun. - Sep.*

Material: Grass and leaf strips are used to make the nest. The nest is not suspended from a point like the Baya Weaver's nest, but it is woven around several twigs or grass blades. The roof dome is thus broad based. The entrance tube is also shorter. The nest is deceptive. Yellow flower petals are seen in the nest wall.

Parental care: Male builds, female inspects, incubates. Polygynous male also helps later.

Eggs: 2-4, white. Incubation 14-17 days. 20.3 x 14.3 mm.

Call: Vocal when nesting. Call *Tilili, tilili, Chit, chit, chichi, chichi, kiti, kiti*. Various notes are given by the male when the female approaches the nest. Has specific alarm calls.

Ecological notes: Partial to swampy and marshy grass strewn terrain. A colony of immature streaked weavers practicing nest building on trees was seen for a few years near Chiplun, Ratnagiri. An active colony has not yet been discovered.(Jayant Kanade)

Cultural notes: Un-noticed bird.

Status: Uncommon.

Black-throated Weaver.

Related species:
***B**lack-breasted / throated Weaver (P.benghalensis-1961) Yellow crown with a dark brown breast band in the male. Female has a yellow supercilium and neck patches, yellow chin, buff-brown breast. Both with streaked back. One small colony in Mumbai. Scarlet or orange coloured petals are often found on the nest. Nesting behaviour similar to the Baya Weaver.*

Male.   **Patteri Sugran**

Description: The fulvous breast is streaked with black. Some males have light streaks. The breeding male has a yellow crown, which is streaked brown in the female and non-breeding male. The yellow supercilium is prominent, and it loops around brown ear-coverts to form a collar. Gregarious. The nest differs from the typical tubular nest of the Baya Weaver. Nests located in swamps are overlooked.

Streaked Weaver adult male & Immature practicing nest weaving.

Female.　　　　　Sugran, Gawlan, Vinkar, Baya.

Baya

Passeridae.

Ploceus philippinus(1957,1958)

Sugruhakarta,Suchimukh,Peetamunda

Size: 150 mm.　　M / F : Alike　　**R, LM**

Distribution : Throughout the region upto about 1000 m.

Description: Breeding male has yellow crown, dark brown throat and ear-coverts. Non-breeding male and female are sparrow like with streaked back. Non-breeding males also build nests. The fresh nest looks green, eventually turning brown. Gregarious. Often seen with warblers, munias, white-eyes, etc.

Nest-site: Colonies on branches projecting over water, old wells, ravines, are at a lower height. Higher up on *palm s*pp., Electric, telephone wires, or thatched roofs. *May - Sep.*

Material: Sugarcane, jowar, bajra, paddy-leaf strips, grass blades, palm shreds. Mud is plastered on the inside of the nest dome ? to make it cool. A pebble is found in many nests at the point of suspension ? to tie the first knot.

Parental care: Male alone builds, female incubates. Male polygynous. Some nests are multistoried, with only the lower chamber open and others closed; a few nests are fused laterally.

Eggs: 2-4, white. 20.3 x 14.5 mm.

Call: Very noisy when breeding. Male chirps anxiously when the female approaches the nest for inspection, calling *Cee, chee, chit, chit, churr.* Wings are flapped in excitement to attract the female. Alarm calls given.

Ecological notes: Baya causes damage to jowar, bajra, paddy, maize, but also eats pest insects. Unused Baya nests are parasitized by Whitethroated Munia through side hole. Spotted Munias and field mice use them for roosting. Predation by snakes and wanton boys is common. Dry nests burn easily and burning nests can be hurled beyond the fireline by wind.

Cultural notes: Baya, capable of performing skillful and intelligent tricks is unfortunately still a popular pet bird. The bird and the nest, feature in literature and visual art.

Status: Common.

Male.
Nest colony in a well, a rapidly vanishing habitat.

Female Baya Weaver taking a mantice to feed the chicks.
Insects form substantial dietary supplement in addition to grain & grub.

*Baya (M) on the nest. The male builds the nest &
the female having found it to her satisfaction mates with the male.*

*The female is seen with a dragon fly in the beak.
This is food for the nestlings*

Baya nests are parasitised by White-throated Munias. Mice are also known to use old Baya nests.

Snakes can predate on Baya's chicks.

Male Baya inspecting leaves for nest material.

This rare photograph taken by Salim Ali shows V. C. Ambedkar holding a six storied Baya nest.(JBNHS 75,1980)

A series of Baya nests is seen on wires running across a well. Male Baya is seen on the wire.

Baya nests on a branch overhanging a water body.

Grey-headed Starling

Chestnut-tailed Starling, Greyheaded Myna

Sturnidae

Sturnus malabaricus (987, 988)

Shyeta Shari

Size: 210 mm   M / F : Alike   **R**

Distribution : The entire region up to 1200 m. Race *blythii* common Belgaum south and rare northwards.

Nest-site: Secondary holes in trees, usually barbet and woodpecker nests are used. *Mar. - Jul.*

Material: Lined with feathers, soft grass, twigs.
Parental care: Both.
Eggs: 3-5,blue-green. 23.8x18.2 mm.
Call: Song and call. Vocal when feeding.
Ecological notes: Frugivorous and insectivorous. Cause seed dispersal. Also feed on nectar and help cross pollination. Flocks suddenly take off, abandoning feeding apparently for no reason, circle the fruiting tree and then settle again to resume feeding.
Cultural notes: The Sanskrit name **Shyeta Shari** means a red-white myna.
As per the Yajurved, this myna was offered in sacrifice to the goddess of learning- Saraswati!
Status: Occasional.

Rakhi-dokyachi Myna, Pawai Myna

Description: Chestnut-tailed Starling with black wing tips, grey-brown back and rufous belly. Affects lightly wooded areas and plantations. Keeps to the canopy rarely coming down to fruiting bushes. Gregarious and vocal. Cling to branches in all manners.

Asian Pied Starling.

Chestnut-tailed Starling.

Related species:**A*sian Pied Starling (Sternus contra-1002) Orange orbital skin, yellow bill of this pied starling help identification. Common in Thane district of Kokan, since a pair is said to have escaped in 1939! Seen perched on wires near marshy swamps. The nest is a globular dome placed on branches or poles. Song is melodious. Call is a liquid whistle.*

Brahminy Starling

Brahminy Myna, Blackheaded Myna

Sturnidae

Sturnus pagodarum (994)

Shankara, Gaurika

Size: 220 mm   M / F : Alike   **R**

Distribution : The entire region.

Bhangpadi Myna, Bamani Myna

Description: Female has a shorter black crest. On cold mornings, during rains and at the time of breeding, the birds fluff their feathers erect the crest giving them an untidy appearance. These intelligent birds have adapted very well to the urban habitat.

Nest-site: Natural tree hollows, holes in wall. Letter-boxes, pipes, broken lamp shades. *Mar. - Jul.*

Material: Lined with leaves, grass, sticks, paper, plastic, aluminium foils, silvery paper.

Parental care: Both. Incubation 16-18 days.

Eggs: 3-4, pale blue. 24.6x19 mm.

Call: A variety of chatters, chuckles, chirping sounds like *Chrr, chrr, krrr, krrr, krak, chir.* Noisy when in pairs or flocks. Mimics bulbul calls.

Ecological notes: Is known to eat poisonous fruits of Yellow Oleander (*Thevetia nerifolia*). Cross pollinator of *Erythrina, Salmalia, Butea, Capparis, Madhuka* spp. Seed dispersal of *Lantana, fig* spp. Insect controller.

Cultural notes: Quickly accepts and adapts to offered nest boxes. Confiding bird.

Status: Common.

This Myna successfully nested in an unused letter-box. Incidently the name on the letter-box is that of a tree !! Audumbar, the name of the tree on this letter box is Ficus glomerata.

Brahminy Starling on the nest.

Rarely seen partial melanism in a Brahminy Starling & a Jungle Babbler.

Rosy Starling

Rosy Pastor

Sturnidae

Sturnus roseus (996)

Madhusarika

Size: 230 mm   M / F : Alike   **WM, PM**

Distribution : The entire region up to the highest altitude.

Nest-site: Colonial nest-holes in earth banks are seen in SE Europe, SW Asia, where locusts are abundant, hence the nest-sites are shifted, depending on the availability of food. *May - Jun.*

Material: Lined with feathers, twigs, leaves.

Parental care: Both.

Eggs: 3-5, faint blue.

Call: Feathers ruffle and crest stands up when calls like chuckles, cries, whistles, warbles are uttered. Vociferously forage on winter crops and fruiting trees. The cries of the farmers guarding their fields, mingle with the bird calls!

Ecological notes: Cause damage to winter crops, but also eats locusts, hence play a mixed role. Cross pollinate flowering trees. Cause seed dispersal. Fat birds on return migration are hunted for flesh, in the North-West.

Cultural notes: According to an Iranian legend, the water from a certain well attracts the pastors to locust infested areas, and then rids the people from the scourge of the locust plague. Considered helpful in Afghanistan. In our areas, the fields are guarded against the clouds of pastors, which are driven away but not killed.

Status: Common.

Bhordi, Gulabi Myna, Palas Myna

Description: Rose-pink and black, crested-starling. Immatures-crestless and brown. Flocks arrive in September and depart by April. They like to keep flowering trees to themselves, threatening intruding birds by wing flapping and loud cries. Rest at noon, on field side acacia trees, mango groves or even in city gardens. During this sojourn, they noisily chatter, as if to plan some future raid on the crop in an unattended field! Roost with parakeets, mynas, pigeons at night.

Br.

Non-Br.

Immature.

Salunkhi, Myna

Description: White wing patch prominent in flight. Yellow pre-orbital skin is bare. The habit of indulging in noisy fights with the other mynas, at times mounting one another, has given it the apt name of **Kalahapriya**, one who likes to fight! They use the nests of the woodpeckers, parakeets, etc. Predation of chicks by crow, coucal and grey hornbill observed. A pair marauding the nestbox of another pair of the same species, evicting the chicks from this nestbox by holding them in the beak and later not even using the emptied nestbox themselves, is recorded! (Amit Pawashe)

Jungle Myna.

Common Myna

Indian Myna

Sturnidae

Acridotheres tristis(1006)

Kalahapriya, Chitranetra

Size: 230 mm   M / F : Alike   **R**

Distribution : Throughout the region but uncommon in the Ghats.

Nest-site: Natural tree holes, under the roofs, in wall holes, embankment holes, hollow pipes, drains. At times multiple holes are seen in a single dead palm stump. *Apr. - Aug.*
Material: Lined with roots, leaves, grass, rubbish, paper, plastic, cloth, feathers.
Parental care: Both.
Eggs: 4-5, blue. Two broods are common. 30.8x21.9 mm.
Call: A noisy bird with varied calls, squabbles, chatters and even pleasant notes. *Keek, kreek, Chirr, chirr* uttered. A typical whirring noise is heard, when the bird takes to wings.
Ecological notes: Seen on garbage dumps. Frequently use plastic for the nest. An incidence of suffocation, when the plastic bag held in the beak turned over the bird's head, when in flight, is recorded. Partial to flowering trees and often follow the farmers, for the insects, as they plough the fields.
Cultural notes: Features in literature. Symbolic pair of Raghu-Myna is immortalized by poets.
Status: Common.

Related Species: *Jungle Myna(A.fuscus-1010) Now seen in urban areas. Less confiding than Common Myna. It is more grey and lacks bare yellow pre-orbital skin. Iris is bluish. Less omnivorous than Common Myna. Common in our area. Takes over nests of owlets, parakeets, barbets.*

Jungle Myna - nominate race.
The bluish iris & tuft at the base of bill are typical.

*Hill Myna
(E. P. Gee, 1942,
BNHS collection)*

*Common Mynas sometimes
get electrocuted and die.
This is an urban hazard.*

Jungle Myna & Spotted Deer.

*Jungle Myna with
yellow iris (A.fuscus fuscus)*

Common Mynas often use plastic as nest material.

Common Myna with a grass-hopper.

*To minimise such risks from nesting near electrical equipment
artificial nests can be offered to Mynas.*

Common Mynas find such unusual but hazardous nest sites.

Brahminy Starling is an inquisitive bird.
Here it is seen inspecting a Baya's nest.

Sri Lanka Hill Myna. *Endemic.*

A flock of Rosy Pastors in worn plumage.

Southern Hill Myna is often kept as a pet.

Starlings: Bill straight, pointed.
Wings short, triangular & lack mirror.

Mynas: Bill decurved. Wings rounded & with mirrors.

Chestnut-tailed Starling.
(Sturnus malabaricus malabaricus)

Southern Hill Myna

Hill Myna, Grackle

Sturnidae

Gracula indica(1016)

Purushvaak Shari

Size: 250 mm   M / F : Alike   **R**

Distribution : Western Ghats up to 1700 m.

Rare in Maharashtra.

Nest-site: In a solitary secondary tree hole or in one amongst many holes on a tree near the forest edge or in a clearing by the road. *Feb. - May.*

Material: Lined with feathers, leaves, bark, etc.

Parental care: Both. Said to pair for life.

Eggs: 2-3, blue, red spotted. 31.6x23 mm.

Call: High pitched alarm call. A range of calls, screeches, chuckles, whistles, murmurs and songs. The calls change with locality. Do not mimic calls of birds in the wild!

Ecological notes: Effects cross pollination, being an ardent nectar seeker. (*Erythrina, Grevillea, Salmalia, Butea, Bombax* spp. *Helicteres isora*). In Bihar, castor crop pest *Ophiusa melicerte* was seen in the stomach contents (SA Handbook). The trade in these talking birds was so flourishing and chick poaching so rampant, that now, it is rare.

Cultural notes: Still a popular cage bird, owing to the astonishing ability of reproducing human voice (Sanskrit name- **Purushvaak Shari**) and bird calls.

Status: Rare.

Pahadi Maina

Description: A strong footed, jet black bird with orange-yellow bill, yellow naked fleshy nape and eye-wattles. White wing band is seen in flight. Noisy flocks on fruiting ficuses with barbets, green pigeons and hornbills. Keep to the canopy and do not come to the ground, except to take a drink. The Grackle is an American bird and is a misnomer for the Hill Myna. Insects and lizards eaten. Winged insects are hawked clumsily.

Common Starling.

Related species: *Bank Myna (Acridotheres ginginianus-1008) Black-capped,blue-grey myna with a tufted forehead, red orbital patch and pink tipped squarish tail. Gregarious and terrestrial. Northern part of our range till Thane up to 800 m. Colony exists around Pune and in Goa (?escaped birds). Attend grazing cattle, ploughing farmers. Colonial nesting in holes in earth banks, walls.*
*Common Starling (S.vulgaris-997) WM to our Northern range in Gujarat. 200 mm. Glossy green, black. White spotted. Beak yellow. Fields and marshes.

Bank Myna.

Haldya, Kanchan, Aamrapakshi

Description: Male is golden yellow, female is dull green. Both have a black eye stripe and wing bands. Often seen in mango groves, hence the name Aamrapakshi. These arboreal birds are noisy. Commonly seen chasing one - another in a typical undulating flight. They drink flower-nectar by clinging to the flowers upside down or in other aerobatic manner.

Black-headed Oriole

Related species:
Black-headed Oriole (O.xanthornus-959) More common in Kokan as compared to the Golden Oriole. In the Ghats up to 1700 m. Chicks have a streaked black neck.
Black-naped Oriole (O.chinensis-954) WM up to 1000 m. It is recorded in Phansad reserve near Dighi, dist.Raigad (S.Pande, P.Mestri).

Eurasian Golden Oriole

Golden Oriole

Oriolidae

Oriolus oriolus(953)

Haridraav

Size: 250 mm. M / F : Dimorphic **R, LM**

Distribution : The entire region up to 1800 m.

Nest-site: Intricately woven, deep hammock like cup in *Mangifera indica, Tectona grandis, Artocarpus heterophyllus* trees, often with the nests of drongos and green pigeons. *Apr. - Jul.*
Material: Bark, twigs, fibres, cobweb. Lined with feathers.
Parental care: Both. More than one clutch.
Eggs: 2-3, white, spotted. 29.3 x 20.3 mm.
Call: Tinkling, fluid whistles *Peeloo, peeloo-loo.* Harsh *Cheeh, cheeh, Whee-oo.* A musical call, easily identifiable in the woods.
Ecological notes: Chicks born in Kokan and Ghats migrate to the Deccan in winter, when they may be found exhausted and victimised by crows. Medical aid, if promptly, given is usually beneficial.
Cultural notes: Folk-lores are woven around this beautiful yellow coloured bird. One name of the Oriole in Kokan is Balanteen-In puerperium, derived from the custom of applying turmeric powder to post-partum women.
Status: Occasional.

Golden Oriole in flight.

Black-naped Oriole.

Golden Oriole in a hammock nest.

Adult & Juvenile Golden Orioles on fruiting trees.

Juvenile Black-headed Oriole.

Immature. **Kotwal, Bangda, Kolsa, Govinda**

Description: Upright stance. Perches on a wire or a bare branch with the deeply forked tail hanging down. Has a white rictal spot. In agricultural areas, it is encountered riding on the back of cattle, making frequent sallies after insects, when it often alights on the ground.

Related species:
*Ashy or Grey Drongo (D.leucophaeus-965). This slim winter visitor keeps to well wooded country. The tail is deeply forked, iris crimson and diagnostic, white rictal spot is absent. Mimics the calls of various birds. Nests in the Himalayas and NE India.

Grey Drongo.

Grey Drongo predating on a Sunbird. An unworthy act for a "Police Bird" !!

Black Drongo

King Crow

Dicruridae

Dicrurus macrocercus (962, 963)

Bhrunga, Angarak, Goprerak

Flimsy nest of a Drongo.

Size: 310 mm   M / F : Dimorphic   **R, WM**

Distribution : The entire region up to 2100 m. The larger sized race *albirictus* is rarely seen in Maharashtra in winter.

Nest-site: The flimsy cup is placed 4-12 m. up in a fork of a thorny tree like *Acacia nilotica*; also in *Dalbergia sisoo*, *Mangifera indica*, etc. trees. *Apr. - Aug.*
Material: Twigs, grass, fibres. Unlined.
Parental care: Both. The nests are ferociously guarded against intruders. Nests are rarely parasitized by Koel and Drongo-Cuckoo.
Eggs: 3-4, pink-white. Spotted, brown-blotched. 27.1x19.8 mm.
Call: Utters harsh loud calls, some like those of a shikra. Noisy *Chee, titi, chichuk, cheece, Cheece.* Musical calls uttered. Duets during breeding.
Ecological notes: Pollinator and pest controller. Devours winged insects in midair sallies, often up to late midnight, around halogen street lamps. Smaller birds sometimes nest on the same tree with the drongo. Flocks on flowering forest trees, at the edges of forest, grassland and field-fires,to catch insect 'sizzlers'! Predates on chicks of small birds.
Cultural notes: Well known for aggressive behaviour and fearlessly attacking birds of prey, when nesting, thus offering protection to the mild mannered birds and is known as the Kotwal-from its policing habit!
Status: Common.

Black Drongo on the nest with three chicks.

White-bellied Drongo

Dicruridae

Dicrurus caerulescens (967)

Dhawal

Size: 240 mm   M / F : Alike   **R, LM**

Distribution : The entire region up to 1500 m., in hilly areas.

Nest-site: In a bamboo clump or in an open forest, 3m. and above. *Mar. - Jun.*
Material: Grass, twigs, fibres, rootlets.
Parental care: Both build the nest. Incubation and feeding by both the parents.
Eggs: 2-4, pinkish, spotted. 23.6x17.8 mm.
Call: Songster and mimic. Pleasant whistles and a few harsh calls. Noisy.
Ecological notes: Mainly insectivorous, but also takes nectar. Prefers hilly areas, keeping to shady patches. Often pursues smaller birds and robs food from them.
Cultural notes: The drongos being excellent vocalists,were dedicated to **Vak**, the goddess of Speech, in Rgved.
Status: Uncommon.

Pandhrya-potacha Kotwal

Description: Black above with white belly, vent and under tail-coverts. Fond of hawking insects in air in aerobatic sallies launched from trees. After eating a few insects, takes a few sips of fresh flower nectar, as if to quench thirst, and then utters a soliloquy. Crepuscular. Seen with ioras, flycatchers, minivets, tailorbirds, woodshrikes, whose calls it mimicks to perfection. This drongo is a beautiful forest bird.

Nesting White-belied Drongo.

Immature Black Drongo.

Look Alike: *The immature ***B**lack Drongo (D. macrocercus) may be mistaken for the whitebellied Drongo. The former has less white on the belly, and the under tail-coverts are streaked grey-white and not uniform white.*

Makhamali Kotwal

Bronzed Drongo

Dicruridae

Dicrurus aeneus (971)

Vachalaha, Dhumyat

Size: 240 mm M / F : Alike **R, LM**

Distribution : Up to 2000 m. in Ghats, Khandala southwards. Rarely in Kokan and Malabar.

Nest-site: Nest is suspended hammock-wise on a dry fork of *Dalbergia, Terminalia* spp., *Tectona grandis* or bamboo stems. *Mar. - Jun.*
Material: Leaf strips, bark, twigs, fibres, plastered with copious amount of cobweb, hence white.
Parental care: Both. Intruding birds of prey are boldly attacked by the parents.
Eggs: 2-4, pinkish, spotted. 21x16 mm.
Call: Normally silent. Capable of excellent mimicry. Whistles and fluid musical calls.
Ecological notes: Move with mixed hunting parties of insectivorous birds. Take flower nectar and cause cross pollination. Very territorial and keep to the same area for years!
Cultural notes: Locally it is called Makhamali (silky), due to the glossy plumage. In Nepal the bird is known as Chaptia, after it's flat beak!
Status: Rare.

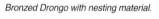

Bronzed Drongo with nesting material.

Description: Small drongo with flat bill and gently forked tail. A bronze-green gloss is seen in sunshine. Rump and belly grey and iris red-brown. Nests are seen near forest water holes. The bird may be encountered in groves close to forest edges. A shy bird, unlike the other pugnacious drongos.

Haircrested Drongo.

***Related species: *** *Spangled or Haircrested Drongo (D.hottentottus-973) The upcurled tail end with a shallow fork is a good field guide. The hair-crest is not seen without binoculars. Spangled blue-black wings are appreciated in direct sunlight. In forested areas up to 1400 m. Takes nectar with the long, curved bill. In Kokan the nest is seen on palm fronds also. Call is metallic. Mimics bird calls.*

Racket-tailed Drongo

Dicruridae

Dicrurus paradiseus (977)

Bhrungaraj, Satyavaak

Size: 600 mm.   M / F : Alike   **R, LM**

Distribution : Patchily and locally throughout the region up to 1500 m.

Nest-site: The see-through cup is placed 5-15 m. up, on a horizontal branch or the base of a palm leaf stalk. *May - Jul.*

Material: Twigs, grass, tendrils, bark.

Parental care: Both.

Eggs: 3-4, white, red-brown. 27.8x20.2 mm.

Call: Loud fluid *Kit, kit, kit,* heard from before the dawn till late night. Mimics Shikra, Crested Hawk-Eagle, Serpent-Eagle, Koel, Pitta, woodpecker, Grey Hornbill, Jungle Babbler, Magpie Robin, Shama, etc. and squirrel calls and human whistles to perfection. An instance has been reported, when a tame elephant in the jungle camp, promptly came in response to the mahouts whistle, which in fact was mimicked by the drongo!

Ecological notes: Cross pollinator of *Salmalia, Bombax ,Erythrina* spp. Termites, moths, beetles, butterflies, grasshoppers and larvae in the tree trunks are eaten. Seen with treepies, blackheaded orioles, woodpeckers and barred jungle owlets. Prefers moist deciduous forests.

Cultural notes: Well known due to the long trailing tail feathers. Was a popular pet, being the master of mimicry, hence the Sanskrit name **Satyavaak**.

Status: Uncommon.

Greater Racket-tailed Drongo with moulted streamers.

Mahabhrungaraj, Pallavapuchha Kotwal

Description: The head is tufted. This is an important aid for identification, since the long tail feathers may be shed during breeding, probably to make incubation easier. The tail is long and spatulate at the ends. When flying, it appears as if the bird is being chased by a pair of bumble bees. Fluid, metallic calls are uttered in flight when birds chase each other.

Greater Racket-tailed or Ceylon Crested Drongo (D.p.lophorinus)

Look Alike: *Lesser Racket-tailed Drongo (D.remifer-972) is not seen in our area. D.paradiseus when in moult may be difficult to identify, but the tufted head is a useful aid.

Tad-Pakoli

Description: Bluish-bellied, slaty grey, short-tailed bird. Bold and aggressive when nesting. While gliding, the bird seizes winged insects in its wide gape, which are then transferred to the toes, shredded and eaten in flight. Flight on the pointed wings is slow sailing and elegant. When perching huddled together on a bare branch or wire, the tail is waggled laterally.

Look Alike: *May be confused with swallows if not carefully observed.*

Ashy Woodswallow

Ashy Swallow-Shrike

Artamidae

Artamus fuscus(982)

Talchatak

Size: 190 mm.   M / F : Alike   **R**

Distribution : The entire region up to 2100 m.

Nest-site: Loose unlined cup in the angle of leaf stalks of palmyra, coconut, fantail palms. *Mar. - Jun.*
Material: Leaves, grass, sticks.
Parental care: Both.
Eggs: 2-3, greenish-white, spotted, ends broad. 23.4 x 17.1 mm.
Call: Harsh *Chek, chek.* Song is given by the breeding female. Mimics jungle myna and redwhiskered bulbul.
Ecological notes: Recorded to eat pungent butterflies of the genera *Danais* and *Euploea.* Devours flower nectar. Keeps near Tad-Palm groves.
Cultural notes: None recorded.

Indian Treepie

Rufous Tree Pie

Corvidae

Dendrocitta vagabunda(1031,1033,1034)

Krushakut, Karayika

Size: 460-500 mm.   M / F : Alike   **R**

Distribution : Race *parvula* common up to 2000 m., *pallida* in Maharashtra, *vernayi* Nilgiri south.

Nest-site: Well hidden untidy nest at a considerable height in a tree. *Feb. - Jul.*
Material: Grass, rootlets, leaves, thorny twigs.
Parental care: Both.
Eggs: 4-5,white, streaked red-brown. 27 x 21.1 mm.
Call: Loud noisy tinkling metallic sound- *Bob-bo-link.* Varied calls of harsh to pleasant in quality. Also *Kokli, Good-day,* etc.
Ecological notes: Omnivorous. Eats weaker birds and carrion. Scavenges on prey of carnivora.
Cultural notes: Considered to be a bird of augury! Other Sanskrit names **Bhash** and **Kuraivak** indicate the birds bark-like and whining sounds.
Status: Rare.

Takachor, Bhera

Description: White wing patches are seen in flight. Seen in mixed hunting parties. Very noisy during courtship when jumping display and mutual feeding are noted.

Andaman Treepie (D.bayleyi-1040) Endemic to Andaman Is.

White-bellied Treepie.

Indian Treepie with the wing of Purple Moorhen.

White-bellied Treepie.

Related species: *****White-bellied Treepie (D.leucogastra-1036) Chestnut back and white hind-neck (sooty in Indian Treepie). Southwards from Goa in the heavy rainfall areas up to 1500 m. Often seen in the company of Racket-tailed Drongo. Endemic.*

Mice feature in the diet of the crow. Here a field mouse is seen nesting in the mud nest of a Red-rumped Swallow. (Page 213)

Crows often draw our attention towards animals in distress. This civet stranded on a lamp pole was later rescued.

Some Corvids :

Indian Treepie.

Carrion Crow of the Himalayas.

Crow pecking a live injured Grey Plover.

Sri Lanka Blue Magpie. Sri Lankan *Endemic*.

White-bellied Treepie. *Endemic* to Western Ghats.

House Crow

Indian House Crow

Corvidae

Corvus splendens (1049,1050)

Vayas, Grushakak, Dhwanksha

Size: 430 mm.   M / F : Alike   **R**

Distribution : Race *splendens* up to 2100 m.

Race *protegatus* Palghat south.

Kawla

Nest-site: *Mangifera indica, Ficus* spp., *Tamarindus indica, Ailanthus exelsa, Casurina litorea* trees. Ledges on houses, electric poles, pylons, roofs, platforms. *Feb. - Jul.*

Material: Sticks, dry twigs, coir, metal-wires, coat hangers, foils, plastic, paper, etc. Adorned with metal cups, colourful cloth, necklaces, straps,spoons, petty articles, taken from houses or picked from garbage dumps.

Parental care: Both. Apparently monogamous. Rears parasitic koel chicks. Reported to have killed a koel chick, some times koel chicks are pushed out of the nests by the crows, when noticed to be different.

Eggs: 4-5, blue-green, speckled. 25 x 20 mm.

Call: *Kaw, kaw.* At noon, soft *Krr, kree* from a shady branch. A separate call for almost every mood and to instruct the chicks.

Ecological notes: Nest parasitized by the koel. Scavenger. Indicator of insanitation. Attacks heronries, baya colonies, kills unfit birds, and also attacks humans if nests are disturbed. 153 injurious insects have been found in the stomach contents of the crow. Urban bird.

Cultural notes: Falsely believed to be one-eyed-**Ekaksha**. Associated with human death, and is important in the post-death ritual of 'Pind-daan' which indicates the peaceful departure of the soul.

Status: Common.

Description: Grey neck, nape, upper breast, otherwise black. Conjugal. Intelligent, adaptive, Cunning, social, inquisitive, wary, scavenging, omnivorous. Enjoys teasing other birds and animals. Known to catch fish from water, in flight. Pirates food from others.

The White Crow: The white crow is an albino. Such crows when sighted generate much interest and awe, since association of the crow to black colour is strong.

Crows lay several eggs in the nest.
House Crow feeding its own chick.

Crows often molest owls in day light.
House Crow & Dusky Horned Owl.

House Crow scavenging on a Lesser Flamingo carcass.
Crows have taken up the job of the vultures.

Partial Albino House Crow.

Mating of House crows is a rarely seen event. As per a superstition this sighting means near death to the observer !

Jungle Crow

Large-billed Crow

Corvidae

Corvus macrorhynchos (1057)

Vankak, Krushakak

Size: 480 mm.   M / F : Alike   **R**

Distribution : The entire region up to 2300 m.

Domkawla

Nest-site: *Ficus* spp.,*Mangifera indica*, *Palm* spp.,*Casuarina litorea, Holoptelea integrifolia, Dalbergia latifolia, Albezia lebbek* tree forks. Also on poles, pylons. *Feb. - Jun.*
Material: Sticks, wool, paper, plastic, choir, twigs.
Parental care: Both. Incubation more by the female. Chicks fed by both the birds.
Eggs: 3-5, blue-green, speckled.38 x 28.1mm.
Call: Deep, hoarse croaking *kaws. Kaa, kaa* when alone, with the neck extended and tail moving up and down.
Ecological notes: A chick lifter and hence a menace to poultry. An indicator of carnivora kill hidden in the forest undergrowth. Scavenges with vultures, treepies. Nest is brood-parasitized by the Koel.
Cultural notes: The crow is never killed or molested.
Status: Common. Increasingly seen in urban areas due to large scale destruction of the forests, its former habitat.

Description: Jet-black with thickset bill. Flocks are seen with house crow at garbage dumps in urban areas, on TV antennas, around carrion and kills in the forests, and to pay homage to their dead kin. Like the house crow, indulges in various flight patterns, and is often mistaken for some other avian species when seen in silhouette.

With the decline in vulture population the jungle & even house crows are efficiently scavenging such road traffic collision kills. This caracas of a jackal was scraped to the bones by crows in a few hours.

Rehabilitation & Conservation :

Orphaned Bonelli's Eaglets can be successfully rehabilitated by relocating them in nests of foster parents. If the eaglets are found outside their nests due to molestation from kids, they can be replaced in their nests within 24 hours. Injuries like penetration of twigs during their fall from nests can be treated.

The cave in Burnt Island. Poachers had erected bamboo frames for removing nests of Edible-nest Swiftlets. These were removed, poachers arrested & cave protected.

Geography of the Region

- *Sanjeev B. Nalavade*

The Sahyadris or the Western Ghats is the high rising border of the Deccan Plateau. It is the most conspicuous physiographic feature of Peninsular India. It commences in the Nandurbar district of northern Maharashtra south of the Tapi valley. From there it runs southwards as a sinuous line more or less parallel to the western coast for about 1600 km. up to Kanyakumari. The Ghat range overlooks the narrow Kokan and Malabar lowland to the west and is separated from them by a line of long cliffs. The steep slope towards the coast is regarded as an indication of faulting and subsequent erosion activity. As a result, the Western Ghats act like a barrier between the low lying coastal belt and the upland plateau. The average height of the crest line is 920 m. in the northern section, Kalsubai (1,646 m.) being the highest peak in this part. Up to south Maharashtra the range is composed of lava with a typical Deccan trap topography with a steep escarpment to the west and step like arrangement to the east.

From south Maharashtra to the Nilgiris the range is composed of heterogenous assemblage of granite and greisses. In Karnataka, the Ghat range runs closer to the coast with an average elevation of 1220 m. Several peaks exceed 1500 m. with Kudremukh (1892 m.), Pushpagiri (1714 m.), Kottebetta (1638 m.) among them. Here, a wide area traversed by the Western Ghats and lying between 940 m. and 1250 m. is called 'Malnad'. The prominent eastern offshoot is the Baba Budan range famous for iron ore mining.

Nilgiri Hills or the Blue Mountains which is an eastern offshoot of the main Western Ghat range joins the latter near Gudalur in south Karnataka. It rises over 2000 m. with Doda Betta near Udhagamandalam or Ooty (2637 m.) as the highest point of the range.

South of the Nilgiri Hills, the Palghat gap is a prominent break in the continuity of the Western Ghat ranges. The gap is about 25 km. wide at the narrowest and acts like a pass with many road and railway lines passing through it. A complex system of lofty hill ranges is found south of Palghat Gap. Together called as the South Western Ghats, the main ranges from the region are Anaimalai Hills, the Palni Hills and the Cardamom Hills. The highest point from the area is the Anaimudi Peak (2,695 m.) which is also the highest peak in Peninsular India.

The coastal belt considered in this book stretches from Surat in South Gujarat (21 degrees N) to Kanyakumari (8 degrees N) in the south. This more or less narrow stretch is almost 1,600 km. long and varies in width from 12 km. to 80 km. The coastal plain is broader around Surat but narrows down towards the south.

The coastal lowland from Maharashtra, Goa and Karnataka is generally described as Kokan whereas further south up to Kanyakumari it is called Malabar. Kokan, 50 km. to 80 km. wide is a rocky and rugged country of elevated plateau and low lateritic hills and detached ridges interspersed by creeks and embayments.

In Goa and Karnataka, the coast is less rugged and wide estuaries are seen. The Malabar Coast is narrower in the north and south and wider in the middle. Low lateritic plateaus and low hills occur east of the alluvial coastal belt. The coast here is characterized by a number of narrow lagoons called backwaters in this part of the country. These are excellent waterways. Sandy beaches coastal sand dunes, mud flats and wave cut platforms provide extensive feeding arena for thousands of waders, which throng here, in winter.

The Western Ghats act like a water divide between the two water systems, one directed towards the Bay of Bengal and the other to the Arabian Sea. The Ghat country is the source region of most of the major rivers of the Peninsula all of which are east or south-east flowing. The important east flowing rivers with their sources in the Ghat country from north to south are the Godavari, Bhima, Krishna and her tributaries (Ghataprabha, Malaprabha, Tungabhadra), Kaveri and her tributaries (Kabini and Bhavani), The west flowing rivers, numerous in number are obviously shorter and swift flowing with narrow gorge like valleys and water falls near their sources with tortuous rocky courses. Some of the major west flowing rivers are the Tansa, Ulhas and Vashisthi in Maharashtra, Mandavi and Zuari in Goa, Kalinadi, Sharavathi and Netravathi in Karnataka and Ponnani and Periyar in Kerala. Many Ghat rivers near their sources have been dammed into reservoirs for irrigation and power generation.

Climate:

The typical climate in this whole region is tropical monsoon type. The climate along the coast is more or less equable with high temperatures throughout the year. The north-south orientation of the coastline and the adjoining Western Ghat ranges are the deciding factors for the climate. The region has three distinct seasons as elsewhere in the country, namely, the summer, the rainy season and the winter. A rise in the mean temperature in March indicates the arrival of hot season. April and May are the hottest months both in the coastal lowland and in the Hills. It is hot and humid along the coast whereas it is cooler and pleasant up in the hills. The Western Ghat experiences low temperature of 20 C and 27 C in April. The high mean temperature is around 30 C along the coast. Some parts along the eastern margins register temperature above 40 C. Light sea breeze is experienced up to the foothills and even up to the western slopes of the Western Ghats.

Pre-monsoon rains called 'mango showers' are experienced all along the region especially in the south. These showers are beneficial for coffee and mango. They also help in bringing down the hot summer temperature by providing cool relief to the otherwise sultry uncomfortable weather, especially in coastal regions.

Forest clearing for cultivation.

The rainy season commences with the onset of southwest monsoon (SWM), which reaches coastal Kerala by the end of May. The normal date of onset is 29" May in Kerala, 3" June around Karwar and 7" June in Mumbai. The Arabian Sea branch of the SWM confronts the Ghat range with the resultant orographic rains, which are heavy in the coastal plain along the western slopes and the crestline of the Western Ghats. The rainfall is between 2000 to 4000 mm. all along the coast and above 4000 mm. in the Ghat region. Agumbe in south Kanara is aptly called the 'Cherapunji of South India' with more than 7500 mm. of rainfall. High rainfall, high humidity and longer wet season have given rise to dense evergreen luxuriant vegetation in this region. Many hill slopes are under tea, coffee, spices or rubber plantation. With the onset of the monsoon there is a general fall in

the mean monthly temperature by 3 C to 4 C especially in the Ghat country. The mean July temperatures are 25 to 27.5 in coastal lowland, below 25 C in the Ghat country and 27.5 to 30 C north of Mumbai. Because of the rain shadow effect, there is a drastic fall in the rainfall amount, east of the Ghat region. This decline is especially noticeable in the north. Mahabaleshwar located on the crest of the Ghat receives more than 6000mm. Panchgani located hardly 20 km. east of Mahabaleshwar receives only 1750 mm. and at Wai 10 km. further to the east rainfall drops down to 800 mm.

July and August are the rainiest months all along but 80 % of the annual rainfall is received in the four months from June to September. In the northern and central portion, the rainy season is of four months. In South Western Ghats the rainy season is longer because of early arrival and late departure of the monsoon and also due to rains received from the North East monsoon. Here the rainy season is almost of seven months.

The retreating of monsoon from Maharashtra and Karnataka allows the mean temperature to rise again, responsible for what is called as October heat period. This is the season when sometimes-cyclonic storms from the Arabian Sea affect coastal region.

Cool season starts from November-December. Low temperature, low humidity, clear skies and bright sunshine characterize this period of pleasant weather. It is the best season for bird watching. The winter migratory birds double the joy of bird watching in these months. Along the coast, there is a fall in the temperature in the north, with little change in the south. The mean January temperature is 25 C and above south of Goa, 22 to 25 C in South Kokan, and 20 to 22 in north Kokan and South Gujarat. The Ghat region is quite cool with mean temperature between 15 and 20 C. During early hours, the winter days in hill stations like Udhagamandalam and Kodaikanal experience temperatures below 5 C.

Destruction of the vegetation for mining.

Avian Geography of the Western Ghats

- Sanjeev B. Nalavade / Satish Pande

*Lanner/Saker Falcon.
Not recorded in our region.
Can be mistaken
for Laggar Falcon.*

Avian geography of the Western Ghats and the West Coast is closely related to the biogeography of the Indian subcontinent. The World is divided into six zoogeographical regions, Neotropical, Australian, Nearctic, Palearctic, Ethiopian and Oriental. Most of India along with South China, South East Asia and North West Islands of Indonesia fall in the last category. A small portion of Northwest Himalayas falls under Palearctic region. The Indian Subcontinent has been further subdivided into a number of zoogeographical subdivisions by Blanford (1870, 1904), Smith (1931), Mahendra (1939). Rodgers and Panwar's scheme of Indian Biogeographic Zones is : 1. TransHimalayan 2. Himalayan 3. Indian Desert 4. Semi-Arid 5. Western Ghats 6. Gangetic Plane 7. North-East India 8. Islands 9. Coasts.

The present day distribution of animals, especially birds in the Indian Peninsula has its roots in the geological history of the region. The major geological events those have shaped the present day distribution, diversity and composition of the fauna of the Peninsula can be chronologically stated as:

1) Fragmentation of Pangaea (*All Earth Greek*), the Super-Continent that existed millions of years ago into two major parts, Laurasia in the north and Gondwanaland in the south with the intervening Tethy's Sea. Indian subcontinent was part of the Gondwanaland, which also contained South America, Africa, Antarctica, Australia and Madagascar Island.

2) 200 million years ago the ancient landmass of Gondwanaland fragmented further and the Indian plate started drifting to the north and northeast direction from its original position in the southern hemisphere around 40 degrees south.

3) On the way to the north, some where near the present day Mauritius, eruption of lava took place giving rise to the formation of Deccan Plateau and other hill ranges of peninsular India.

4) Block fracturing of the Western parts of the Peninsula and marine subsidence (sinking) of the fractured fragments in the Arabian Sea gave rise to the scarps-line of vertical cliffs,of the Western Ghats and formation of Kokan-Malabar lowland.

5) Gradual denudation of the Peninsular Hill ranges took place lowering down their general heights while still in the northward journey in the Palaeozoic and Mesozoic eras

(205 to 135 million years ago)

6) As the Peninsular landmass closed in the Asian landmass, the intervening Tethy's Sea began to disappear progressively westward from the east.

7) The physical contact between the Indian landmass and its Asian counterpart gave rise to squeezing and folding of sediments giving rise to Tertiary Mountain areas the mighty Himalayas.

8) The lagoons and marshes formed as the vanishing relicts of the Tethy's sea were filled with the sediments brought from the Himalayas and the Peninsular Hills giving rise to the Great Indian Plains.

9) The first physical contact of the Peninsular India with the Asian plate in the east (Assam Gateway) allowed the influx of Indo-Chinese and Malayan fauna to the Peninsula. In due course of time, Tethy's sea disappeared from the west thus establishing contacts with Asia in the northwest allowing the Ethiopian and Palaearctic faunal elements to enter the Indian Peninsula. The formation of the Assam gateway is one of the most important phases in the biogeographical, particularly avifaunal, evolution of the Indian Subcontinent.

10) Pleistocene glaciation on the Himalayas forced many northern plants and animals to take refuge in the south. After the retreat of pleistocene glaciers from the Himalayas, the conditions in the south became warmer, many animals moved towards the higher parts of the South Indian hills where they are now confined.

11) Advent of man in India is another important event.

The two most important events that have shaped the ecology and biogeography of the Subcontinent are: a) the Himalayan uplift, which changed the geology, climate and bio-diversity composition, and b) the anthropogenic impact operational for the last five thousand years but severe during the past four centuries, leading to widespread conversion, degradation and disappearance of natural habitats including forests.

Climate is a secondary factor in the evolution of the present day biogeographical characters of the Subcontinent. The monsoon climate characters are of recent origin. Presently the role of rainfall is comparatively more important than that of the temperature.

Some peculiarities of the present distribution of our avifauna are:

- There is a concentration of faunal elements including birds in the small areas of south Western Ghats, Northeast India, the Himalayas and Andaman and Nicobar Islands-the Hotspots identified by Birdlife International.
- This concentration and isolation is in the form of relict or refugial islands. The south Western Ghats is not a center of radiation but is at the receiving end. These concentrations are the places where original forest-cover has not been fully destroyed. The fauna has retreated to these favourable places.
- The refugial islands have created a marked discontinuity in the faunal distribution (disjunct distribution). Many bird species have two populations- one confined to the Southwest India and the other to the Northeast, with a huge intervening gap where these birds are either absent or patchily distributed. Some of the Western Ghats bird species,mostly forest birds, with such disjunct distribution are Great Pied Hornbill, Brown-Backed Needletail, Common Goldenbacked Woodpecker, Great Eared Nightjar, Speckled Piculet and Rufous Woodpecker
- Altitudinal zonation is significant in the Himalayas. It is also observed to a lesser extent in the Peninsula in the lofty mountains in the Southwestern Ghats. Some plants, reptiles and mammals like Nilgiri Tahr show altitudinal affinity. Few altitude specific birds are: Black Bulbul, Nilgiri Pipit, Nilgiri Wood Pigeon, Nilgiri,Grey-breasted and Wynaad Laughing Thrushes and White-bellied Treepie.
- There is northward impoverishment of flora and fauna as one moves from the southwest to the north along the Western Ghats, as is evident by these figures pertaining to forest birds: Kerala 230, Kodagu 205, Goa-Amboli 182, Mahabaleshwar-Koyna 168, Nashik Ghats 142 (Nalavade,1997, per.com.)

The Western Ghats and the West Coast has a complex assemblage of birds with different affinities. Many bird species noted here are more or less restricted to the entire tropical region and represent the earlier bird fauna. These may be called Gondwana autochthonous elements (Trogons, Orioles).

- The biodiversity in the Western Ghats is limited compared to that seen in the NE India because of its location at the receiving end, lesser degree of speciation and absence of some avian families.
- Some guilds of birds unique to the tropical rainforests are absent in the Western Ghats. Some of these are Curassows, Toucans, Manakins, Antbirds, Tyrrant Flycatchers and Birds Of Paradise. Many tropical forest avian species are poorly represented in the Western Ghats (Pittas,Trogons,Sunbirds).

Many birds from our region have their origin in the Indo-Chinese and Malayan regions. Some bird species with Malayan affinities are Indian Pitta, Spiderhunters, Broadbilled Roller, Tit-Babblers and Blue-bearded Bee-eater. Red-legged Falcon has a Chinese affinity. There is a striking similarity in bird fauna of Malabar and Chota-Nagpur plateau. Birds common to these areas and Sri Lanka are Darkfronted Babbler, Indian Swiftlet, Crimsonfronted Barbet and Frogmouth. African and Madagascar affinities include birds like Yellow-throated Sparrow, Spotted Grey Creeper, Lesser Florican, Sirkeer Cuckoo and Lesser Cuckoo. West Asian affinity is evident through Sandgrouses and Larks. The winter migratory birds that come all the way from the Himalayas, Central Asia and even East Europe mainly represent the Palaearctic element. Many cranes, storks, ducks and waders belong to this category.

Regarding the avifauna of the Western Ghats, the Indo-Chinese and Malayan influence is prominent. The suggested routes or highways of faunal movement from the Northeast India to the southwest Ghats are many:

*From the Northeast along the Himalayas to the west upto the Doab of the Ganga and Yamuna rivers and thence southward through the Aravali ranges. *From the Northeast long the Eastern Ghats. *From the Northeast via Rajmahal Gap, Satpuda range to northern Western Ghats and thence southward to the southwest hills (Satpuda Hypothesis by S. L. Hora).

It is possible that a large area of the Peninsula was under evergreen forests in the past when the overall climate was warmer and the forest birds, which are now confined to the Southwest and the Northeast, were found continuously over a larger area. The gradual change in climate and increasing dryness converted the evergreen forests into deciduous or scrub land. Birds adapted to the evergreen forests in the intervening areas retreated to the remaining humid and favourable southwest Ghats. The barrier of the Deccan Plateau prevented further influx of the birds from the northeast source pool and hampered the eastward movement of the forest birds, acting like a climato-vegetational barrier.

No endemic bird families are found in India (R. Daniel, 1997). Bird fauna of the Western Ghats has only 16 endemic species comprising hardly 4 % of the species of the region. In spite of this, the Western Ghats plays an important role in the avian geography of the Peninsula since it alone supports 70 % of its avifauna, south of the Satpudas.

Vegetation of the Western Ghats

- S. D. Mahajan

The Western Ghats region comprising the main north-south range and its numerous offshoots on the western and the eastern sides sustains rich and diverse types of prime vegetation. It may be designated as an important Floral Reserve characterized by a wide range of floristic patterns, high degree of endemism as well as tremendous and ever increasing human interference from the prehistoric period. Geology, latitudinal and altitudinal differences, climatic, edaphic (soil related) and biotic factors have been responsible for the development of a myriad of vegetation. The average annual precipitation as well as the humid hydro-period increase as one moves from north to south and decrease from west to east in the rain-shadow area in general.

The rich and diverse vegetation of the Western Ghats, according to a moderate estimate, includes over 7,000 species of flowering plants, about 10 % of which are likely to be arboreal (trees and large shrubs). These are the ones, which offer abundant fruits to the avifauna in this region. In addition to this, the flora consists of a few thousand species of Cryptogams (non-flowering plants). More than 200 species are endemic and an equal number of plants are highly endangered. It is feared that some species are becoming extinct every year and many more become critically endangered due to loss of habitat as a result to deforestation and destruction of natural ecosystems. Based on the basic work done by Champion (1936) and Champion and Seth (1968), later corroborated and accepted in principle with some modifications by many workers, the vegetation of the Western Ghats,particularly in relation to its association with the avifauna may be classified as under :-

1. Evergreen or tropical rainforests along elevated ravines, tablelands and crest line receiving high rainfall.
2. Semi-evergreen or mixed moist deciduous forests.
3. Moist deciduous forests
4. Dry deciduous forests.
5. Scrub and thorn forests.
6. Sholas, Savannah and Montane forests.
7. Grasslands and pastures.
8. Littoral swampy halophytic vegetation
9. Woodlands and orchards and,
10. Cultivated fields, agricultural land.

1. Evergreen Forests in the Western Ghats are mostly restricted in pockets, higher plateaus, steeper slopes and in deep valleys along the main mountain range such as in the Silent Valley (Kerala), Anmod (Goa), Kodachadri and Anashi

National Park (Karnataka), Radhanagari Sanctuary, Koynanagar and Mahabaleshwar area(Maharashtra).

The least disturbed and loftier forests representing the climatic climax designated as the virgin forests, are found mainly in the southern Western Ghats. Those present in the Sahyadri region are relatively stunted and tending to degrade into semievergreen or mixed forests. This is a high rainfall area, with annual rainfall averaging more than 2,500 mm. with higher relative humidity and better soil moisture during major part of the year.

Thousands of Devarais or Sacred Groves, ranging from meagre 1 hectare to over 100 hectares scattered all over, along or near the main mountain range deserve a special mention here. Many of these sustain almost undisturbed or pristine evergreen vegetation and often serve as a refuge for arboreal birds and mammals. The evergreen rainforests of the lower latitudes and altitudes exhibit well defined stratification, the top canopy *dominants* (growing upto 50 m) being *Dipterocarpus, Canarium, Olea, Eleocarpus, Palaquim, Vateria, Garcinea, Myristica, Persica,Cinnamomum, Poeciloneuron, Mesua, Diospyros, Alstonia, Mangifera* and *Holigarna*. The vegetation is characterized by canopy density between 90 and 100 percent, particularly due to abundance of second canopy trees, youngsters of taller dominants, lianas (extensive perennial woody climber),vines, thickets of cane, bamboo brakes in the openings and along edges and rich sciophytic vegetation (shade tolerant) including saprophytic (thriving on dead organic matter) fungi, autotrophic algae, parasitic fungi and spermatophytes (seed bearing plants). Epiphytic (growing on another plant without obtaining nourishment from it) mosses, liverworts, ferns and orchids as well as terrestrial ferns abound in number and diversity. In the high plateau and slopes along the crest-line in the Sahyadri range (e.g. Mahabaleshwar and Bhimashankar areas) the evergreen vegetation is reperesented by *Syzygium, Terminalia, Actinodaphne, Memecylon,Garcinea, Litsea, Nephelium, Sideroxylon, Mallotus* and *Ficus* spp.

The forest openings created by natural calamities or human interference develop secondary vegetation of a variety of shrubs, undershrubs, suffruticose (herb with woody main stem) perennial herbs and ephimerals (short lived seasonal plants) forming dense thickets. Impenetrable thickets of 'Karvi' (*Carvia callosa*), 'Waiti' (*Mackenzia integrifolia, Syn: Strobilanthes perfoliatus*) over slopes and those of Pandanus along streams are of common occurrence. Wetlands of various dimensions are not uncommon in lowlands. These develop luxuriant hydrophytic vegetation, predominantly formed of amphibious hydrophytes (aquatic). Being a storehouse of magnificient diversity and richness of flora and fauna, these rainforests represent the 'Gene Pools' of the country. They can be aptly called the Paradise of birds, especially because of the abundance of fruits, nectar and insects.

2. Moist Deciduous Forests:

Moist deciduous forests occurring in flatter areas and over gentle slopes of hills and hillocks are perhaps the most extensive type of vegetation on the either side of the Western Ghats. The average annual rainfall ranges between 1500 and 2500mm. The forests are simpler in structure with or without less evident stratification as compared to that in the evergreen forests.

Most of the trees are basically mesophytic (terrestrial plants growing in normal conditions) and shed their leaves to avoid adverse effects of transpiration during 3-4 drier months in the winter season. The canopy density ranges between 60 and 80 percent in less disturbed forests. Parasitism and epiphytic commensalism is relatively poor, while seasonal undergrowth is richer in comparison with that in the rain forests. Mudumalai, Bandipur, Bhadra and Dandeli Wildlife sanctuaries in Tamilnadu and Karnataka sustain luxurient moist deciduous forests. Similar, but poorer and highly disturbed forests occur all along the length of the Sahyadri over the western and eastern slopes characterised by similar moisture conditions. Water holes or perennial water body is a common feature of the scene. The dominant and co-dominant species are those of *Terminalia, Anogeissus, Albizzia, Dalbergia, Lagerstroemia, Tectona, Gmelina, Adina, Mitragyna, Grewia, Butea, Bombax* and *Pterocarpus*. Other commonly occuring arboreal species are *Chloroxylon sweitenia, Madhuca indica, Spondias pinnata, Garuga pinnata, Emblica officinalis, Careya arborea, Boswellia serrata, Acacia catechu, Vangueria spinosa, Bridelia retusa, Morinda citrifolia* and *Schleichera oleosa*.

3. Semi-evergreen/Mixed Moist Deciduous Forests:

The classification of vegetation, being man-made and arbitrary, is not rigid in the sense that intermediate types occur between the adjacent ones. Gradual merging as indicated by intermixing of plant species belonging to evergreen and deciduous types is of common occurrence. Trees such as *Anogeissus, Butea, Bombax, Lagerstroemia, Mitragyna* and *Terminalia* from the moist deciduous forests often invade disturbed rain forests. This vegetation is considered as semi-deciduous, mixed moist deciduous or semi-evergreen. This situation is more common in the Sahyadri region (Maharashtra and Goa) having a

relatively prolonged dry season. Examples of this may be seen around Bhandardara, Chandoli Sanctuary (Maharashtra) Doodhsagar forest (Goa), Kemangundi (Karnataka) and Peechi Wildlife Sanctuary in Kerala.

4. Dry Deciduous Forest:

Dry deciduous forests are spread over areas receiving low rainfall, annual average ranging between 500 and 1000 mm. The trees are smaller and shorter in height as compared to that characteristic of moist deciduous forests. They remain leafless over a long period (5-8 months) during the prolonged dry season. Ephemeral herbs and twines abound during the favourable season. Dominant and sub-dominant tree species are *Tectona grandis, Acacia chundra, Anogeissus latifolia, Bombax ceiba, Dalbergia latifolia, Lannea coromandelica, Holoptelea integrifolia* and *Zizyphus xylopyra*. Other commonly occurring trees and shrubs include *Bauhinea racemosa, Cassia fistula, Sterculia urens, Gmelina arborea, Dolichandrone falcata, Acacia leucophloea, Maytenus rothiana, Mimosa hamata, Woodfordia fruticosa, Zizyphus oenoplea, Boswellia serrata, Albizzia lebbeck* etc. Many plant species are common in moist and dry deciduous forests, which very often unnoticeably merge into each other. Dry deciduous forests occur over hills and hillocks in the rain shadow in the East, they also occur as a result of degradation of the moist deciduous type elsewhere.

5. Scrub Forests / Thorn Forests:

Scrub forests mainly dry deciduous scrub or thorn forests exist in the semi-arid zone in the eastern foothills before the Western Ghats completely merge with the Deccan Plateau in the Sahyadri region. In the southern Western Ghats area of Tamilnadu and Kerala secondary bio-edaphic evergreen and semi-evergreen scrub forests also occur at places, but in general such forests exist in regions receiving scanty and irregular rainfall (usually 500 mm or less). The vegetation consists of stunted, much branched xeromorphic (morphologically adapted for scarcity of water) shrubs and under-shrubs, many of which are armed with thorns (Pointed, hard, protective modification of stem), spines (modified leaf or its part) or prickles (pointed projections from the external surface of stem, leaf, flower, fruit). Extensive thorn forests are found in Maharashtra in Ahmednagar, Pune, Satara, Sangli and Kolhapur districts; in Karnataka in Belgaum and Dharwad districts and in Biligirirangan hills in southern Tamilnadu. *Acacia, Zizyphus* spp., Cacti, *Euphorbia, Asclepiadacae* members and grasses dominate the vegetation.

6. Sholas Montane and Savannah Forests:

a. Sholas: The Nilgiri hills, Annaimalai Tinneveily and Palani hills and other South Indian hilly regions exhibit extensive grassy downs over higher slopes and tablelands, interspersed with densely forested gorges with evergreen vegetation. Such vegetation is known as the Sholas, which have affinities with the high altitude ranges of Assam, Meghalaya and Arunachal Pradesh. Several avian species hence occur only in these two disjunct areas. The sholas are basically evergreen forests of closed canopy comprising some temperate element growing in high valleys (above 1,000 m.) interspersed with grasslands over steep slopes. The canopy is poorly differentiated into layers. The floristic composition is similar to that of lofty evergreen forests at lower altitude in the adjacent areas. Hill grasses and members of *Asteraceae, Acanthaceae, Scrophulariaceae* and *Fabaceae* are abundant.

b. Montane Forests (Sub-tropical Hill Forests) :
As the name indicates, these forests are restricted to higher hills particularly the Anaimalai, Palani and Nilgiri hills between 1,000 m and 1,600 m. They also occur to a fair extent at Mahabaleshwar, Bhimashankar, Kalasubai and over some hill forts in Maharashtra and are composed of a mixture of evergreen and temperate species. Occurrence of *Rhododendron* spp. *and Berberies aristata* on the tableland of Udhagmandalam (Ooty) in Tamilnadu as well as several fern species at Mahabaleshwar are noteworthy in this connection.

c. Savannahs: This is supposed to be a mixture of luxurious grassland and arboreal vegetation. Trees are few and far between, many of which are deciduous, while scattered fire hardy shrubs and tall coarse grasses are common.
The characteristic species of the type are *Phoenix humilis, Adina cordifolia* and *Lagerstroemia parviflora*. According to most forest ecologists, there are no true savannahs just as no true grasslands in India. They can be considered as relict communities that may be only biotic or bio-edaphic and not climatic. The scattered lofty trees are in fact remnants of high forests of the past. The Nilgiri 'Subtropical Hill-Savanna' is chiefly distributed over slopes of Nilgiri and Palani Hills between 1,000 and 1,700 m. in the area with well distributed annual rainfall of more than 1,500 mm.

7. Grasslands:
True grassland ecosystem hardly exists in the Western Ghat region as also elsewhere in our country. The grasslands, wherever present, are formed as a result of degradation of the original forest type due to human interference. Once formed, biotic and often edaphic factors help in their perpetuation over extensive areas. The seral (plant communities in intermediate stages of succession) communities thus get established due to continued intervention and pressure of human activities.
The grasslands are classified as under:

a. Xerophilous: From pre-historic times, these occur in semidesert conditions along with thorn forests over extensive terrain in the Deccan Plateau region of Maharashtra and Karnataka. Common grasses belong to the species of

Eragrostis, Aristida, Heteropogon, Apluda, Chloris, etc. A host of xeromorphic herbs like *Acanathaceae, Euphorbiceae, Asteraceae* and *Fabaceae* make their appearance in the favourable season.

b. Mesophilous: Present in small patches in degraded evergreen and deciduous forests all along the Western Ghats. *Andropogon, Digitaria,Pennisetum, Themeda, Panicum, Setaria* are some of the commonly occuring grasses here. The grassy downs of the Nilgiris (Savannah) are temperate variation of this type.

c. Hygrophilus: Also known as wet meadows exist patchily in Kokan and Mawal area and in the south exist in high rainfall areas and along wetlands.
Arundo,Hygrorhiza,Saccharum,Erianthus and *Phargmites* are well known examples.

8. Littoral, Swampy Halophytic Vegetation:
Also called Mangrove forests or Deltoid forests are restricted along the mouths of all the larger west-flowing rivers, the remaining coastal area having mostly rocky or sandy beaches. Dabhol creek in Ratnagiri district, Kali Nadi creek near Karwar in North Karnataka and Mandavi creek in Goa sustain fairly good mangrove forests, though in patches, on soft tidal mud flats. Plants in halophytic vegetation (growing in salinity) to which the mangroves belong exhibit xeromorphic characters as a result of adaptation to physiological drought like conditions. They are hardy, slow growing, much branched often with thick leaves and pneumatophores.
Vivipary (seedlings develop on parent trees) is observed in some species.
Rhizophora mucronata, Sonneratia apetala and *Avicennia officinalis* dominate the mangrove vegetation. Other noteworthy species are *Brugeria conjugata, Ceriops candolleana, Acanthus illicifolius, Aegiceros majos* and *Exoecaria agallocha*. However, much of this vegetation is lost or is being lost due to human interference and negligence during the last few decades. What is left may be seen in narrow strips and small patches. Two species of palms namely *Nipa fruticans* and *Phoenix padulosa*; a gregarious sand binding grass *Spinifex squarrosus* and a convolvulaceae creeper *Ipomea biloba* occur in the intertidal zone near mangrove vegetation. This ecosystem is supposed to be crucial as breeding ground for many animals including some fish, crustaceans, molluscs and birds.

9. Woodlands and Orchards :
The age-old practice of cultivation of Coconut (*Cocos nucifera*) and Betel-nut or Areca (*Areca catechu*) in the well-watered valleys and plains in the Western Ghats and Coastal lowlands has created a rather stable, though secondary and manmade ecosystem. It is often coupled with cultivation of shade tolerant fruits or spices like Pineapple (*Ananas sativus*), Cardamom (*Ellateria cardamomum*), Black pepper (*Piper nigrum*).

Extensive Cashewnut and Mango orchards over ever increasing area is a common feature in Kokan and adjoining region. Other notable fruit trees are Jack-fruit, Bread fruit, Chiku (*Achras sapota*), Kokam (*Garcinea indica*) and Jam (*Syzygium jambos*). Tea (*Thea sinensis*), Coffee (*Coffea arabica*) plantations occur over extensive areas of Nilgiri, Palani and other hills of South Western Ghats since the beginning of the nineteenth century. All these and other similar plantations have become a significant sector of the vegetation of the Western Ghats.

Cultivation of arboreal species under afforestation projects run by the State Forest Department, social forestry, non-government voluntary organizations, as well as private bodies and individual farmers have developed woodlands of various dimensions. Though there is still much scope in such tree farming, as the rate of deforestation is alarming, a significant vegetation cover has been and is being formed due to this activity. Extensive monocultures of Teak (*Tectona grandis*) over hill slopes in moderate rainfall areas, extensive Suru (*Casuarina equisetifolia*) cultivation along sea shores, and large scale commercial plantation of various species of *Eucalyptus* all over the length and breadth of the Western Ghat region are of common occurrence. The British colonists introduced pine (*Pinus* spp.) and *Eucalyptus* spp. in the Palani and Nilgiris in the early decades of the nineteenth century. Since then, numerous exotic species have been and are being introduced and grown successfully, e.g. Silver Oak (*Grevillea robusta), Gliricidia sepium,* Rubber (*Hevea braziliensis*), Australian Acacias (*Acacia Auriculiformis, Acacia menzium*), Subabhul (*Leucena leucocephala*). Some of these species have been naturalized and exhibit natural regeneration forming fairly dense patches in the course of time. Many indigenous species like *Dalbergia sissoo, Bombax ceiba, Azadirachta indica, Dendrocalamus strictus, Bambusa arundinacea, Gmelina arborea* are also used for commercial purposes and/or energy plantation.

Besides this, a host of ornamental trees, shrubs and vines, both indigenous and exotic are cultivated on a large scale adding to the greenery and beauty of the landscape. The importance of mixed plantation and ecoplantation against monoculture has been realised by the concerned authorities as is evident by increasing number of such plantations all over the area.

10. Cultivated Fields, Agricultural Land:
A fairly large percentage of the total area is under agricultural practices from pre-historic time showing a trend of increase during the last century. The chief rain-fed Kharif crops are Rice (*Oryza sativa*), Jowar (*Sorghum vulgare*), Nachani (*Eleusine coracana*) and a host of pulses, minor cereals and oilseeds. Area under irrigation is gradually increasing associated with large-scale cultivation of cash crops like sugarcane (*Saccharum officinarum*), Cotton (*Gossypium* spp.), Groundnut (*Arachis hypogea*) spices as well as hybrid, high yielding varieties of food grains like Jowar, Maize and Wheat. Ever increasing trend of cultivation of fruit plants (Grape, Strawberry, etc.), vegetables (Onion, Potato, Cabbage and leaf vegetables) and floriculture and even export of these commodities is changing the cropping pattern in recent years. Shifting cultivation is still practised in hilly area, resulting in the loss of the original vegetation followed by impoverishing the land by soil erosion. Several noxious exotic weeds, namely Congress grass (*Parthenium hysterophorus*), Besharam (*Ipomea carnii*), Ranmari (*Eupatorium triplinerve*) and water Hyacinth (*Eicchornia crassipes*) have spread far and wide. Many localities especially on eastern margins have been turned to non-arable land due to water logging, salinity and excessive use of chemical fertilizers. Constructing dams and bunds, some of which have evolved into extensive manmade wetlands, has created innumerable reservoirs.

Nilgiri Tahr in the Nilgiri mountains.

Wonderful Birds

- *Satish Pande*

It is believed that birds evolved from reptiles in the Triassic period of the Mesozoic era, some 250 million years ago. One of the earliest known bird is the fossilized *Archeopteryx lithographica,* which was discovered embedded in limestone at Solnhofen, Bavaria in 1861. Avian reptilian lineage is evident from their nucleated red blood cells, reptilian type of early embryological development, peculiar atlanto-occipital articulation, egg laying habit and periodic moulting of feathers and beak sheaths (Puffins). Their feathers appear to be modified scales, since in certain owls, eagles and sandgrouses leg scales are replaced by feathers. It is interesting to note here that the intriguing egg-laying poisonous mammal, the Platypus (*Ornithorhynchus anatinus,*Ord:Monotremes), residing in Australian and Tasmanian waters, has an enigmatic bird-like beak, reptile like claws and body covered not with feathers but fur!

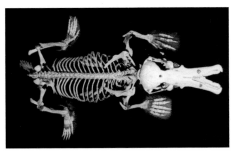

Birds are feathered bipeds. Most birds can fly, few like the ostrich, emu, cassowary, penguin cannot. Bird bones are pneumonised to reduce weight to aid flight. Flightless birds have heavier bones. In aquatic diving birds skeletal pneumonisation is almost absent to reduce buoyancy. Birds have well-developed air sacs in cervical, thoracic and abdominal regions. The air sacs are separate from the lungs. Birds have variable respiratory rates. The avian lung has two components, the primitive paleopulmo- where the air-flow is continuous and unidirectional and the complex neopulmo- with branching bronchi. In cormorants, ducks, gulls, owls, cranes and buzzards the neopulmo is less developed. Birds have diverse types of syrinx situated at tracheal bifurcation, an apparatus which is useful for vocalization. Their larynx does not have vocal cords. Syrinx is simple in Suboscines (primitive songbirds) which have monotonous songs and is complex in Oscines (New world songsters) which have musical songs. Many cranes, geese and swans have a long looped trachea. This is accommodated either in the excavated keel or in the subcutaneous region. The long tracheal loops have a wider diameter to reduce the air-flow resistance. These wind-pipes may also contribute to the quality of bird sound (Trumpeter swans).

To make flight possible, keel of the sternum is very well developed to provide insertion for the strong pectoralis flight muscles. If man was to fly, his sternum would protrude five feet in front of the chest to support the equally strong pectorals. In flightless birds the sternum is flat, with the exception of the penguins, who 'fly' under water with the aid of supra-coracoid muscles. The fused clavicles or the wish bone enhance bird flight by allowing unhindered shoulder joint movements and by providing attachment to flight muscles. Wing is the modified forelimb with various lengths of humeri, radiaocarpals and carpo-metacarpals in different species. The humerus is short, has a shallow articulation at ball and socket shoulder joint in humming birds, which are capable of complex wing movements. Fast flying birds have narrow pointed wings (Falcons, swifts). High-speed gliders like gulls have long wings with tapered non-slotted ends. Large, slow soarers like storks and eagles have broad, long, slotted wings. Most smaller flight birds have elliptical, small, slotted wings. The slots are moved with the alula (thumb) and this helps in controlling speed and maneuvering flight. The Handley-Page 'slotted wing' safety device used in airplanes is based on this principle.

To keep their weight less, birds have a rapid food transit time through their short intestines, which varies from an hour and a half to three hour. Birds excrete frequently. Carnivorous birds have a shorter length of bowel than frugivorous and herbivorous birds. Birds like parrots that eat dicotyledonous seeds, have a ridged palate. Seeds are held between the upper beak and the palatal ridge and are broken with the lower beak. Grainivorous birds, which eat dry food, have well functioning salivary glands. Others like swifts develop enlargement of salivary glands during breeding and the sticky saliva is used for building their nests.

Birds have taste buds on the tongue, bill, near choanal orifices or on the roof of oropharynx. Birds dislike salty and bitter food and relish sweet and sour taste. Many but not all birds have crops for storing food which they gulp in a hurry. But birds like gulls store food in the oesophagus. Food is digested in the stomach which is made of the proventriculus and the gizzard or ventriculus. Flesh eating birds have a simpler stomach while fruit, grass and insect eating birds have a complex stomach where the gizzard is used for grinding food. Grit in the form of gravel, pebbles, quartz is swallowed by the birds to assist grinding of food and trichuration. This is commonly seen in ostriches, crows and pigeons. In carnivorous birds, bones, hair, teeth, claws are not digested but are formed into pellets and are vomited after each feed. Crows, herons, thrushes and owls are seen to spit the pellets in this fashion. Vultures subsist on rotting flesh of dead animals and they appear to be resistant to food borne infections like botulism. Digestion of food is carried out by digestive juices secreted by the stomach, liver, bowel and the coecum. Birds drink fresh water, but few like the hornbills can go without it, meeting this requirement through juicy fruits. Few marine birds can drink saline water and the salt is excreted through the salt glands located near the nostrils.

Birdbrain is made of unconvoluted cerebral hemispheres, large rhombencephalon and evolved corpora striata. Birds have sharp reflexes & a keen sense vision. The nostrils are variously shaped and are placed anywhere on the beak or cere while a few birds like gannets do not have nares and they breath through the angle of mouth. The size of olfactory bulb varies from species to species. It is large in storm petrels, which locate their nests in darkness aided by the sense of smell, and it is small in swifts. The avian eye balls are large, less mobile and capable of tremendous acuity, binolcular vision and adaptation. Nocturnal birds have an excellent day and night vision. Avian ears are covered with contour feathers, they lack pinna and hence are inconspicuous. The beaks are variously shaped to meet food and other requirements. Carnivorous birds have hooked strong beaks, sunbirds have long bills for sucking nectar, nightjars have short bills and a wide gape, while storks, cormorants, herons use their strong beaks for attacking intruders. The outer covering of the bill in certain birds moults in a reptilian fashion. The beaks have sensory organs that help aquatic birds like curlews find food in squelch and mud. Flycatchers have sensory bristles or vibrissae over the base of the bill for sensing aerial insects.

Prehensile claw of a Pitta for perching.

While the upper limbs are modified into wings, avian lower limbs are also modified. The leg muscles are well developed to enable rapid take off. Most birds have four toes with three facing forwards and one backward (Anisodactyl). This arrangement is useful for perching and grasping. The toes may be webbed or padded to aid paddling in water as in cormorants and ducks. Cuckoos, barbets and woodpeckers have two forwards and two backward pointing toes (Zygodactyl) which are useful for climbing. In owls and ospreys one backward pointing toe can move forwards. In most swifts all four toes point forward (Pamprodactyl) and hence they are incapable of perching and these birds cling to vertical surfaces. In the tree swift one toe is reversible and it can perch. In kingfishers the toes are partially fused (Syndactyl). Some birds like certain woodpeckers, kingfishers, lapwings have three toes (Tridactyl). In the latter all three digits point forwards, hence lapwings cannot perch. Ostriches have two toes, first and fourth being absent. The large third digit is adapted for running. In tall birds like storks, to be able to maintain balance, the tibio-tarsus and tarso-metatarsus are equally long. The femur is hidden within the abdomen and the so-called thigh is actually the tibio-tarsus which bends forwards at the tarso-metatarsus. This explains why the bird 'knee' bends forwards and not backwards like in humans. The digits are equipped with claws that vary in size and strength. Birds have up to 60 vertebrae and in hornbills the first two vertebrae, atlas and axis are fused to support the large head. To compensate the immobility of eyes, bird necks exhibit significant movements. Frogmouths and bitterns have relatively mobile eyes.

Birds are warm-blooded animals with body temperature between 38 and 44 degrees. The avian four chambered hearts are triangular with a double (arterial and venous) circulatory system. Smaller birds have bigger hearts and in sunbirds the heart weighs 2 percent of the body weight. Birds have rapid heart rates and high metabolic rates for supplementing flight requirements. Their body is covered with feathers, which are modified epidermis. The feathers are thinnest in the axillae. The feathered patches (pterylae) and bare areas (apteria) are uniquely distributed in different species. The insulating feathers maintain temperature. Air sacs eliminate excess heat through internal perspiration. Birds lack sweat glands. The feathers are of three basic types. The contour feathers (pinnae) cover the body. They are modified into flight feathers or tail feathers, latter aid flight navigation. Down feathers lie beneath contour feathers and these non-conducting underclothes maintain body temperature. Filoplumes, wires and vibrissae are hair like modified feathers. Old and diseased feathers are shed or plucked by birds and new take their place. Some birds like ducks moult flight and tail feathers all together but the replacement of body feathers is more gradual. They are briefly incapable of flight and hence they gather in the center of large remote lakes till new feathers grow. Female hornbills moult in the security of the nest.

Feathers give a uniform contour to their body. Flying birds have 9-12 primary remiges which are attached to the anatomical hand. In most birds carpal remiges are absent. Secondary remiges are attached to the arm. They are arranged in two patterns, diastataxy and eutaxy depending on the presence and absence of space between the fourth and fifth feathers. Short coverts protectively cover remiges. Each feather has a central axis, which is cylindrical at base-calamus and tapered distally-rachis. It has serial branches-barbs, which in a composite manner form vanes. Barbs have parallel branches called barbules. Their hooks lock the flat distal barbicels. This framework forms a strong, flexible and light feather. In some species chicks are born featherless (nidicolous). In some they are born with down (nidifugous). Their first plumage is the juvenile plumage, which may be retained into adult life (babblers). In others after one or more moults (winter and summer plumages) the adult plumage is attained (eagles, pigeons). This knowledge helps their field identification. Feathers are ornamental and colourful, and humans till date prize 'aigrettes' of egrets and peacock feathers. In the past, quills were used for making arrows and for writing. Colour of the feathers varies with surroundings. Birds in hotter areas are drab coloured and in wetter and cooler areas are colourful.

This may depend on the amount of incident ultraviolet rays, which in turn affect the degree of pigmentation. Birds spend much time preening their feathers in order to be air worthy. Some birds preen each other. Many birds take dry baths with hot sand to get rid of the feather mites and ticks. Some birds oint the feathers with plant oils, others have powder feathers. Many birds take a daily water bath. Aquatic diving birds apply preen gland oil to their feathers to resist water logging.

Birds are either resident or migratory. Birds migrate over various distances either by day or night. They fly over the tallest peaks and cross the widest oceans. Most of our migrants come from the temperate and palaearctic regions in the north (ducks, storks), from the east (Red-legged Falcon), and also from the west (Lesser Cuckoo). Few are passage migrants, which are on their southerly course. Their migratory routes are poorly understood. From the Northwest they come along the Indus plains, from the Northeast they fly over the Brahmaputra while marine birds heading south fly along the coasts. Palaearctic migrants cross the Himalayas and they often use Himalayan valleys. Interestingly, these narrow passes where avian migrants bottleneck, were known to our ancestors. In *Meghdoot* by Kalidas (fourth centuary AD) these passes (Randhra) are called Hamsa-Randhra and Kraunch-Randhra, those used by geese and cranes respectively. Migration is studied by bird-ringing, a technique invented by Mortensen in Denmark.

Food compulsions, duration of sunlight and breeding instincts initiate migration. There are winter, summer, monsoon, local and altitudinal migrants. Visual landmarks, position of the sun, moon and stars, magnetic fields, gravitational pull, earth's spin and inclination, sound of waves, inter-flock communications, genetic knowledge are their navigational tools. They may fail to reach the desired destination due to storms, cyclones, poor visibility, distracting night lights from cities on their migratory routes and failing strength due to exhaustion. Many birds do not migrate and are endemic to certain areas and others prefer to stay at certain altitudes where-ever they occur (Black Bulbul, Nilgiri Wood Pigeon). The Western Ghats is an endemic bird area, which is home to 16 endemic bird species.

Birds communicate through vocalization and body language. During breeding the spectacular plumage helps in attracting the partner. Avian males in general are more attractive. Cranes perform elaborate dances. Most birds perform nuptial displays. Gonads enlarge during breeding. Most birds have a left ovary and left oviduct, with the exception of 16 avain orders, prominent amongst which are vultures and falconides, which have paired gonads. Eggs from the ovary are collected in the oviduct and the shell gland forms shell. A diseased shell gland causes egg shell brittleness. Birds breed seasonally while some do so year round. Eggs vary in size, shape and design in different species. Tropical birds lay less eggs while temperate birds lay more eggs, where daylight hours are longer in summer when they breed. With the exception of ostrich, larger birds like eagles, vultures, lay one egg every year or in alternate year. Smaller birds lay smaller sized but proportionately larger eggs. Hens lay an egg every day. Most birds lay a determinate clutch every year, which if destroyed, another clutch may be laid. In a clutch an egg is laid every 24 hours but all eggs may hatch simultaneously, since the chicks communicate from within the eggs shell and decide the pipping time. Birds incubating their eggs transiently lose the belly feathers and develop a warm incubation patch, which is rich in capillaries. Megapodes do not incubate the eggs themselves but do so by burying the eggs in a mound of decaying leaves and rubbish. Chicks break the hard egg-shell with the egg tooth which is lost soon after pipping. In larger birds like condors two eggs may be laid 4 to 5 days apart and Kiwis do so at an interval of a week to two months. Parasitic cuckoos do so in a few minutes for obvious compulsions. Chicks leave their nests immediately after birth (megapodes, plovers) or remain in the nest for upto 8 months (albatross).

Birds attain breeding maturity from the age of 6 weeks (Japanese quails) to 8 years (albatross). Most birds do so in 11 months. Male birds lack a well-developed phallus. They have a common opening of urogenital tracts called the cloaca. Their cloacae are ridged and these are of three types; intromittent (ratites, ducks) non-intromittent (fowl, buntings) and intermediate types (vasa parrots). Males have a single testis, placed medial to the kidney. During mating cloacae are approximated and fertility depends as much on healthy gonads as on good undisturbed perching places. Swifts mate in flight. Cloacal injuries and surgeries can cause infertility. There can be a considerable interval, at times up to one year, between copulation and egg laying.

Eggs fail to hatch if they are infertile. Fertile eggs may not hatch if they get very wet or -

Cloacal pecking prior to mating in Yellow-billed Babblers.

- dehydrated or if the eggs get infected. Egg infection occurs through the shell if the nest hygiene is poor or if the shell is weak. Shells are microporous for breathing. Infection can also be acquired in the oviduct. Most embryos die in shell during the early part of incubation, when antibodies do not pass to the embryo. Even herbivorous birds feed worms and insects to their chicks. Many birds maintain nest cleanliness by eating or removing the fecal sacs. Martins, swallows, hornbills eject faeces outside the nests. Eggs and chicks in filthy nests are disease prone and can be infested with lice, ticks, mites. Orphaned chicks die if unattended but they can be artificially reared. Interestingly, while rearing orphaned chicks if foster parents of other species are used, the chicks are known to face pairing problems later.

Life span of birds varies from a decade (small passerines) to seventy years (royal albatross). Larger and captive birds live longer. Birds face threats due to habitat loss, overexploitation of natural resources, changing methods of traditional forest management, shifting cultivation, hunting, trapping and abandonment of religious protection accorded through tradition. 42 avian species found in the Kokan and the Western Ghats now feature in the Red Data List. Death comes to aves through aging, disease and by falling prey to others. Insecticides, pesticides, fertilizers, polluting gases and fumes, wood preservatives, molluscicides, oil spills, tannins and alkaloids in barks are all poisonous to birds. Birds are useful as controllers of insect pests, devourers of vermin, as scavengers, pollinators of flowers, dispersers of seeds and suppliers of guano. They impart joy through graceful flight, melodious songs and exquisite beauty and fire the imagination of artists through their sublime existence. To lesser humans, they supply eggs, flesh and bird-nest soups! But they also spread disease. The ticks and mites, which infest them carry germs. Several of these organisms are human pathogens. Bacterial, viral, rickettsial, fungal, retroviral, arboviral, adenoviral, protozoal parasitic, haematozoal, helminthal parasitic, arthropodal ectoparasitic, etc. types of diseases afflict birds. Lice, mites, fleas, flies, bugs, ticks, gnats, mosquitos, midges can transmit blood parasites from one bird to another bird or animal. Kyasanur Forest Disease is an example of wild birds getting parasitized by Ixodid Ticks-Acarina (JBNHS, 69-1,55-78). Contamination of drinking water due to bird droppings can be major health hazard. Some pests travel inter-continentally, lodged in the plumage of migrating birds. Both wild and pet birds should be handled with aseptic precautions to minimize risk of cross infection and infestation. Psittacosis is a well-known bird related human disease. Birds are also known to suffer from hypertension, atherosclerosis and diabetes. In this aspect they are similar to humans !

Bird Behaviour

- *Anil Mahabal*

Barn Swallow.

Emlen (1995) has rightly indicated that behaviour is the 'overt expression of the coordinated life processes of an animal including the means by which the animal maintains its relation with the environment.' To be able to achieve this one must have physical equipment like sense organs. Birds possess an acute sense of vision, hearing, balance and equilibrium, olfactory power, taste, touch and so on. Besides these birds exhibit highly complex instinctive behaviour regarding nutrition, aggression, reproduction, social relations, sleep and care (Van Tyne and Berger, 1965).

Song Behaviour: Throughout the animal kingdom birds are outstanding for the variety of sounds they make and for the versatility of their songs. It is nature's gift to bird-life. Some examples from the region are Hill Myna, Shama, Blackcapped Blackbird, Malabar Whistling Thrush, Magpie Robin & Red-whiskered Bulbul. Many species of drongos and shrikes are well known as mimics. Birds communicate and express through various calls, not only within pairs or species but also at inter-species level. Hence an alarm call or a warning signal emitted by a bird is information of the approach of a predator to other birds and animals. Peacocks, mynas, herons, lapwings, babblers & treepies are some alarm birds. Likewise, communal noise made by Indian Mynas in the evening at their communal roosts has a social significance. Further, few bird species are vocal throughout the year whereas many have attractive vocalizations during the breeding periods only. Others like the Openbill Stork are incapable of emitting sound and they communicate by bill clapping sounds.

Feeding Behaviour: All animals are directly or indirectly dependent on plants for nourishment. But birds as a group seem to feed mainly on animals and a few birds utilize plant food directly. But the chicks of the vegetarian birds also require animal matter in the initial period of their life. Birds have a two-way relationship with plants. Plants influence their life processes and birds in their turn influence the natural history of plants.

This concept of 'Ornithobotany' was discussed by Salim Ali and later by Sharma (1995) in Indian Weaver birds. There are various interesting aspects of food behaviour such as finding food, food preference, seasonal variations in food supply, symbiotic and social feeding habits, solitary and gregarious feeding, water requirements and mass-drinking flights in various species of sandgrouse in arid and semi-desert areas. Food is considered as a limiting factor and it is thought that the development of territorial behaviour has resulted to provide a substantial food supply during the nesting period.

Reproductive Behaviour: Reproduction is an integral part in the life of every animal for the procreation of progeny and in turn survival of the species. Breeding and nesting behaviour are exhibited in a diverse pattern in the various orders of birds. Breeding season, territoriality, defense, type of courtship displays, pair formation, mating, polygamy (mating of male with several females), polyandry (mating of one female with several males), nest-building, nesting density and dispersion of nests, synchronization in breeding in colonies or otherwise, nidification, eggs and clutch size, incubation, nestling, sex-ratio, mortality, nest predation and breeding success, social relation between parents and young ones and various other environmental factors related to this breeding behaviour are essentially fascinating. Much study needs to be done to fathom these aspects of avian life.

Social Behaviour: There are several phases in the social activities of birds, since sociality is a universal phenomenon among birds. Gregariousness like swarms of bees, school of fishes, herds of deer and grouping in number of animals, birds too have a tendency to flock formation. Flocking behaviour is seen at inter-specific and intra-specific level. Birds of different species come together in various contexts. Mixed feeding flocks during the day (hunting parties of insectivorous birds-Yellow-browed Leaf-Warbler, tit, tree-creeper and frugivorus birds like barbet, bulbul, myna, koel & hornbill), mixed breeding colonies (herons, egrets, cormorants), mixed flocks while migrating (geese, swans, teals, ducks) and mixed communal roosts during the night (mynas, egrets, crows, parakeets, pastors). This gregariousness in birds has evolved possibly for protection from predators and improved efficiency in getting food. Populations are the basic units of evolution. All over the world populations and their dynamics have been studied in various bird species. This has led to analysis and assessment of the density and numbers of birds, structure of the populations and to calculate the biomass, the energy balance in various habitats, and the movement of energy in different levels of the food chain. Studies have been made to estimate the global bird population. It is estimated to be at a magnitude of 200 to 400 billion individuals. In the Western Ghats the population and communal roosting habits are studied in Indian Myna, Rosy Pastor, Rosy-ringed Parakeet and Pariah Kite, with aspects like seasonal fluctuation in the population and flock sizes, arrival and departure time at roosts, communal displays, direction of routes. (Mahabal, Anil, pers. com.) There is a relationship between roosting behaviour and feeding dispersion in birds. Hence flock feeders and solitary feeders are both seen roosting either solitarily or communally.

Grey-bellied Cuckoo.

In several species this habit may change from one season to another season. Likewise there is a correlation between breeding colonies and feeding dispersel in many bird species. The importance of such assemblages of birds serves as 'information-centers' for food finding. Ward P. and Zahavi (Ibis 115, 1977) have put forth this novel hypothesis.

Species like nightjars, night herons, owls, frogmouths are nocturnal. They spend the day in some shadowy roost and become active in late evening. They spend the night in getting food and performing social activities. It is a well-established fact that various behavioural patterns have a genetic control favoured through natural selection. Devastation of nature and degradation of natural habitats have threatened many bird species across the world, from 1111 species in 1994 to 1186 in 2000, a shocking 12% rise of all the global bird species [IUCN Bulletin vol.32, (No.3):2001:8]. There are many ways to save these species from total extinction, one of them being captive breeding-aviculture. This has been done with pheasants and ducks in India and their progeny successfully reintroduced and released in the wild. The rapid growth of urban population and the expansion of urban space at the cost of agricultural land and forested areas lead to environmental problems, habitat loss and deterioration of the quality of animal life. In this context it has become very crucial to study the behavioural adaptation in bird-life including their migration behaviour and distribution pattern in general.

Migration Behaviour: Bird migration is the periodic bi-directional movement and it is probably as old as flight itself. Birds migrate over long distances, often extra-limitally during the winter months. These are the winter migrants. Birds,

which move locally over short distances, are local migrants and they are usually the resident birds of that area. Birds also exhibit altitudinal migration and this is prominently seen in Himalayan birds (Crested Bunting, Brown Rock Chat, Rosy Minivet, Punjab Redvented Bulbul, Northern Tree-Pie, Baybacked Shrike, etc.) and to a lesser extent in the Western Ghats birds also (Scarlet Minivet, Nilgiri Laughing Thrush, Malabar Parakeet, Nilgiri Pipit,etc.). This up and down movement from plains to high mountains and back is seen during winter and summer months. Besides this, resident birds of Himalayas (Himalayan White-tailed Nuthatch, Himalayan Greenfinch, Firebreasted Flowerpecker, Himalayan Tree-creeper, Simla Yellow-browed Tit, etc.) show winter-summer vertical migration within this ecosystem only. In general, migration is an adaptation and an effective solution to the problem posed by seasonally changing environment such as harsh climate, scarcity of food and non-availability of suitable breeding grounds. Due to the ability of flight birds can move to distant favourable places.

Distribution Pattern: Birds are distributed in six zoogeographical regions all over the world namely Nearctic, Neo-tropical, Palaearctic, Ethiopian, Australian, and Oriental. The range and distribution pattern of several bird species is still not clearly understood. The correct interpretation of avian zoogeographical distribution is dependent upon taxonomic concepts, hence the need for correct classification. No two- bird species have exactly the same distribution. The zoogeographic regions are further divided into subregions, provinces, biogeographic zones, biomes on the basis of peculiarities of vegetation type, ecology, climate, physiography and soil for analyzing the distribution of the flora and fauna. The location of the Western Ghats conforms to the Oriental region, Indian subcontinent, Southwestern province, Western Ghats biogeographic zone, which is further divided into coastal Malabar and the Western Ghat mountains. Within biotic provinces there are several biotopes and habitats. With this context it is easier to assess the fascinating aspect of bird behaviour and help us to understand and act when conservation efforts are required. Ample scope exists to study every minute aspect of bird community and behaviour, the distribution pattern in relation to ecological niche, with respect to the various ecosystems in the Indian scenario. The Western Ghat complex, a global Hotspot rich in terrestrial biodiversity is one such area for study.

Tail feather of Koel (F)

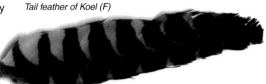

Important Bird Areas (IBA) Programme

- Asad R. Rahmani

India comes in top ten of world's biodiversity richest nations. Its immense biologial diversity represents about 7% of world's flora and 6.5% of world's fauna.

It embraces 10 biogeographic zones and 26 biotic provinces. There are more than 600 species of amphibians and reptiles, nearly 1,300 species of birds and 359 species of mammals in India. Among the larger animals, 173 species of mammals and 130 of birds, 15 of reptiles are considered endangered.

Birds are good indicators of the state of our environment, spatial biodiversity and consequently, sustainability. Birds are an important source of revenue through bird-watching tourism. As elsewhere in the world, birds have great significance in many Indian cultures. Birds also provide an excellent means to create awareness of nature and the environment among young people. Moreover, places where a wide variety of birds are found tend to have a wide variety of other forms of life. Birds are thus good indicators of diversity of plants and animals. How effectively we are conserving the world's birds is a means of assessing how successful we are in safeguarding ecosystem's functions and biodiversity as a whole the world's ecospace.

Question arises as to why should we care about birds? Of the 1,186 threatened 1,175 (99%) are at risk from human activities such as logging, agriculture, hunting and trapping. Around a third of the world's seabird species are at risk, mainly because of over fishing. Habitat loss and degradation threaten 1,008 (85% of all) species. A total of 367 (31% of all) threatened birds are directly exploited in some way mainly through hunting for food or for trapping for the cage-bird trade, but the main cause of extinction in recent times has been invasive species such as rats and cats. Virtually all of these were of island birds that lacked defense against introduced predators It is a serious worry that a quarter of all threatened birds are currently affected in this way. Three-quarters of all threatened species depend on forests, but widespread deforestation continues unabated. Wetland birds are threatened because of drainage of marshes and pollution of river systems. These population declines demonstrate clearly that we are abusing our marine, forestry and wetland environments the very ecospace on which we depend for survival. Species extinction is no longer isolated natural events but the result of major changes in the world's ecosystems. These ecosystems provide vital services such as maintaining global climate patterns, mediating the carbon cycle, safeguarding watersheds and stabilizing soils. For future generations to rely on these ecosystems, we must act now to safeguard them.

Over 1,200 bird species across the world (about 12% of the world's birds) are currently under threat of extinction. The Bombay Natural History Society (BNHS), Birdlife International (UK) and the Royal Society for the Protection of Birds (UK), the three premier organizations have come together to establish the Indian Bird Conservation Network which includes NGO's and individuals who want to contribute towards bird conservation. One of the major aims of the Network is to identify and protect Important Bird Areas (IBAs) throughout the country. The IBA program was launched at the BNHS in March 1999. Some IBA sites could be the last refuges for certain species, and if we lose such sites, the species would be in danger of extinction. IBA sites are selected using scientific and practical methods. Besides the existing protected areas which form the backbone of the IBA network many new sites are also identified. Ideally, each site should be large enough to support self-sustaining populations of as many species as possible, for which it was identified. A site-based approach however, does not suit all birds. The sites-based approach needs to be combined with conservation of the wider environment. The function of the Important Bird Areas (IBA) programme is to identify and protect a network of sites, critical for the long-term survival of wild bird populations.

The following are the criteria to identify any site as an IBA. A site must meet at least one of the criteria described below :

Globally Threatened species: The site regularly holds significant numbers of a globally threatened species, or other species of global conservation concern.

Restricted-Range Species: The site is known or thought to hold a significant component of a group of species whose breeding distributions define an Endemic Bird Area (EBA) or Secondary Area (SA).

Biome-Restricted Assemblages: The site is known or thought to hold a significant component of the group of species whose distributions are largely or wholly confined to one biome.

Congregations: A site may qualify on any one of the four criteria listed below :

i). Site known or thought to hold, on a regular basis, 1% of a biogeographic population of a congregatory waterbird species.

ii). Site known or thought to hold, on a regular basis, 1% of the global population of a congregatory seabird or terrestrial species.

iii). Site known or thought to hold, on a regular basis, 20,000 waterbirds or 10,000 pairs of seabirds of one or more species.

iv). Site known or thought to exceed thresholds set for migratory species at bottleneck sites.

The Indian Bird Conservation Network (IBCN): It is a Network of Indian organisations and individuals who have agreed to collaborate to promote the conservation of birds in India and through them, the conservation of biological diversity as a whole. IBCN is a non-political open network with a mission to help in exchange of information and ideas between the partners or members.

If you want further information, kindly contact :
Bombay Natural History Society
Hornbill House, Shaheed Bhagat Singh Road,
Mumbai - 400 023, India.
E-mail: bnhs@bom4.vsnl.net.in
Tel. : +91-22-2821811, Fax : +91-22-2837615

Indian Ethno-Ornithological Perspective :

- Suruchi Pande

Sanskrit is a unique language encompassing a period of 5000 years. It continues to play an important role in the life, thought and expression of Indian people. Sanskrit is a window to our ancient knowledge. It helps us understand how our ancestors viewed nature in general and bird life in particular. It is even more important today to understand our compassionate heritage. Vedic philosophy was essentially based on good and moral living achieved in harmony with oneself and with nature. Indian thought has always sanctified preservation of biodiversity. Sanskrit literature views birds in various perspectives. Here are a few examples. Many more are cited in the book.

Lexicons: Several synonyms for the word 'bird' can be found in the Sanskrit Koshas (dictionaries). In the Amarkosha, 27 synonyms are listed in this verse.

खगे विहंगविहगविहंगमविहायसः।
शकुन्तिपक्षिशकुनिशकुन्तशकुन्द्विजाः ॥२॥
पतत्रिपत्रिपतगपतत्पत्ररथाण्डजाः ।
नगौकोवाजिविकिरविविष्किरपत्रयः ॥३॥
नीडोद्भवा गरुत्मन्तः पित्सन्तो नभसंगमाः ।... ॥४॥

-Amarakosha II.5

Philosophical Thought: We come across subtle, philosophical avian references like,

शकुनीनामिवाकाशे पदं नैवोपलभ्यते।
एवं प्रज्ञानतृप्तस्य मुनेर्वर्त्म न दृश्यते ॥२३॥

-Udyogaparva, Yanasandhiparva, Adhyaya 63

As we cannot trace the footprints of a bird flying in the sky, so we cannot fathom the mind of a realised soul.

Use in Rituals: In a special ritual of the Vedic period known as Chayana, performed at the event of Somayaga, a peculiar type of Chiti (altar) was built. It had the shape of a falcon in flight.
As per the Vedic thought the soul of the departed was supposed to be taken to the higher abode on the wings of a bird.

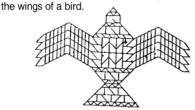

Observation: In the Rgveda, there is a reference to twenty one types of small birds which are said to be capable of digesting poisonous flowers.

It is said,

त्रिः सप्त विष्पुलिङ्गका विषस्य पुष्पमक्षन्।
ताश्चित्र न मरन्ति नो वयं मरामाऽऽरे अस्य
योजनं हरिष्ठा मधु त्वा मधुला चकार मंडल।

- Rgveda I. 191.12

Habits: Ruddy Shelduck is usually referred to as a symbol of conjugal devotion in the Sanskrit literature. But in the Mahabharata, the scavenging habit of this duck (Chakravaka) is mentioned.
It is described as -

अद्य कर्णस्य चक्राङ्गः क्रव्यादाश्च पृथग्विधाः ॥३७॥
शरैश्छित्रानि गात्राणि विहरिष्यन्ति केशव। ...॥३८॥

-Karnaparva, Adhyaya 74

Arjuna says to SriKrishna, 'today Chakravaka and other flesh-eating birds will devour Karna's various organs'. The scavenging habit of the Shelduck is well documented today. The falcons were trained to obey commands and hunt. The art of falconry was practised since long and the sanskrit text Shyainik-Shastra by Raja Rudradev, dating back to the 16 th centuary, describes this art.

Descriptions: Accurate descriptions of birds which are proved to be correct by modern ornithological standards are often encountered. In the drama 'Abhijnanashakuntalam' by Kalidasa, King Dushyanta says,

स्त्रीणामशिक्षितपटुत्वममानुषीषु
संदृश्यते किमुत याः प्रतिबोधवत्यः ।
प्रागन्तरिक्षगमनात् स्वमपत्यजात-
मन्यैर्द्विजैः परभृताः खलु पोषयन्ति ॥

-5.22

Untaught cleverness of women is seen (even) among females other than humans! What then (of) those who are possessed of reason? Indeed, the female cuckoos get their chicks reared by other birds, before they (are able to) fly in the sky.

Migratory routes: References to migratory routes and mountain passes used by birds are encountered in the Sanskrit literature. Here is an example from the Meghaduta by Kalidasa of a route used by birds for crossing the mighty Himalayas.

प्रालेयाद्रेरुपतटमतिक्रम्य तांस्तान् विशेषान्
हंसद्वारं भृगुपतियशोवर्त्म यत् क्रौञ्चरन्ध्रम्।
तेनोदीचीं दिशमनुसरेस्तिर्यगायामशोभी
श्यामः पादो बलिनियमनाभ्युद्यतस्येव विष्णोः ॥ १ ६ १ ॥

-Purvamegha

A Yaksha (a divinty of an inferior order) is asking a cloud to take a message to his beloved.
He says, "passing beyond the wonders on the slopes of the Himalayas, with your form stretched out obliquely you appear like the sable foot of Lord Vishnu intent on putting down Bali. You should go to the north through the gap in the mount Krauncha (Krauncha- a crane), through which cranes and swans (or geese) proceed to the Manasa lake.

Conservation: Unless the importance of nature is imbibed in our consciousness nature shall never be truly conserved. The mantra of conservation is not new to us. It has always been a part of our rich heritage. In the Bhavishya Purana, it is said,

अपुत्रस्य हि पुत्रत्वं पादपा इह कुर्वते। ॥ २७॥

- Madhyama Parva; 4 - Vruksharopana

Trees are like children for those who are childless. In the text 'Kashyapashilpa' it is said that those trees should not be cut for human use where birds take shelter. Similar reference is found in the Bruhatsamhita,

खगनिलयभग्नसंशुष्कदग्धदेवालयश्मशानस्थान्।
क्षीरतरुधवबिभीतकनिंबारणिवर्जितांश्छिन्द्यात्॥ १ २ ०॥

- Adhyaya, 53

The word Khaganilaya means- bird's abode. Where Khaga is a bird and Nilaya is the abode.
Mythology: Right from the time of the Vedas there are references to the Eagle (Garuda). Several other birds also have an important place in Indian mythology. Puranas are the main source of mythology. In Agnipurana, worship of an eagle is described and it is said that it should be performed by reciting a hymn,

ॐ पक्षिराजाय हूँ फट्॥ २८॥

- Agnipurana, Adhyaya 307

In this line, an eagle is praised as the king of birds. The special epithets of an eagle are

पत्रिराड्वैनतेयस्तथा नारायणध्वजः।
काश्यपेयोऽमृताहर्ता नागारिर्विष्णुवाहनः॥ १ ० ॥

- Agnipurana, Adhyaya 269

Eagle the king of birds, born in the family of Sage Kashyapa, son of Vinata, one who obtained nectar of immortality, is an enemy of snakes and carrier of Lord Vishnu.

Abstract Ideas: An abstract idea pertaining to the swan exists in the Sanskrit literature. The swan is believed to be capable of separating milk from water and partake only the milk. For example,-

नीरक्षीरविवेके हंसालस्यं त्वमेव तनुषे चेत्।
विश्वस्मिन्नधुनान्यः कुलव्रतं पालयिष्यति कः॥ १ २ ॥

Bhaminivilasa

O swan! If you are ever disposed to be idle in separating milk from water, who else on earth will fulfil your family vow ?
Because of the swan's abstract capability to sift and eliminate bad from good, the seeker in pursuit of the highest truth is called Hamsa or Paramahamsa.
Use of birds in Indian medicine: Several surgical instruments devised by Sushruta were based on the shapes of bird beaks.
Bird flesh was prescribed to treat certain diseases. In Vrukshayurveda, it is said that trees infested with worms can be treated by smoking them with the mixture of white mustard and pigeon's flesh. Thankfully good pesticides have now replaced these prescriptions!

War: Strategic formations of infantry when fighting wars were named after bird formations (Krаunchavyuha- swans in flight), or after shapes of bird wings. In the text Vaimanikshastra by Maharshi Bhardwaja, a model of an airplane is described as 'Shakunavimanakrtihi' (an airplane like a bird).
Metaphors: Innumerable Subhashitas (quotations) with reference to birds and animals can be seen. These metaphorical quotes criticize human nature, e.g.-

तुल्यवर्णच्छदः कृष्णः कोकिलैः सह संगतः।
केन विज्ञायते काकः स्वयं यदि न भाषते॥

Because of their similar colour, if the crow is in the company of the cuckoos, who will recognize a crow if it does not speak?
Superstitions: In the Rgveda we find references to the owl- Uluka and Khargala. This bird was noted for its fearful cry which suggested ill-fortune. Owls were offered at the horse sacrifice to the forest trees. On the other hand sighting of the bird Khanjanaka (Wagtail) was believed to be auspicious. It is said in Bruhatsamhita,

... सूर्योदये प्रशस्तो नेष्फलः खंजनोऽस्तमये॥ १ ०॥

- Adhyaya 45

If a wagtail is seen during sunrise it is auspicious, if seen during sunset it is not so.
Religion: There are religious concepts about birds. People believe that Kokila-Koel is the symbol of Goddess Parvati. Women respect it and keep a fast during the Kokilavrata. Unfortunately this respectful relation has led to extensive trapping of the Koel so as to enable fasting women to have a glimpse of this bird !
Magico-Religious concepts: Birds were sadly

offered to various deities while performing sacrifices. In the Yajurveda, there is a list of several birds stating their co-relation with particular deities. Birds were used in the practice of black magic. Such references are also found in the Kamasutra and Kautiliya Arthashastra.

Cultural concepts: Peacock is our national bird. There are many mythological stories describing how the peacock acquired its beautiful plumage. Peacock is supposed to be the carrier of Goddess Saraswati, the Goddess of Knowledge. Their corelation is symbolically presented in Tantrism.

Language: It is the uniqueness of Sanskrit language that almost every word bears a strong relationship with the root verb. The tiny multicoloured sunbird is called 'Sinjiraka', where the root verb is 'Shija-Shinja' which means to tinkle or jingle. It refers to the sweet jingling call of sunbirds.

Mysticism: Mystical references to birds are particularly seen in the Upanishadas. For example,

पाशं छित्त्वा यथा हंसो निर्विशङ्कं खमुत्क्रमेत् छिन्नपाशस्तथा जीवः संसारं तरते सदा ॥ २ २॥

Kshurikopanishada

By remorselessly breaking all bonds, like the swan flying in the sky, so also the realized person or the yogi emancipates himself.
In the Nadabindu-Upanishada, the Omkar (Divine sound Om) is compared with the swan.

Nature: There are innumerable references to various birds in the Sanskrit literature where the serene beauty of nature is described. For example,

सारसांश्चक्रवाकांश्च नदीपुलिनचारिणः । सरांसि च सपद्मानि युतानि जलजैः खगैः ॥ ३॥

Valmiki Ramayana, Aranyakanda, Sarga 11
Lord Rama and Seeta saw, *Sarus cranes and Brahminy ducks wandering on the bank of the rivers and lakes bearing lotus flowers and various aquatic birds.*

Birds and Architecture: The skillful nest builder, the Baya weaver, was called Sugruha- one who builds a beautiful house. Another bird Pushkarsaad is mentioned in the Yajurveda. It builds a beautiful nest on floating lotus leaves amidst a pond filled with blooming lotus flowers. In the fire sacrifice it was offered to Twashtra, the artisan of Gods. This bird is none other than the Whiskered Tern.

... कलविङ्क्षो लोहिताहिः पुष्करसादस्ते त्वाष्ट्रा ... ॥

Vajasaneyi Samhita 24.31

Large Hawk Cuckoo (M)
Sarus Cranes - a family.

A Brief History of Indian Ornithology :

- *Satish Pande*

Rgveda (1400 BC) mentions about 20 birds which were sacrificed to deities. *Yajurveda* (1400 BC) describes about 60 birds. It is evident that facts like nocturnal habit of shelducks, parasitic behaviour of cuckoos, human speech imitating ability of parrots and hill mynas were observed by our ancestors. *Sushrutsamhita* and *Charaksamhita* (200 AD) have classified birds into four categories-Pratuda (the peckers), Vishkira (the scratchers), Plava (the water birds) and Prasaha (the birds of prey). Quails have been classified into four types-Vartak, Vartika, Vartir and Laav (Buttonquails). Around 400 AD Buddhist monks traveled widely to spread Buddhism. While crossing the snow-clad Himalayas they used warm clothes with stuffing of soft owl feathers. In order to procure owl feathers, they minutely observed these nocturnal birds. Types of owls, their nesting habits and calls are described in details in many ancient texts. Falconry was a popular royal past-time and *Shyenik Shastra*, a Sanskrit text on falconry was written in the times of Raja Rudradeva of Kumaon in the 16 centuary, copies of which can be seen in some libraries even now. The concept of conservation of nature exists in the Indian thought since ancient times and many such references are seen in the *Ramayan and Mahabharat (400 BC), Bruhatsamhita (500AD), Kautilya's Arthashastra (700 AD).* (Suruchi Pande, Per.Com.). The Sanskrit text *Mrugapakshishastra (1300 AD)* by Hamsadeva (translated in Marathi by Maruti Chitampalli in 1993) lists 127 mammals, 97 birds, but their descriptions are sometimes not accurate.

In 1713 Edward Buckley a surgeon at Fort George, Madras published an appendix to Ray's 'Synopsis Avium et Pisum'. The earliest descriptions of Indian birds can be found in 'A Natural History of Birds' by Eleanyar Albin (1738), 'A Natural History of Unknown Birds' and 'Gleaning on Natural History' by George Edwards (1743-1764). Sonnerat collected (1775) and described bird specimens from the Malabar Coast and Pondicherry in 'Voyage aux Indes Orientalis'(1782). Scientific names to many of these Indian birds were given by Linnaeus, Gmelin, Scopoli and Latham.

Adolph Delessert (1837) collected specimen from Calcutta, Bhutan and Nilgiris and one of the babblers was named by Jerdon as *Crateropus delesserti*. The begining of Ceylon (Sri Lanka) ornithology can be traced to Jolm G. Loten (1752), the Governor of Ceylon. Dr. John Latham wrote 'General Synopsis of Birds' (1781), 'India

Ornithologicus' (1790) and started 'A General History of Birds'(1828) at the age of 81. Hardwick's drawings of Indian birds were published in 'Illustrations of Indian Zoology' (1830-34) by Dr. J.C.Grey. John Gould, chief taxidermist of the Zoological Society simultaneously published the drawings of Himalayan birds by Mrs. Gould, described by N.A.Vigors. Capt. James Franklin collected more than 160 bird specimens from Vinghan Hills and made their drawings for the Asiatic Society, sent later to the Zoological Society. W.H.Sykes described raptors from the Dukhen in the 'Proc. of the Zool. Soc.'(1832). S.R.Tickell published the list of birds from Bihar through the Asiatic Society.

Dr.T.C.Jerdon's, 'The Catalogue of the Birds of the Indian Peninsula' appeared in the 'Madras Journal of Literature and Science' (1834-46) wherein 420 species were described. (Syke's list-236, Franklin's list-156). He completed 'Illustration of Indian Ornithology' in 1846. Hodgson described a new hornbill *Buceros nepalensis* and was first to describe bird migration and altitudinal distribution of Indian birds in 'Physical Geography of the Himalayas' (J. of Asiatic Society of Bengal, 18,1849). 'Materials for the Ornithology of Afghanistan' (JBNHS,44-45,1944-45) by Whistler listed bird specimens collected by Hutton, Blyth and Dr.Griffith in 1845. Ludlow recorded birds from Capt. Pemberton and Dr. Griffith's collection of 126 species in 'Birds of Bhutan' (Ibis, 1937). 'Birds of Asia' commenced by Gould in 1849 was completed by Dr.Boulder Sharpe in1883. Sir Norman Kinnear systematized the enormous collections in the BNHS when he was the first stipendiary Curator in 1907.

The contribution to Indian ornitholgy by T.C.Jerdon, Edward Blyth and Brian Hodgson (1860-72) is significant. T.C.Jerdon wrote *The Birds of India* in three volumes (1862-64). T.F. Bourdillon, H.T.Fulton and H.S.Fergusson studied birds of SW India between 1872-1922 and significantly contributed to the avian knowledge. Allan O. Hume, one of the founders of the Indian National Congress, edited the first bird journal-*Stray Feathers*. In 1876 the first bird list of SW India appeared in it. Rev. S. B. Fairbank and H. A. Terry, G. W. Vidal and A. O. Hume published their lists of birds of Sawantwadi, Konkan and Travancore in the 1880's. E. A. Butler published the tentative catalogue of the Birds of the Deccan and South Marattha country (1880). The Nilgiri Game Association (1879) and The Sacred Heart College at Kodaikanal (1895) contributed to the knowledge of Natural History and birds of the Nilgiri and Palani Hills. By 1888 *Stray Feathers* ceased publication but the newly founded Bombay Natural History Society (1883) started its journal (1886) and this publication is continued till date. Collections of skins of birds of the Western Ghats and around Bombay are preserved in BNHS, Zoology Department of St. Xavier College, Bombay, and Zoological Survey of India, Western Region Station, Pune. It has significantly contributed to ornithology by publishing authentic and first hand observations about Indian birds all these years. It was in the *Journal of Bombay Natural History Society* that J. Davidson published the comprehensive list of Birds of Uttar Kannada in 1898. An important work *The Nidification of Birds of the Indian Empire* was published in 1890 by E. C. Stuart Baker partly based on egg collections of John Stuart. He also wrote *Game-birds of India, Burma and Ceylon* in 1911. The notes of R.M.Betham on bird nesting around Pune were published in the JBNHS between 1899-1904. E. H. Atkin wrote the classical book *The Common Birds of Bombay* in 1905. It was during this period that the first of the series of *Fauna of British India-Birds* was compiled by W. T. Blanford and E. W. Oates which was later revised between 1922-30 by E. C. Stuart Baker. In 1937 J. Williams listed Game-birds in Anaimalai Hills and E. G. Nichols worked on birds of Kodaikanal. Hugh Whistler published the 'Popular Handbook of Indian Birds' during the same period. Ali and Whistler listed 'Birds of Mysore' in a five part note in JBNHS (1942-43). F. N. Betts (Coorg,1929-51), R. S. P.Bates, A. P. Kinloch (Londa), A. M. Primrose (1904,Nilgiris) and Walter Koelz contributed significantly to the existing knowledge of the birds of SW India. The bird skins collected by Walter Koelz are presently in the American Natural History Museum under the 'Koelz Collections'.

Salim Ali and H. Abdulali wrote *The Birds of Bombay and Salsette* between 1936-45. Dr. Salim Ali and H. Whistler surveyed birds of Travancore and Cochin (1935-37), Mysore (1939-40) and Goa (1975, with Robert Grubh). The classic *Handbook of the Birds of India and Pakistan* by Salim Ali and S. Dillon Ripley was published in 1968-75. It still remains a major reference book to our birds, though several birds have now been added to their list of 1200 birds. Ripley's *A Synopsis of the birds of India and Pakistan* was published in 1961. *Birds of Kerala* by Salim Ali was published in 1969. Humayun Abdulali published the comprehensive list of the birds of Maharashtra in 1973. Prof. K. K. Neelakantan published a unique illustrated guide to the birds of Kerala in 1958 in the local Malyalam language. The efforts of Biswaymoy Biswas, J.C.Daniel (Hon.Secretary, BNHS), V. S. and L. Vijayan, Asad R. Rahmani-Director, BNHS (Bustard, Florican), N. G. Pillai, Reza Khan, A. J. T. Johnsingh, Madhav Gadgil

(Project Lifescape), R. J. Ranjit Daniel and Malati Hegde (Information system for the birds of Western Ghats,1986), Hainz Lainer (Birds of Goa), Anil Mahabal (work on population and roosting behaviour of birds), Prakash Gole (*Journal of the Ecological Society*) are noteworthy. *Newsletter for Birdwatchers* edited by Zafar Futehally published for the past 40 years since 1959 needs a special mention for its role in enhancing the study of ornithology and for attracting, inspiring and widely dispersing notes from serious and amature bird watchers. Recent bird books by Ben King (A Fieldguide to the birds of South-East Asia,1975), Martin Woodcock (Collin's Handbook to the Birds of the Indian Subcontinent,1980), Richard Grimett, Carol and Tim Inskipp (1998), fieldguides by Bikram Grewal (993), Krys Kazmierczak (1998) and R. J. Ranjit Daniels (1997) have significantly added to the information on the SW Indian birds, the latter is devoted to this area. Aasheesh Pittie has compiled a comprehensive ornithological bibliographic index to the birds of India (1995). The major reference to Indian Ethno-Ornithology is *Birds in Sanskrit Literature* by K. N. Dave (1985).

'Bird photography in India' by R. S. P. Bates (JBNHS, vol 40,1939) was one of the first papers on this subject. Others by B. T. Phillips (1945), Christina (Birds of an Indian Garden) and Loke Wan Tho (vol 50-4,1952) soon followed. The works of M. Krishnan, T. N. A. Perumal, Krupakar, Senani, H. Hanumant Rao in the field of avian and wildlife photography needs a special mention. After the introduction of high speed flash photography, fast emulsions and smaller camera, avian photography progressed by leaps and bounds but in reality could never be practiced ubiquitously due to the high costs. The work of several bird photographers illustrates this book.

Institutes for the study of avifauna in South Western India :
1. Bombay Natural History Society, Mumbai
2. Zoological Survey of India, Western Region Station, Pune 3. Center for Ecological Sciences, Indian Institute of Science, Bangalore 4. Salim Ali Center for Ornithology and Natural History, Coimbtore 5. Department of Environmental Biology, Shivaji University, Kolhapur
6. Department of Environmental Sciences, Bharati Vidypeeth, Pune 7. Wildlife Institute of India, Dehradun (Western Ghat projects) 8. Jawaharlal Nehru Center for Advanced Scientific Research, Jakkur, Bangalore 9. Kerala Forest Research Institute 10. Nilgiri Wildlife and Environment Association, Udhagamandalam. 11. Department of Zoology, University of Calicut, Kozhikode.

Bird Remains from Indian Archaeological Contexts : A Bird's Eye View

- Pramod P. Joglekar

Introduction

Birds evolved during the Mesozoic Era (approximately 140 million years ago) and radiated to a large extent during the Cenozoic Era (Carrol 1997). In this article, we are restricting to a relatively short period of time, i.e. the Pleistocene Epoch. This is because the archaeological context of the birds, in relation to the human activity is only from the deposits of the last 2 million years (Pleistocene).

Animal remains are found in form of fossilized or semi-fossilized skeletal members such as the bones, teeth, cartilage, horns and antlers. Palaeozoology includes study of animal remains from all geological periods. Archaeozoology/zooarchaeology is an interdisciplinary branch of modern archaeology. It is concerned with humans and their environment in general and humans and animals in particular. Essentially it focuses on human relationship with animals, including those living on land, in the water and the air (Thomas and Joglekar 1995). In archaeozoology, essentially animal remains from last 2 million years are examined. Using these remains of animals, archaeozoology endeavours to understand animal-human interaction in the past that covers all these past humans and the associated animals. In the last 2 million years many species of humans (*Homo*) have emerged, of which only one subspecies (*Homo sapiens sapiens*) survives today. The other (now extinct) human forms such as *H. habilis** (1.6-2.0 million), *H. rudolfensis** (1.7-1.9 million), *Homo erectus** (0.3-1.6 million) and *Homo sapiens neanderthalensis** (30,000-100,000 years) were engaged in fishing, hunting, trapping and scavenging (Joglekar 2000). Obviously, all these human forms had been closely associated with animals of all kinds.

Remains of skeletal elements of animals are found at an archaeological site essentially because the animals play a variety of roles in human society. Humans evolved as carnivorous predators dependent on their physical and mental prowess to kill other animals for food. For over last 2 million years humans are relying on animals for food. Even in present-day humans have retained some of the ancestral behavioural traits that reflect the original hunting-gathering life (Joglekar 2001, manuscript). Since the domestication of animals (approximately last 10000 years) animals like dogs and cat are important as pets, while animals such as donkey, ox, horse and camel serve as beasts of burden.

Due to all such varied roles that animals play in human life, their skeletal parts are deposited in the habitational sediments at archaeological sites. The archaeofaunal assemblages contain remains of food refuse as well as natural burial of animals. Human activities like butchering, cooking and consuming meat portion leaves specific and distinct marks on the skeletal elements. It is possible for a trained archaeozoologist to easily distinguish between natural burials and food refuse (Thomas and Joglekar 1995).

The Nature of Evidence

Evidence for use of plants and animals in the past comes from two distinct, yet related sources-
(1) actual skeletal elements recovered from archaeological sites during systematic explorations and excavations, and
(2) visual records of plants and animals in the form of rock/wall paintings, sculptures, seals, coins and other forms of pictorial representations made in the past.

Archaeozoological work begins with sorting and classification of the skeletal elements recovered from archaeological sites. The remains after initial sorting are cleaned, numbered and stored on the site for future laboratory analysis. In archaeozoological laboratory these are identified and then processed using a computer-based system to obtain routine as well as information necessary to answer specific questions related to the general archaeological research design. The primary work of species identification is carried out using general morphological keys prepared to classify mammals, birds, reptiles, fish, and invertebrate animals like molluscs. However, as one progresses below the family level, identification becomes difficult and hence often metrical isolating criteria are required (Joglekar 2000 in press).

For bird species identification, many morphological and metrical identification keys are available, primarily due to efforts of archaeozoologists in countries such as France, Spain, Germany, Belgium and Holland. Using the osteological distinguishing markers, birds are identified up to the genus level. However, identification to the species level becomes extremely difficult as the bones obtained from archaeological contexts are fragmentary and many a times are devoid of very minute osteological characters. Therefore, in archaeozoological reports, one finds bird identification halted at the genus level. Particularly, separation of domestic and wild bird forms (e.g.

chicken and jungle fowl) requires consideration of morphometric variability artificially induced in the domestic stock as well as sexual and other types of morphological variation seen in wild populations across the Indian subcontinent.

Though many animals including birds are found in ancient Indian literature, art, sculptures, coinage and paintings (Joglekar 2000a), these are primarily so far been dealt from artistic historical perspective. Several scholars have worked on animals in Indian art and literature (for details see Kadgaonkar 1993,1995). This subject has not yet been treated from bio-historical point of view. Therefore, the methodology of obtaining biological, ecological and environmental information from the pictorial representations of animals in various art forms is yet to be formulated. Birds appear in a variety of Indian religious and cultural contexts. They symbolise the carnal emotions and the body itself. They are a dominant theme in Indian sculptures and therefore, often are found in sculptures as iconographic expressions of Hindu, Buddhist and Jain gods, goddesses and their vehicles. For instance, peacock is the vehicle of "Jins" like Trimukha, goddess such as Mahamanasi and Bala and Amitabh Buddha (Joshi 1979: 348-369). These birds as vehicles of their deities or demigods are likely to be shown in various iconic forms. Obviously, their representations are expected to be more stylistic than naturalistic. The tradition of manuscript painting existed before Jahangir (early 17 century A.D.), but during his period Mughal court encouraged many artists to specifically paint birds and animals. Some of the excellent paintings of birds include those of Himalayan Blue-throated Barbet (c. 1615 A.D.), falcon (1619 A.D.) and dipper (*Cinclus cinclus*) painted in 1620 A.D. by Nadir ul' Asr Ustad Mansur (Verma 1999).

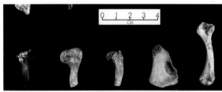

Bird bones found at Inamgaon-a Chalcolithic site (1500-700 BC) in Maharashtra (after Panwalkar-1995).

Considering the vastness of the subject and huge data available on birds in ancient India, this article is aimed to provide only the glimpse of the birds found in Indian archaeological contexts. Hence, in this article selected information has been used regarding occurrence of bird bones and evidence in other forms such as terracotta figurines, toy birds, bird on coins, birds on royal emblems and seals. Also birds referred in ancient religious as well as secular texts, folk tales and fables provide valuable information. However, this kind of information has not been utilised in the present article. Primary aim of this article is to introduce to a general reader the findings from archaeolozoological studies conducted in India during last few decades. For a more curious reader the list of original references will give additional information.

Birds in the Pleistocene (before 10000 B.P. - Before Present)

One of the major issues in Indian palaeozoology is related to ostrich (*Struthio camelus*) which is a large flightless bird. This bird is at present distributed in the Ethiopean zoogeographic region. In the fossil form ostrich (*Struthio asiaticus*) is known from the Lower and Middle Siwaliks of Northwestern India. However, it is completely unknown till the Late Pleistocene (Upper Palaeolithic) dated to about 45000-20000 B.P (Badam and Sathe 1990). This is enigmatic since this large bird is now totally extinct from the Indian Subcontinent. Some scholars have even expressed doubt regarding identification of ostrich in the Upper Palaeolithic context (Tiwari 2000). Tiwari (2000) has provided a rock painting from Sahara, Africa where three ostriches are depicted. He has stated that such figures are not found in Indian rock paintings and suggested those identified as ostriches may be of fowls or other similar birds.

Ostrich egg-shells were found at Chandresal (District Kota, Rajasthan), Nagda (District Ujjain, Madhya Pradesh) and Ramnagar (District Mandsor, Madhya Pradesh). Radiocarbon dates obtained for the samples of egg shells from these sites indicate that these belong to the early Upper Palaeolithic phase and approximately in the time bracket of 40000-31000 B.P.(Giriraj Kumar 1990). The Upper Palaeolithic levels at Patne (Maharashtra) dated to about 25000 ± 200 B.P. have yielded many fragments of ostrich eggshell. These have been engraved in simple cross-hatching pattern (Sali 1980, 1984). More than thousand ostrich egg shell fragments were found near Kota in Rajasthan from Upper Palaeolithic contexts. Ostrich eggshells seem to have been very useful item of the Upper Palaeolithic inhabitants, since these shells have been found at many (more than 40) sites in Rajasthan, Madhya Pradesh, Uttar Pradesh and Maharashtra (Badam and Sathe 1990). It has been suggested that the strong empty ostrich eggshells were used as water containers and were engraved to decorate these bowls. In the Luni basin, Rajasthan ostich eggshells have been found from the context dated to 9800 ± 700 B.P. (Mishra *et al.* 1999).

Birds in the Protohistoric Period (10000-3500 B.P.)

The rock art of these hunters and gatherers gives a fairly detailed description of their hunting techniques (Neumeyer 1993: 67). For example, a hunting scene depicted at Lakhajoar, Madhya Pradesh shows a hunter using a bow and arrow to shoot down a bird (possibily a peacock) perched on a tree. At Bhimbetka, Madhya Pradesh, 24 bird drawings have been found from nine rock shelters. The birds identified from the prehistoric period are blue magpie, cattle egret, spoonbill, pochards and peacock (Mathpal 1984:134). The peacock is seen depicted in many rock shelter sites of the early hunters and gatherers in India, e.g. Vikram Khol (Orissa), Malegatti (Karnataka), Bhadrauli (U.P.), Bori East, Mahadeo Hills, Madhya Pradesh . Many of the early bird depictions are related to rituals. For example, a rock painting from Gupha Masir, Madhya Pradesh has a bird on top of a human body. The meaning of this painting and the relevance of a bird on top of a human figure is unclear. Possibly this bird is connected to some spiritual aspect of death. Similarly, another painting of a later period from the same place shows two fowls (possibly domestic) in some ritual context.

Domestic fowl depictions are known from the third millennium B.C. and it has been suggested that the Red jungle fowl was its ancestor (Wood-Gush 1959). It is not possible to know whether the fowls belonged to wild *Gallus gallus* (Red jungle fowl), *Gallus sonnerati* (Sonnerat's jungle fowl), or the domestic variety. Both species of wild fowls have overlapping range (Delacour 1965) and it is difficult to separate these from the domestic fowls (Tiwari 2000). An early representation of peafowl is seen at Mahadeo Hills.

At the Mesolithic site of Mahadaha (Pratapgarh District, Uttar Pradesh) out of 5648 fragments, only 41 were of bird bones (Joglekar *et al.*, unpublished report 2001). The proximal portion of the first phalanx of a crane has been identified from the Eastern Area in layer 1. The size of this bone indicates that it belonged to the Sarus Crane (*Grus antigone* or the Eastern Common Crane (*Grus grus*). The latter migrates into India from north and these birds are seen in many Indian States between October and March. Heron bones have been found from the Eastern Area. These include the coracoid and the first phalanx. All these bones are charred and fragmentary. The species of heron is probably *Ardea cinerea* (Grey Heron) or *Ardea purpurea* (Purple Heron), which inhabits ponds, lakes and all types of water bodies. It is distributed throughout Indian Subcontinent. Five bones of duck or a teal have been found from the Eastern Area. These bones may be of the Common Teal (*Anas crecca*) that is a common visitor from Europe during the winter. A charred femur fragment (MDH 632) found from grave no. VIII sealed by layer 2 may be of the Whiteheaded Stifftail Duck (*Anas leucocephala*). This duck generally visits lake sites in North India during winter. Apart form these, three categories of birds have been recognized and are described as small (Sparrow sized), medium (Crow sized) and large (Peafowl sized). The bones are and too

fragmentary to be identified specifically. At the Mesolithic site of Damdama in the same region of Mahadaha, bones of Red Jungle fowl (*Gallus gallus*) have been found.

Bone remains of small birds like Heart-spotted Woodpecker are rarely found.

Several birds are represented in the skeletal remains from various Harappan sites. These include peacock from Kuntasi and Shikarpur in Gujarat; *Grus* sp. from Shikarpur; and domestic and/or jungle fowl from several sites like Harappa and Mohenjo-daro (Pakistan). From the faunal remains at Rupar, Bhola Nath (1968) has identified *Francolinus francolinus* (black partridge). Also a few bird bone were not identifiable from the Late Harappan levels at Kuntasi

The Chalcolithic-Neolithic period has yielded evidence of domestic and/or jungle fowl from a good number of sites. Peacock bones have been found from Chalcolithic sites of Inamgaon, Kaothe and Balathal. *Anser indicus* has been reported from Inamgaon and Kaothe. *Ardea* sp. (Heron) bones were found from Kaothe only. Shah (1975) has identified a few bones of pigeon from the site of Dhatva in Gujarat. However, she has not mentioned the species. Similarly, Clason (1979) has identified one bone as of a duck like bird from Kodekal in Karnataka, where it was not possible for her to determine the species. Thomas (1992a) and Pawankar (1995) have identified a few bones of bustard *Ardeotis nigriceps* from the Chalcolithic sites of Tuljapur Garhi and Inamgaon in Maharashtra. This is the only skeletal evidence of the giant bird in Indian archaeological context.

During the Chalcolithic period, several birds are seen portrayed on the pottery. These include peacock, crane, swallow, heron and duck. Of these birds, peacock is seen at many sites. Cranes are on the pottery at Navadatoli Phase II (Madhya Pradesh) and Songaon Phase IIb (Maharashtra). Heron is seen only during Phase V at Daimabad (Maharashtra). The ducks are depicted at Navadatoli Phase III and Inamgaon

Phase III (Maharashtra). Datta (2000) has identified a painting as of vulture from Navadatoli Phase I. However, the depiction is not clear and could be a representation of a large bird like peacock or a bustard.

Nilgiri Laughing Thrush, a bird from wet habitat, hardly leaves any fossil material after death.

Peacock depictions are seen in Indian sculptures from very early times. They are painted on pottery from Harappa, Mohenjo-daro, Chanhu-daro and Lothal (Kadgaonkar 1993) and are seen in various forms and at various places all over India.

A funeral urn found at Harappa shows a composite animal having peacock's head and the dead person's spirit is shown in the belly. This figure had obvious ritual significance, but we do not know its exact meaning now. Peacock is associated with the burials of Chalcolithic period at Daimabad (Maharashtra). Perhaps not only peacocks, but birds in general were related to death rituals and human spirit was equated to a bird from very early times.

Birds in the Historic Period (from 3500 B.P.)
Birds identified from the historic period at Bhimbetka rock shelter site are peafowls (male and female) and fowls (chickens: hens and cocks) besides pochards (Mathpal 1984:134). Peacocks are seen in many art forms throughout the ancient and medieval period art.

The eagle (*Garuda*) appears on the *torana* at Sanchi (Madhya Pradesh) during the Early historic period. *Hamsa* the Indian goose starts appearing as a common decorative motif from the Mauryan period (3rd century B.C.) where generally the depiction is natural.

A few owl like terracotta figurines have been found at Harappa and at Inamgaon. The purpose of these figurines is not known. Kadgaonkar (1995) suggests that these birds were important to agricultural communities and hence may be incorporated in the rituals. Similarly, isolated owl figurines have been discovered at Vaishali (Bihar) and Kaushambi (U.P.) in historical levels. However, the owl is rarely seen in Indian art. This bird got

associated with the goddess Chamunda and it appears in iconography as her vehicle from the 5-6 th century A.D. in post-Gupta period (Kadgaonkar 1995: 30-31).

Concluding Remarks
The data available regarding the birds in ancient period comes from two distinct sources- actual skeletal remains and the visual representations of bird in various art forms. However, both these data sources present skewed pictures of the ancient avifauna of the Indian Subcontinent. Though information available about birds since the prehistoric times is not complete, one can make a few observations about the birds in the past.

At present one finds several hundred species of birds still existing in India.. A variety of birds are depicted in the art of prehistoric as well as historic period. But considering the diversity of birds today, the depicted species are almost negligible. It is also interesting to note that many of the birds that are depicted in pictorial from are not represented in the skeletal record. Similarly, a number of birds whose bones are found are not depicted on pottery or are not seen as terra cotta figurines. The mode of creation of these two kinds of archaeological records (i.e. skeletal and pictorial) is varied and hence this apparently mismatching picture of the bird in archaeological contexts emerges. Both these methods of gathering information about bird are in developing stages. Archaeozoological work depends on a good and authentic comparative collection of bird skeletons. One needs at least one skeleton of both sexes of each species, and if possible from various geographical areas and subspecies. Obviously this requires sustained efforts of several decades. Though at Deccan College, Pune, efforts are going on to build such a collection over last two decades, the bird skeletal collection is not adequate. Therefore, it is necessary to note here that identifications done so far are not final (anyway nothing can be all time final in any scientific discipline). As our understanding of Indian bird osteology grows, it may be possible to develop finer identification keys. At a number of protohistoric and historic sites, many bird bones are kept aside as "unidentified" or just identified into broad categories as "duck", "crane" or "pigeon". These elements could be identified in the future with more advanced morphological and metrical keys.

Threatened Birds of India :

S.No./EnglishName/ScientificName/Category

1. Pink-headed Duck
Rhodonessa caryophyllacea   EX?
2. White-rumped Vulture   Gyps bengalensis   CR
3. Long-billed Vulture   Gyps indicus   CR
4. Slender-billed Vulture   Gyps tenuirostris   CR
5. Himalayan Quail   Ophrysia superciliosus   EX?
6. Siberian Crane   Grus leucogeranus   CR N
7. Jerdon's Courser   Rhinoptilus bitorquatus   CR
8. Forest Owlet   Athene blewitti   CR
9. White-bellied Heron   Ardea insignis   EN
10. Oriental Stork   Ciconia boyciana   EN
11. Greater Adjutant   Leptoptilos dubius   EN
12. White-headed Duck
Oxyura leucocephala   EN N
13. White-winged Duck   Cairina scutulata   EN
14. Great Indian Bustard   Ardeotis nigriceps   EN
15. Bengal Florican Houbaropsis bengalensis   EN
16. Lesser Florican   Sypheotides indica   EN
17. Nordmann's Greenshank   Tringa guttifer   EN N
18. Rufous-breasted Laughingthrush
Garrulax cachinnans   EN
19. Spot-billed Pelican
Pelecanus philippensis   VU
20. Lesser Adjutant   Leptoptilos javanicus   VU
21. Lesser White-fronted Goose
Anser erythropus   VU
22. Baikal Tea l   Anas formosa   VU
23. Marbled Teal
Marmaronetta angustirostris   VU
24. Baer's Pochard Aythya baeri   VU N
25. Pallas's Sea-eagle   Haliaeetus leucoryphus VU
26. Greater Spotted Eagle   Aquila clanga   VU N
27. Imperial Eagle
Aquila heliaca   VU N
28. Nicobar Sparrowhawk   Accipiter butleri   VU
29. Lesser Kestrel Falco naumanni   VU N
30. Nicobar Scrubfowl
Megapodius nicobariensis   VU
31. Swamp Francolin   Francolinus gularis   V U
32. Manipur Bush-quail
Perdicula manipurensis   VU
33. Chestnut-breasted Partridge
Arborophila mandellii   VU
34. Western Tragopan
Tragopan melanocephalus   VU
35. Blyth's Tragopan   Tragopan blythii   VU
36. Sclater's Monal   Lophophorus sclateri   VU
37. Cheer Pheasant   Catreus wallichi   VU
38. Hume's Pheasant   Syrmaticus humiae   VU
39. Green Peafowl   Pavo muticus   VU
40. Sarus Crane   Grus antigone   VU
41. Black-necked Crane   Grus nigricollis   VU
42. Hooded Crane   Grus monacha   VU
43. Masked Finfoot   Heliopais personata   VU N
44. Sociable Lapwing   Vanellus gregarius   VU N
45. Wood Snipe   Gallinago nemoricola   VU
46. Spoon-billed Sandpiper
Eurynorhynchus pygmeus   VU N
47. Indian Skimmer Rynchops albicollis   VU
48. Pale-backed Pigeon
Columba eversmanni   VU N
49. Nilgiri Wood-pigeon
Columba elphinstonii   VU
50. Pale-capped Pigeon Columba punicea   VU
51. Dark-rumped Swift Apus acuticauda   VU
52. Rufous-necked Hornbill

Aceros nipalensis   VU
53. Narcondam Hornbill
Aceros narcondami   VU
54. Yellow-throated Bulbul
Pycnonotus xantholaemus   VU
55. Nicobar Bulbul   Hypsipetes nicobariensis   VU
56. Grey-sided Thrush   Turdus feae   VU N
57. Rusty-bellied Shortwing
Brachypteryx hyperythra   VU
58. White-bellied Shortwing
Brachypteryx major   VU
59. White-browed Bushchat
Saxicola macrorhyncha   VU
60. White-throated Bushchat
Saxicola insignis   VU N
61. Marsh Babbler Pellorneum palustre   VU
62. Rusty-throated Wren-babbler
Spelaeornis badeigularis   VU
63. Tawny-breasted Wren-babbler
Spelaeornis longicaudatus   VU
64. Snowy-throated Babbler
Stachyris oglei   VU
65. Jerdon's Babbl Chrysomma altirostre   VU
66. Slender-billed Babbler
Turdoides longirostris   VU
67. Black-breasted Parrotbill
Paradoxornis flavirostris   VU
68. Grey-crowned Prinia Prinia cinereocapilla   VU
69. Bristled Grass-warbler Chaetornis striatus   VU
70. Broad-tailed Grassbird
Schoenicola platyura   VU
71. Kashmir Flycatcher   Ficedula subrubra   VU
72. White-naped Tit   Parus nuchalis   VU
73. Beautiful Nuthatch   Sitta formosa   VU
74. Green Avadavat   Amandava formosa   VU
75. Yellow Weaver   Ploceus megarhynchus   VU
76. Dalmatian Pelican   Pelecanus crispus   CD N
77. Oriental Darter Anhinga melanogaster   NT
78. Painted Stork   Mycteria leucocephala   NT
79. Black-necked Stork
Ephippiorhynchus asiaticus   NT
80. Black-headed Ibis
Threskiornis melanocephalus   NT
81. Lesser Flamingo   Phoenicopterus minor   NT
82. Ferruginous Duck   Aythya nyroca   NT
83. White-tailed Eagle   Haliaeetus albicilla   NT N
84. Lesser Fish-eagle   Ichthyophaga humilis   NT
85. Grey-headed Fish-eagle
Ichthyophaga ichthyaetus   NT
86. Cinereous Vulture   Aegypius monachus   NT
87. Red-headed Vulture   Sarcogyps calvus   NT
88. Nicobar Serpent-eagle Spilornis minimus   NT
89. Andaman Serpent-eagle Spilornis elgini   NT
90. Pallid Harrier   Circus macrourus   NT
91. White-cheeked patridge
Arborophila atrogularis   NT
92. Satyr Tragopan Tragopan satyra   NT
93. Tibetan Eared-pheasant
Crossoptilon harmani   NT
94. Little Bustard   Tetrax tetrax   NT
95. Houbara Bustard   Chlamydotis undulata   NT
96. Great Snipe   Gallinago media   NT
97. Asian Dowitcher
Limnodromus semipalmatus   NT
98. Beach Thick-knee
Esacus magnirostris   NT
99. Black-bellied Tern   Sterna acuticauda   NT
100. Andaman Wood-pigeon
Columba palumboides   NT
101. Andaman Cuckoo-dove
Macropygia rufipennis   NT

102. Nicobar Pigeon   *Caloenas nicobarica*   NT
103. Nicobar Parakeet   *Psittacula caniceps*   NT
104. Long-tailed Parakeet
     Psittacula longicauda   NT
105. Andaman Scops-owl   *Otus balli*   NT
106. Andaman Hawk-owl   *Ninox affinis*   NT
107. Ward's trogon   *Harpactes wardi*   NT
108. Blyth's Kingfisher   *Alcedo Hercules*   NT
109. Brown-winged Kingfisher
     Pelargopsis amauropterus   NT
110. Malabar Pied-hornbill
     Anthracoceros coronatus   NT
111. Great Hornbill   *Buceros bicornis*   NT
112. Brown Hornbill   *Anorrhinus tickelli*   NT
113. Yellow-rumped Honeyguide
     Indicator xanthonotus   NT
114. Andaman Woodpecker   *Dryocopus hodgei*   NT
115. Nilgiri pipit   *Anthus nilghiriensis*   NT
116. Firethroat   *Luscinic pectardens*   NT N
117. Chestnut-backed Laughingthrush
     Garrulax nuchalis   NT
118. Grey-breasted Laughingthrush
     Garrulax jerdoni   NT
119. Rufous-throated Wren-warbler
     Spelaeornis caudatus   NT
120. Wedge-billed Wren-babbler
     Spenocichla humei   NT
121. Giant Babax   *Babax waddelli*   NT
122. Rufous-vented Prinia   *Prinia burnesii*   NT
123. Long-billed bush-warbler
     Bradypterus major   NT
124. Rufous-rumped Grassbird
     Graminicola bengalensis   NT
125. Black-and-rufous Flycatcher
     Ficedula nigrorufa   NT
126. Nilgiri Flycatcher   *Eumyias albicaudata*   NT
127. Andaman Drongo
     Dicrurus andamanensis   NT
128. Andaman Treepie   *Dendrocitta bayleyi*   NT
129. Andaman Crake   *Rallina canningi*   DD
130. Nicobar Scops-owl   *Otus alius*   DD

Key :	Category	No.
EX	Extinct (possible)	2
CR	Critical	8
EN	Endangered	10
VU	Vulnerable	58
CD	Conservation Dependent	1
NT	Near Threatened	53
DD	Data Deficient	2

N = Non-breeding visitor.
35 species are endemic to India

Reference: BirdLife International. (2001)
Threatened Birds of Asia: The Bird Life International
Red Data Book.
Cambridge, UK: Bird Life International.

Darter.

Water Hyacinth-an environmental threat.

Jerdon's Courser. *Critical. Not in our area.*

GLOSSARY

Affinity Structural or behavioural resemblance due to a common ancestor or family relationship.
Albinism Complete or partial absence of pigment.
Axillaries Feathers of the arm-pit junction of the body & underwing.
Bristles Sensory rictal feathers usually in barbets, flycatchers.
Carpal Bend of the wing. Patch in this area is carpal patch.
Casque Growth on the beak of some hornbills.
Cere Fleshy, bare skin at the base of the beak containing nostrils - usually in raptors.
Culmen Superior ridge of the upper mandible.
Coverts Small feathers covering flight & tail feathers.
Dimorphic Different plumages, usually in male & female.
Eclipse plumage Temporary plumage after breeding - usually in ducks.
Endemic Confined to an area, indigenous.
Family Group of related genera.
Feral Wild populations once escaped from captivity.
Filoplume Hair like feather (e.g. in breeding egrets).
Fledging Time of leaving the nest. A stage of feather growth capable of first flight.
Flight feathers Primary, secondary, tertiary & tail feathers - also remiges.
Fringes Margins of feathers.
Gape Corner of the mouth where mandibles meet.
Genus Group of closely related species.
Gonys Angle at the bottom of lower mandible (in gulls & terns).
Gorget Chest band.
Gular Related to the throat.
Gular pouch Loose skin extending from the throat (e.g. Pelicans).
Immature Fully grown but not adult. Juvenile. May last for 2-4 years.
Iris Peri-pupillary, coloured, dynamic circular membrane. (Pleural - Irides)
Jizz An overall impression useful for bird identification.
Lore Part between the eye & the beak.
Melanistic Dark plumage due to excess melanin pigment.
Mandible Beak.
Mantle Part between back & scapulars.
Mask Dark area covering the eye and ear coverts (Shrikes).
Mesial stripe Central throat stripe.
Mirror White spots on the outer primaries of gulls.
Mixed hunting party Several avian species moving together in a forest for feeding.
Moult Seasonal shedding of feathers.
Morph Normal plumage variant in a species (e.g. pale, dark).
Nidification Related to nests, nesting habits.
Nominate First named subspecies with the same name for species & subspecies.
Nuchal Related to neck.
Ocelli Eye-like shiny feather spots (e.g. peafowl).
Oology Study of eggs (exterior, markings, etc).
Passerine The largest Order of sparrow like, small to medium sized perching song-birds.
Pectoral Breast region.
Pelagic Mainly living on the open sea.
Plumage Related to feather.
Polygynous Male mating with more than one female.
Polyandrous Female mating with more than one male
Primary projection Distance from tip of longest tertiary to wing tip, in a folded wing.
Race Different looking, geographically separated populations of a species-Subspecies.
Rictal Related to gape.
Rump Base of tail & lower back.
Species Distinct, isolated group of organisms capable of successfully breeding with each other.
Speculum Glossy, rectangular, coloured wing panel in ducks.
Streamers Long projections from tail feathers.
Subadult Prolonged immature stage prior to adulthood (e.g. raptors & gulls.)
Talon Claw of a bird of prey.
Taxonomy Scientific classification of living beings.
Vent Area adjoining the cloaca & separate from undertail-coverts.
Trousers Loose tibial feathers.
Vinaceous Colour of red wine.
Wattle Bare fleshy skin on part of the head (e.g. lapwings, fowl)
Wing-bar Differently coloured bar on coverts or on base of flight feathers.

Red-headed Bunting.

Blue-bearded Bee-Eaters.

Partial Albino Red-vented Bulbul.

Pied Falconet   Collared Falconet
(Microhierax melanoleucos) (M. caerulescens)
These small falconets are seen in NE Indian states.

White-spotted Fantail Flycatcher.

Index of Scientific Names of Birds :

Index of Sanskrit Names of Birds :

Karalphal Bak 129 / करालफालबक

Karandav 97 / कारंडव

Karayika 323 / करायिका

Karnikar 228 / कर्णिकार

Karvaan 126 / करवाण

Kashikani 124 / काशिकाणी

Kashthakut 201 / काष्ठकूट

Kashthashuka 153 / काष्ठशुक

Kathaku 190 / कठाकु

Katukwan 108 / कटुक्वाण

Kaulik 305 / कौलिक

Kaushik 168 / कौशिक

Kesari 7 / केसरी

Khajak 32 / खजाक

Khakamini 49 / खकामिनी

Khanjanak 217 / खञ्जनक

Khanjanika 105, 106 / खञ्जनिका

Khanjari 250 / खञ्जरी

Kharashabdakurar 131 / खरशब्दकुरर

Kharchatak 253 / खरचटक

Kharpidwa 252 / खरपिद्वा

Khilkhila 100 / खिलखिला

Kikideewi 183 / किकिदीवि

Kiriti 296 / किरीटी

Kohasa 50 / कोहासा

Konth 168 / कोंठ

Kooj 207 / कूज

Kotth 168 / कोट्ठ

Krishnakraunch 91 / कृष्णक्रौंच

Krukal 144 / कृकल

Krukalika 170 / कृकलिक

Krusha 212 / कृश

Krushachanchu 16 / कृशचञ्चु

Krushakak 323 / कृशकाक

Krushakut 323 / कृशकूट

Krushnaangshuka 154 / कृष्णांगशुक

Krushnachud 226 / कृष्णचूड

Krushnapakshi 249 / कृष्णपक्षी

Krushnika 248 / कृष्णिका

Krushnottaamng 154 / कृष्णोत्तमांग

Kshatrak 185 / क्षत्रक

Kshemankari 49 / क्षेमंकारी

Kshiprachala 127 / क्षिप्रचल

Kshiprashyen 76 / क्षिप्रश्येन

Kshudragrudhri 48 / क्षुद्रगृध्री

Kshudravalguli 283, 294 / क्षुद्रवल्गुली

Kubbha 164 / कुभ्भ

Kuhukashtha 158 / कुहू-काष्ठ

Kuhumukh 158 / कुहूमुख

Kukkutak 85 / कुक्कुटक

Kulalkukkut 164 / कुलालकुक्कुट

Kulingaka 63 / कुलिंगक

Kumarishama 248 / कुमारीश्यामा

Kumbholuk 169 / कुंभोलूक

Kumud 47 / कुमुद

Kunal 103 / कुणाल

Kurar 91 / कुरर

Kurari 133, 141 / कुररी

Kurubahu 254, 255 / कुरुबाहु

Kuruwaak 254, 258 / कुरु वाक

Kushitak 124 / कुषीतक

Kusitaangi 257 / कुषितांगी

Kutidushak 211 / कुटिदूषक

Laav 90 / लाव

Laav Gairik 82 / लावगैरिक

Laav Paansul 84 / लावपांसुल

Laghad 75 / लग्गड-लग्घड

Laghukurari 137 / लघुकुररी

Langan 76 / लंगण

Latukika 300 / लटूकिका

Latushak 235, 236 / लटूषक

Latwa 236, 282 / लट्वा

Lohaprushtha 180 / लोहपृष्ठ

Lohaprushthakank 49 / लोहपृष्ठकंक

Lopa 90 / लोपा

Maalaya 57 / मालाय

Madgu 11 / मद्गु

Madhujivha 86 / मधुजिव्हा

Madhuka 232 / मधुक

Madhukar 290 / मधुकर

Madhup 293 / मधुप

Madhupkhag 289 / मधुपखग

Madhusarika 311 / मधुसारिका

Mallikaksha 44 / मल्लिकाक्ष

Manjul Datyuha 99 / मंजुलदात्यूह

Manjulitak 198 / मंजुलितक

Manjuliyak 196 / मंजुलियक

Marul 97 / मरुल

Matrunindak 192 / मातृनिंदक

Matsyaranka 18, 185 / मत्स्यरंक

Matsyrankashyen 73 / मत्स्यरंकश्येन

Mayuraghni 70 / मयूरघ्नी

Meghanulasin 89 / मेघानुलासिन्

Meghrao 96 / मेघराव

Mrunalkanth 33 / मृणालकंठ

Mugdhabalak 25 / मुग्धबलाक

Nagashi 57 / नागाशी

Nakta kraunch 21 / नक्तक्रौंञ्च

Nakuti 211 / नाकुटी

Nandan 278 / नंदन

Nandimukh 39 / नंदिमुख

Nandimukhi 41 / नंदिमुखी

Naptruka 174 / नप्तृक

Nasachchinna 39 / नासाच्छिन्ना

Neel latwa 276 / नीललट्वा

Neelaang saras 91 / नीलांगसारस

Neelachatak 234, 275, 280 / नीलचटक

Neelachhawi 234 / नीलछ्छवी

Neelagreeva 244 / नीलग्रीव

Neelkapot 146 / नीलकपोत

Padmapatrashayika 139 / पद्मपत्रशायिका

Padmapushpa 206 / पद्मपुष्प

Pakshagupta 233 / पक्षगुप्त

Pandavika 239 / पांडविका

Pandushama 239 / पांडुश्यामा

Paniyakak 9 / पानीयकाक

Pankajit 73 / पंकजित

Pankakeer 111 / पंककीर

Panktichar Kurar 91 / पंक्तिचरकुरर

Paraavat 146 / पारावत

Paroshni 302 / पारोष्णी

Paryantika 229 / पर्यंतिका

Patangika 262 / पतंगिका

Patragupta 233 / पत्रगुप्त

Patri 60 / पत्रि

Pechak 168 / पेचक

Peetamunda 306 / पीतमुंड

Peetanetra 259 / पीतनेत्र

Peetatanu 230 / पीततनु

Peetpaad 109 / पीतपाद

Phutkari 263 / फुटकारी

Pichhabaan 67 / पिच्छबाण

Pik 158 / पिक

Pikaang 206, 220 / पिकांग

Pingachakshu 170 / पिंगचक्षु

Pingalaksha / पिंगलाक्ष

Pingalika 19 / पिंगलिका

Pinga 119 / पिंग

Pippal 196 / पिप्पल

Plav 10 / प्लव

Index of Marathi Names of Birds :

Black Necked Crane (Grus nigricollis) Is seen in Ladakh.
(Photo Courtsey-o/c ZSI, Solan - HP.)
The recently discovered colony of Blue-tailed Bee-Eaters
near Amravati in Eastern Maharashtra (Raju Kasambe, per.com.)

ACKNOWLEDGEMENTS

Photographer Credits :
Bird photography is challenging, demanding, tiring and requires unlimited patience and skill. Rare opportunities and hard work make great photographs. We are therefore particularly grateful to the contributing photographers. This book was possible due to the wholehearted support and contributions from several leading and amateur bird photographers from India and other countries. Many excellent photographs are digitally altered for the sake of illustration and not quality. While acknowledging, page numbers are given in bold type. Illustrations are numbered in clock-wise direction. Stamps and feathers are not numbered. Their credits are given separately. First photograph in the right column is numbered one. When two or more photos are placed side by side in one row, the right sided is a, then b, c, etc. towards the left. In case of change from this pattern, necessary clarifications are given for that page. Four examples are given :

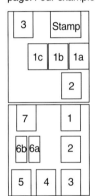
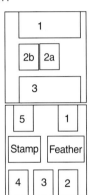

List of abbreviations of the names of photographers is given below in alphabetical order.
Full names of all others are mentioned in the acknowledgement section :

AA-Abrar Ahmad. **AAZ**-Ashfaq A. Zarri. **AD**-Anil Damle. **ADh**-Atul Dhamankar. **AM**-Anil Mahabal. **AP**-Amit Pawashe. **AR**-Asad Rahmani. **AZ**-Anant Zanzale. **BC**-Bharat Cheda. **BJ**-Bill Johnson. **BP**-Banda Pednekar. **C&R**-Chaitra M. R. & Rajesh B. P. **CB**-Chandrashekhar Bapat. **CDS**-Carl D'Silva. **CF**-Clement Francis M. **CKS**-Chan Kai Soon. **DA**-Digital Artwork. **DY**-Dilip Yardi. **EJL**-Rev. E. J. Lott. **GW**-Gehan D'Silva Wijeyeratne. **HP**-Hira Punjabi. **JHD**-John Henry Dick. **JM**-Jijo Mathews. **JSRP**-Jean Sebastien Rosseau-Piot. **KHT**-Kim Hyu Tae. **KJ**-Kishor Joshi. **KP**-Kiran Purandare. **KSR**-K. S. Rajshekhara. **KVP**-K. V. Pandit. **MB**-Milind Bendale. **MD**-Mike Danzenbaker. **MM**-Mahesh Mahajan. **MPN**-M. P. Nagendra. **MVD**-Mukund Deshpande. **NL**-Navendu Lad.(First person to photograph both the rediscovered Indian birds-Jerdon's Courser & Forest Owlet.) **NS**-Niranjan Sant. **PB**-Pramod Bansode. **PJ**-Prakash Joglekar. **PKu**-Prabhakar Kukdolkar. **PML**-P. M. Lad. **RgK**-Raghunandan Kulkarni. **RK**-Raju Kasambe. **RN**-Rishad Naoroji. **RP**-Raja Purohit. **RPr**-Rajesh Pardeshi. **RS**-R. S. Suresh. **RuK**-Rudd Kampf. **PN**-Pramod Nargolkar. **RVT**-Rajkumar Vijaykumar Thondaman. **SBN**-Sanjeev Nalavade. **SB**-S. Balachandran.

SF-Steven Falk. **SGN**-S. G. Neginhal. **SI**-Shrikant Ingalhalikar. **SK**-Shreesh Kshirsagar. **Sku**-Sudhakar Kurhade. **SN**-Sattyasheel Naik. **SPl**-Sachin Palkar. **SPr**-Subhash Puranik. **SP**-Satish Pande. **SR**-Satish Ranade. **SS**-Sanjay Shegaokar. **ST**-Saleel Tambe. **TNAP**-T. N. A. Perumal. **TP**-Tony Palliser. **TSU**-T. S. U.de Zylva. **VC**-Vijay Cavale. **VK**-Vishwas Katdare. **VT**-Vinay Thakar. **VM**-Valeri Moseikin. **WH**-William Hague. **YB**-Yashodhan Bhatia.

P1: CF-1, Merek Kosinski-2, 3. **P2**: SP-1, 3. PML-2, 7, CF-4, 5. Merek Kosinski-6. **P3**: TP-1, 3. JHD-2, 4. **P4**: MD-1. Frank O'Connor-2, Dick Newell-3, TP-4, 5, JHD-6a,b, 7a, b. **P5**: BJ-1a, b, TD-2, TP-3, 4, JHD-5a, b. **P6**: KJ-1, 3, SP-2, PML-4. **P7**: RVT-1, SGN-2, 4, NS-3, CF-5. **P8**: TD-1, Ido Tsurim-2, 4, JHD-3. **P9**: ST-1, SP-2, 4, SGN-3, KJ-5. **P10**: NS-1, 3, 6, RVT-4, b, SP-4a, CF-6. **P11**: SK-1, ST-2a, NS-2b, 3, SP-4. **P12**: Hemant Shinde-1a, 2a, TD-3, JHD-4a, b, c, Tetsu-5. **P13**: ST-1, CF-3, VK-2, SP-4a, PML-4b. **P14**: CF-1a, 4, SP-2, 3. **P15**: ST-1, 2, RVT-3, CF-4, NS-5a, b. **P16**: NS-1, 2, 3, ST-4, **P17**: SK-1, SP-2, 4, 5, NS-3. **P18**: SP-1, SK-2, ST-3, KJ-4. **P19**: NS-1, ST-2, SP-3, SGN-4. **P20**: SP-1, 2, 3, 4, 5. **P21**: SP-1, 2, TSU-4. **P22**: PB-1, 3, Ansar Khan-2. **P23**: GW-1a, JHD-1b, Paul Gale-2a, b, PB-3, CB-4, 5, Peter Jones-6. **P24**: SP-1a, SK-1b, CB-2, NS-3, MPN-4, CF- 5. **P25**: SP-1, 4, CF-5, NS-3, CDS-2. **P26**: CDS-1, SP-2, 4, RaK-3. **P27**: RP-1, Rak-2, KJ-3, MM-4, CDS-5a, SP-5b. **P28**: SP-1, 2, NS-3, MM-4, GW-4. **P29**: NS-1, SP-2, 5, RP-3, KJ-4. **P30**: ST-1, SP-2, 3, 4. **P31**: NS-1, DA-2, KJ-3, NS-4, 5a, ST-5b. **P32**: SP-1, 3, ST-2, SGN-4. **P33**: SP-1, 3a, b, BC-2, KJ-4. **P34**: SP-1a, b, 3a, KJ-2, 4, 5, KVP-3b. **P35**: PML-1, SP-2, 3, NS-4, PB-5. **P36**: DA-1, SP-2a, b, 3, SK-4. **P37**: RP-1, ST-2, John Edwards-3, SP-4. **P38**: ST-1, SP-2, EJL-3, SP-4. **P39**: SP-1, NC-2, Anil Kashyap-3, 4. **P40**: BC-1, EJL-2, DA-3, SP-4, NS-5. **P41**: NS-1, SP-2, MM-3, 4, EJL-5, JSRP-6. **P42**: SF-1a, NS- 2b, SP- 3b, RaK-1b, JRRP-2a, EJL-3a, KJ-4, BJ-5. **P43**: NS-1, SP-2, 5, ST-3, EJL-4. **P44**: EJL-1, 3, SP-2, CB-4, ST-5. **P45**: Lawrence Poh-1, RVT-2, CKS-3, EJL-4, CDS-5, NS-6. **P46**: NS-1a, 4, 5, SP-1b, C. Kolhatkar-2, Devendra Singh-3. **P47**: KSR-1, SP-2, PML-3, KJ-4, NS-5. **P48**: ST-1, 5a, b, SN-2, MD-3, RP-4. **P49**: SP-1, 2, 5, ST-3, NS-4. **P50**: 1, 2b, 4, SP-2a, HP-3. **P51**: VM-1, VK-2, 4, VC-3, SP-4. **P52**: SP-1, 2, CF-3. **P53**: NS-1, 2, 3, AZ-4, SBN-5, CF-6. **P54**: DA-1a, 4, SBN-1b, SP-2, 3, NS-5. **P55**: SP-1, AM-2, MM-3, Suresh Pardeshi-4. **P56**: SP-1a, 2a, SPl-1b, c, NS-2b, C&R-2c, 3, BP-2d, CF-4 (flight). **P57**: SPl-1, MPN-2, NS-3, PN-4. **P58**: SP-1, NS-2, ST-3, CF-4. **P59**: NS-1, 2, 7, 8, 9, SKu-3, ST-4, 5, AP-6. **P60**: PML-1, JHD-2a, b, 4a, AD-3, ST-4b, NS-5. **P61**: NS-1, 2a, 3a, P.Mestri-2b, MD-3b, Mike Shipman-4. **P62**: RN-1, KP-2, SP-3a, JM-3b,c,, NS4, 5. **P63**: CF-1, SP-2, EJL-3. **P64**: PN-1, CKS-2, SP-3. **P65**: KJ-1, NS-2, SP-3a, SBN-3b. **P66**: NS-1a, b, 2a, b, 3a, b, AP-4. **P67**: NS-1, 3, HP-2, DA-4. **P68**: NS-1, 3, 4, SP-2, Job Joseph-Rufous-bellied Eagle-flight. **P69**: KJ-1a, Lean Yen Loong-1b, NS-2a, b, 3. **P70**: CDS-4, AZ-3, SP-1, 5. **P71**: NS-1, 6, SP-2, 3a, 5, GW-3b, RN-4. **P72**:SP-1, 4, Satoshi Maenishi-1, CF-2, GW-3a, SP-3b, 4. **P73**: ST1, 2, 3, Jugal Tiwari-4, SBN-5. **P74**: BC-1, CF-2, VM-3, SP-4, Yves

Adams-5. **P75**: NS-1, CB-2, SK-3, AC-4. **P76**: SBN-1, Prashant Kanvinde-2, SP-3. **P77**: NS-1a, b, George McCarthy-2, JSRP-3. **P78**: RN-1, DA-2a, George McCarthy-2b, CF-3a, b, c, AP-4a, SBN-4b, NS-4c. **P79**: PML-1, SP-2, 4, NS-3. **P80**: AA-1, SP-2, 3, KJ-4, NS-5, RP-6. P81: SP-1, 2, SK-3. **P82**: DA-1, Erach Bharucha-2, Rajat Bhargava-3, EJL-4, SI-5. **P83**:Ajit Kulkarni-1, RK-2, 3, 5, 6, 7, SP-4. **P84**: ST-1a, RK-1b, 2, 3, 5, AD-4. **P85**: SPr-2a, b, 3, SB-4. **P86**: CF-4, PML-1, GW-2, NS-3. **P87**: AA-1, NS-2, ST-3, PML-4, DA-4. **P88**: MVD-1, 2, KJ-3, SP-4. Art work & seals. **P89**: RP-1, SP-2, RVT-3, DA-4. AZ-5. **P90**: Deepak Joshi-1, RS-2, NS-3, RK-4a, b, SP-5. **P91**: PML-1, SP-2a, b, VT-3, ST-4, NS-5, MB-6. **P92**: VT-1, YB-2, NS-3, MVD-4, MB-5. **P93**: RP-1, NS-2. **P94**: CKS-1, S. Harishchandra-2, Premsagar Mestri-3, Paul Gale-4, SP-5, SK-6, KHT-7, ST-8. **P95**: P. N. Papanna-1, Herman Van Oosten-2, CKS-3, Bas van De Meulengraf-4, Shamita Harishchandra-5. **P96**: SP-1, 3, KHT-2, DJ-4, ST-5. **P97**: NS-1, DA (Prabhat Films-Poster)-2, CF-3, 4, BP-5. **P98**: BP-1a, DA-1b, NS-2, 3. **P99**: SP-1, ST-2, 4, NS-3. **P100**: AR-1, RP-2, PML-3. **P101**: PML-1, 4, AD-2, Sandeep Labade-3, CF-5, SP-6, AR-7, 8. **P102**: SP-1, BP-2, JHD-3a, b, PB-4a, b, CF-5, PML-6. **P103**: NS-1, 2, Pramod Pawashe-2, SP-4, Amol Warange-5. **P104**: Peter Draper-bottom, KJ-centre, SP-top. **P105**: PML-1, PB-2, EJL-3, 4b, JHD-4a, SN-5. **P106**: RuK-1, EJL-2, Adrian Webb-3a, NS-3b, 5, GW-4. **P107**: RS-1, SP-2, DA-3, ST-4. **P108**: Adrian Webb-1, SP-2, 5, NS-3, AP-4. **P109**: RP-1, Yves Adams-2, SK-3. **P110**: SBN-1, NS-2, ST-3a, 4, 5, NS-3b. **P111**: ST-1, Michael Gelinas-2, GW-3. **P112**: JHD-1, Dave Curtis-2, SF-3, George & Lindsey Swann-5, SP-4, 6. **P113**: PML-1, ST-2, SP-3, SB-4, NS-5, SP-6. **P114**: RuK-1, MD-2, BJ-3, PML-4. **P115**: KJ-1, SB-2, 3, YB-4, Naoto Kitagawa-5. **P116**: ST-1, NS-2, 3, PML-4. **P117**: KJ-1, RVT-2, YB-3, 4.. **P118**: YB-1a, 2, ST-1b, KJ-3, EJL-4, 5. **P119**: PML-1, SP-2, ST-3, PJ-4, GW-5. **P120**: ST-1, EJL-2, SP-3, 5, PML-4. **P121**: EJL-1, 3, 5, SF-2, KJ-4, J. Judge-6, SP-7. **P122**: JHD-1, RuK-2, CF-3, PML-4, RVT-5. **P123**: PML-1, SP-2a, b, KJ-3, 4a,b, SP-5. **P124**: ST-1, SP-2, GW-3, KJ-4, 5. **P125**: YB-1, 3, KJ-2, MVD-4. **P126**: NS-1, P. N. Papanna-2, CF-3, 4. **P127**: SP-1, RK-2, PN-3, RP-4, PB-5, YB-6. **P128**: NS-1, JHD-2, Ido Tsurim-3, 4, SP-5. **P129**: SP-1, 2, 3, 4. **P130**: James Flynn-1, JHD-2, Bruce Craig-3, BJ-4, TP-5. **P131**: PML-1 (bottom), 2, SP-3 (flight), KJ-4. **P132**: SP-1, 2, 3, 4, 5, 6. **P133**: NS-1, 3, 4, 5, SP-2. **P134**: KJ-1, Lok Wan Tho-2, 6, WH-3, PML-4, SP-5, RuK-7. **P135**: SP-1, PML-2, CDS-3, Khalid Rafeek-4, 5. **P136**: SP-1a, b, 2, 3. **P137**: PML-1, SP-2, 3, 4, SB-5. **P138**: WH-1a, ST-1b, c, 2, SP-3 (bottom). **P139**: PML-1, CF-2, C&E-3, GW-4. **P140**: SP-all. **P141**: SP-1, 2, C&E-3, Dave Brokeman-4. **P142**: WH-1, SP-2a, 3b, 4, BP-3a, DY-2b. **P143**: KJ-1, 2, 3, 5, 6, SP-7. **P144**: KJ-1, AD-2, RK-3, SN-4, PML-5. **P145**: ST-1, 5, SP-2, 6, MB-3, GW-4. **P146**: EJL-1, PK-2, MVD-3, SP-4. **P147**: ST-1, SP-2, 3, CF-4. **P148**: PK-1, MM-2, SP-3, NS-4, CF-5. **P149**: ST-1, TSU-2, SP-3. **P150**: SPI: 1, RVT-2, 4, SP-3, CF-4, WH-5. **P151**: CF-1, KJ-2a, AP-2b, Nazim Siddiqui-3. **P152**: RVT-1, 3, ST-2. **P153**: ST-1, GW-2, RP-3, SP-4. **P154**: MVD-1, SP-2,

PML-3, AP-4, 5. **P155**: SPI: 1, NS-2, 3a, BC-3b, JHD-4a,(sketch), RVT-4b. **P156**: SB-1, CKS-2, 3, SP-4. **P157**: SP-1, 2, 4, NS-3. **P158**: KP-1, NS-2. **P159**: CF-1, 3, RVT-2a, b. **P160**: KK-1, CKS-2, SP-3, VC-4. **P161**: RVT-1a, VC-1b, Suppalak Kladbee-3a, SP-2a, b, c, 3b, SI-4. **P162**: CKS-1a, b, SP-1c, SI-2. **P163**: SR-1, DY-2, RS-3, GW-4a, b. **P164**: SP-1, 2, 6, GW-2, CF-4, RKu: 5, MVD-7. **P165**: DY-1, TSU-2, SP-3, SPI: 4a, Hema Gupte-4b. **P166**: JHD-1, Sanjay Karkare-2a, SPI-2b, NS-3, PML-4, RP-5. **P167**: TNAP-1, NL-2a, SN-2b, NS-3, VJ-4a, CF-4b. **P168**: PML-1, 3b, NS-2, DY-3a, MPN-4. **P169**: SP-1, 6, 8, TNAP-2, SPr-3, SK-5, PML-7. **P170**: NL-1, 3, SN-2, SP-4. **P171**: MVD-1(moth), SP-2, 3, 7, GW-2, PML-5, 8, SR-6, JHD-9. **P172**: NS-1, Parag Dandage-2, TNAP-3, M. P. S. Prasad-4, SP-5. **P173**: NS-1, GW-2, Suresh Elamon-3, Ramchandran-4. **P174**: CB-1, NS-2, Amol Mithari-3, RS-4. **P175**: CF-1, CB-2, Kees Baker-3, Avinash Nangare-4, SPI-4a, JHD-4b. **P176**: Premsagar Mestri-1, SBN-2, SP-3, JM-4, 5. **P177**: SP-1, CDS-2a, b, 3a, b, SP-4. **P178**: SP-1a, 2a, b, 3c, AP-1b, c, JHD-3a, b, VK-4. **P179**: KK-1, PML-2, NS-3, JHD-4. **P180**: NS-1, 4, EJL-2, RVT-3. **P181**: SP-1, 3, 4, Nirmal Kaur-2. **P182**: SP-1, 4, 6, CF-2, VT-5. **P183**: SP-1, 2, 3, NS-4. **P184**: CB-1, SP-2, 6, MVD-3, AZ-4, NS-5, 7. **P185**: ST-1, 2, 6, G. N. Papanna-3, SP-4, JHD-5. **P186**: SP-1, NS-2a, b, SPr-3, RP-4. **P187**: SP-1, PN-2, Arnon T-Sairi-3, 7, ST-4. **P188**: SP-1, Shivaji Jaware-2, RVT-3, Suppalak Kladbee-4, NS-5, Paul Gale-6, 7, ST-8. **P189**: RP-1, TSU-2, SP-3, 4. **P190**: NS-1, 2, 3b, Harish Ingawale-3a, PJ-4. **P191**: ST-1, 2, SP-2, 3, 4. **P192**: GW-1, NS-2a, b, 3, Deepak Joshi-4. **P193**: M. P. S. Prasad-1, SP-2, 6, MM-3, NS-4, DA-5. **P194**: NS-1, 4a, b, 5, SP-2, 3, 4c. **P195**: EJL-1, SP-2, 3. **P196**: SP-1, RP-2, CF-3. **P197**: NS-1a, b, c, 4, CF-2a, b, 3. **P198**: JHD-1, NS-2, GW-3, ST-4, RP-5. **P199**: SP-1a, 2, 5, KJ-1b, 4, NS-3a, GW-3b. **P200**: SR-1, ST-2, 3, Datta Ugaokar-4, PB-5. **P201**: SP-1, 3, 4, NS-2a, b. **P202**: SP-1, SB2, H. S. Ananth-3, TSU-4, Laxmi Joshi-5a, NS-5b. **P203**: H. S. Ananth-1a, b, NS-2, GW-3, TSU-4, RVT-5, PML-6. **P204**: PML-1, 4, NS-2, JHD-3. **P205**: SP-1, ST-2, SGN-3, NS-4. **P206**: SP-1, 2, 5, NL-3, Hema Gupte-4, RVT-6. **P207**: SP-1, 2, 3, MB-4, SK-5. **P208**: Amol Mithari-1, ST-2, 3b, VJT-3a, RP-4, SP-5. **P209**: SP-1, 2, 5, ADh-3, DY-4. **P210**: AP-1, DY-2, CB-3. **P211**: DY-1, ST-2, CDS-3a, b, JHD-4a, b, c. **P212**: SP-1, EJL-2, NS-3, 6, ST-4, CF-5. **P213**: KSR-1, 4, HP-2, NS-3, ST-5. **P214**: SP-1, 4, CKS-2, RVT-3. **P215**: ST-1, SP-2, 3, 4, 5, 6. **P216**: SS-1a, C&R: 1b, NS-1c, 3, SP-2a, b, ST-4. **P217**: SP-1, 2, 4, CF-3. **P218**: Ashley Fisher: 1, SP-2, 3. **P219**: CKS-1, SP-2, 4, 6, 9, 10, EJL-3, Bjorn Johansson-5, SB-7, JM-8. **P220**: SP-1, 4, ST-2a, KP-2b, CF-3. **P221**: SP-1, PML-2, RS-3. **P222**: SP-1, 4, 5, RVT-2, ST-3. **P223**: ST-1, CF-2a, SP-2b, RVT-3, CDS-4a, b. **P224**: CDS-1, ST-2, SP-3a, b, RVT-4. **P225**: SP-1, 2, 4, RVT-3. **P226**: AA-1, 2, CF-3, SK-4. **P227**: SK-1a, CF-1b, AP-2a, b, ST-2c, KJ-3. **P228**: ST-1, 5, SP-2, 3, 4. **P229**: VC-1, CB-2, KJ-3. **P230**: Ulhas Rane-1, CF-2, EJL-3, RVT-4. **P231**: SP-1, 3, GW-2, C&R-4, CF-5. **P232**: KK-1, RSS-2, PB-3, SP-4. **P233**: SP-1, 2b, RVT-2a, ST-3, Vivek Sinha-4. **P234**: RVT-1, ST-2, 5, SP-3, 4. **P235**:

SP-1, 3, 5, CF-2, GW-4. **P236**: SP-1, 2, AP-3. **P237**: AP-1a, b, c, 4, NS-2a, b, 5, SP-2c, d, 3a, b. **P238**: JHD-1a, b, Jugal Tiwari-2, DA-3. **P239**: SP-1, 3, RVT-2, 4.. **P240**: JHD-1a, b, Shamita Harishchandra-2, TSU-3, RVT-4. **P241**: SP-1, 4, 5, RVT-2, ST-3. **P242**: SR-1, GW-2, JM-3, SP-4, ST-5. **P243**: SP-1, 2, 5, NS-3, SPr-4. **P244**: KHT-1, SP-2, RVT-3, ST-4. **P245**: RVT-1, 4, 5, SP-2, MD-3, KHT-6, EJL-7, Tim Loseby-8. **P246**: NS-1a, KVP-1b, SK-2, ST-3a, SP-3b, KJ-4. **P247**: SK-1, 2, SP-3, 4, 5. **P248**: KP-1, SP-2, 4, 5, RS-3. **P249**: SP-1, 2, 3, MB-4, NS-5, Suresh Pardeshi-6. **P250**: SP-1, CB-2, JHD-3, NS-4. **P251**: YB-1a, SP-1b, 2c, NS-2a, b, 3a, b, c, d. **P252**: CF-1, 4, NS-2, SP-3. **P253**: SP-1, 2, NS-3. **P254**: NS-1, GW-2a, AAZ-2b, CDS-3a, b, RVT-4, EJL-5. **P255**: SB-1, YB-2, SR-3, SS-4. **P256**: SR-1, GW-2, Spr-3, Venkat Swamappa-4. **P257**: SP-1, 2, 3, NS-4. **P258**: GW-1, CF-2, SPr-3, 4, ST-5. **P259**: SP-1, 3, ST-2, NS-4. **P260**: EJL-1, SR-2, SKu-3, VC-4, SP-5. **P261**: ST-1, 5, SP-2, 3, SS-4. **P262**: SP-1, 3, 4, RP-2. **P263**: RVT-1, SS-2, CF-3, SP-4. **P264**: SP-1, 3, RS-2, BP-4, CF-5. **P265**: SP-1, 4, JHD-2, JM-3. **P266**: SP-1, 4, 5, CF-2, SB3. **P267**: SP-1, 2, 3. **P268**: SP-1, 2, 3. **P269**: SP-1, 3, NS-2, ST-4, AP-5. **P270**: JHD-1a, b, GW-2, VC-3, SR-4, 5. **P271**: SP-1, Otto Pfister-2, 4, Kees Baker-3. **P272**: C&R-1, TSU-2, ST3. **P273**: EJL-1, SP-2a, CF-2b, 3, AAZ-4, JHD-5. **P274**: CF-1, 5, SP- 2, 3, AR-4a, AAZ-4b. **P275**: ST-1, CF-2, 3, GW-4, SP-5. **P276**: JHD-1, SP-2, ST-3, 4, CB-5, NS-6. **P277**: RVT-1, KK-2a, b, NS-3. **P278**: PML-1, RVT-2, KSR-3. **P279**: KSR-1, CB-2, Lok Wan Tho-3. **P280**: RVT-1, SP-2, SS-3. **P281**: KSR-1, EJL-2, SPl-3, CF-4. **P282**: Joanna Van Gruisen-1, EJL-2, SP-3. **P283**: MM-1, SP-2, PB-3a, EJL-3b. **P284**: NS-1a, SP-1b, 2. **P285**: SP-1, ST-2, 3, NS-4. **P286**: ST-1, Lok Wan Tho-2, Manoj Kulshrestha-3. **P287**: RVT-1, SP-2, 3, GW-4. **P288**: Ram Mone-1, CKS-2, SP-3, 4. **P289**: NS-1, 3, 4, SP-2, 5. **P290**: SP-1, 2, 3, RP-4. **P291**: SP-1, 4, SK-2, ST-3. **P292**: SP-1a, 2, 3, SPl-1b. **P293**: SP-1, 5, NS-2, CF-3, 4. **P294**: CF-1, DA-2, RVT-3, 4. **P295**: SP-1, 3, 4, GW-2, SN-5a, KP-5b. **P296**: SP-1, 4, 5, RK-2, SK-3. **P297**: SP-1, 2, 4, CF-3. **P298**: CDS-1, NS-2, 3, CF-4, SP-5. **P299**: EJL-1a, Suppalak Kladbee-1b, AP-2, ST-3. **P300**: SP-1, 3a, b, KSR-2, ST-4. **P301**: RVT-1, SP-2, 5, ST-3, 4. **P302**: CDS-1, GW-2, NS-3, SP-4, SI-5. **P303**: ST-1a, SP-1b, 2c, 3, CF-2a, PJ-2b. **P304**: ST-1, CDS-2a, CF-2b, Suresh Pardeshi-3, 5, NS-4, SPr-6. **P305**: R. S. Dayanand-1, 2, SP-3, SS-4. **P306**: MB-1, 2, SP-3, 4,. **P307**: KJ-1a, b, 2. **P308**: KJ-1a, b, 3, Harish Ingawale-1c, 2, Salim Ali-4. **P309**: CF-1, ST-2, CF-3. **P310**: SP-1, 3, 4, ST-4, KK-5. **P311**: Vijay Tuljapurkar-1, RVT-2, NS-3, CF-4. **P312**: SP-1, 3, NS-2. **P313**: E. P. Gee-1, ST-2, CF-3, SP-4, 6, 8, RPr-5, Krys Kazmierczak-7. **P314**: SP-1, 2, 5, GW-4, NS-3. **P315**: RVT-1, EJL-2, ST-3. **P316**: CF-1, SP-2, 4, NS-3, CF-4. **P317**: SP-1, 2, 5, ADh-3, CF-4. **P318**: SP-1, ST-2, 5, NS-3, KJ-4. **P319**: CF-1, ST-2, SP-3. **P320**: JHD-1, NS-2, 4, ST-3, Ooi Beng Yean-5. **P321**: RVT-1, GW-2, ST-3. **P322**: SP-1,4, ST--3, Chaiyan Kasorndorkbue-3,5. **P323**: NS-1, SB-2, EJL-3, 5, ST-4. **P324**: SP-1, MVD-2, 3, GW-4, RVT-5, SK-6, CF-7. **P325**: GW-1, Dr. M. G. Kanitkar-2, SP-3, 4. **P326**: NS-1, KJ-2, SP-3, Mukund Jere-4, 5,

6. **P327**: SN-1, SP-2, 3. **P328**: SP-1, 3, 4. 5. VK-2. **P329**: J.Tadphale-1, SGN-2. **P330**: SP-1, SBN-2,. **P331**: SF-1. **P333**: JHD-1. **P336**: NS-1. **P337**: SP-1. **P338**: SP-1. **P339**: RS-1. **P340**: CF-1. **P341**: CF-1. **P342**: CF-1. **P343**: AR-1, SP-2. **P344**: DA-1. **P346**: RVT-1, CB-2, SP-3a, DA-3b. **P347**: JHD-1a,b. **P349**: Rahul Warange-1. **P350**: DA-1. **P351**: RVT-1. **P352**: KVP-1, RVT-2. **P354**: SK-1, Mohan Panse-2, NL-3, NS-4. **P355**: SP-1. **P:359** EJL-1, NS-2, CKS-3a, 3b, CF-4. **P363**: SP-1. **P366**: KJ-1. **P368**: SP-1, CF-Inset. **P372**: SP-1. **P373**: SP-1, NS-2, GW-3, KJ-4, Ashwin Deshmukh-5. **P374**: AR-1. **Flying Page**- (Double spread) : 1-NS, 2-MB.

Feathers : Only those feathers definitely identified are illustrated. We thank : Amit Pawashe, Deepak Joshi, Raju Kasambe, Saurabh Shet, Premsagar Mestri, Chandrahas Kolhatkar, Sanket Shet, Jummabhai, Bharat Mallav, Satish Pande, Nivedita Pande, Smita & Vishu Kumar, Shailesh Joshi, Vishwas Joshi, Atul Joshi, Tejashree Joshi, Ram Mone, Abhay Soman.

Postal Stamps : Those stamps depicting birds from India and other countries are from the valuable collections of : T. C. Jacob, Naresh Chaturvedi, Anil Mahabal, Satish Pande, Rohan Lowlekar, Mukund Deshpande, Vandana Ghormade, Madhura Deshpande.

Coins & Currencies illustrated in the book are from the precious collection of K. V. Pandit.

Symbols given on each page are designed by Satish Pande, Saleel Tambe.

Digital avian sketches (abbreviated as DA) illustrated in the book are based on photographs taken by the authors, references sited in next two paragraphs for Sanskrit text and Archaeology & also the following references : 1. A Pictorial Guide to the Bids of the Indian Subcontinent, Salim Ali & S. Dillon Ripley, illustrated by John Henry Dick. 2. The Book of Indian Birds, Salim Ali, illustrated by Carl D'Silva. 3. Birds of the Indian Subcontinent, Richard Grimmet, Carol Inskipp, Tim Inskipp. 4. A Field Guide to the Birds of India, Krys Kazmierczak, illustrated by Ber van Perlo.

Sanskrit Text : Critical compilation and short listing of Sanskrit names of birds is done by Suruchi & Satish Pande. Shrikant Bahulikar gave guidance. Prasad & Vidya Joshi did type setting of Sanskrit index. References to birds can be found in several Sanskrit texts. Some major available references are: 1. The Hymns of the Rgveda, Tr. Ralph T. H. Griffith & Ed. by Prof. J. L. Shastri, Pub. - Motilal Banarasidass, Delhi, 1999.
2. Shukla Yajurveda (2 Volumes), Ed.-Dhundiraj Ganesh Dixit Bapat, Pub. - Rajasaheb, Sansthan Aundh, Shaka, 1862.
3. Subhashitaratnabhandagaram, Enlarged & Re-Ed. by Narayan Ram Acharya, Pub. - Nirnaya Sagar Press, Bombay, 1952.
4. Vasantarajashakunam, Vasantaraja, Jagadishwara Shilayantralaya, Mumbai, 1940.
5. Sushrut Samhita, Ed. Vaidya J. T. Acharya & Narayan Ram Acharya, Pub. Chaukhamba Orientalia, Varanasi, 1980.
6. Charak Samhita, Ed. by Vaidya Vamanashastri

Datar, Nirnaya Sagar Mudranalaya, Mumbai, 1922.
7. Arthashastra, Kautilya, Marathi Tr. by Prof. R. P.
Kangale, Pub. Maharashtra Rajya Sahitya aani
Sanskriti Mandal, Mumbai, 1982.
8. Brihatsamhita, Varaha Mihira, Tr. By Pt.
Achyutananda Jha, Pub. Chaukhamba
Vidyabhavan, Varanasi, 1959.
9. Valmiki Ramayana (Critical Ed.), Critically Ed. by
Prof. G. H. Bhatt, Oriental Institute, Baroda.
10. Mahabharata (Critical Ed.), Bhandarkar Oriental
Research Institute, Pune.
11. Kalidas Granthavali, Sitaram Chaturvedi, Pub.
Bharat Prakashan Mandir, Aligadh, Samvat 2019.
12. Birds in Sanskrit Literature, K. N. Dave, Pub.
Motilal Banarasidass, 1985.
Indian Avian Archaeology : References to birds in
Indian Archaeology taken in the book are from the
following authors : Kadgaonkar, S., Nath, B.,
Pawankar, S. J., Thomas, P. K., Joglekar, P. P.
Some major reference sources are:
1. Bulletin Of the Deccan College Post-Graduate
and Research Institute. 2. Indian Iconography, Joshi,
N. P. , 1979. 3. Prehistoric Rock Paintings of
Bhimbetka, Central India, Mathpal, Y. 1984. 4.
Bulletin of Indian Museum. 5. Man and Environment.
6. Quartenary Environments and Geoarchaeology of
India. 7. Riddles of Indian Rock-shelter Paintings,
Tiwari, SK, Pub. - Sarup & Sons. 8. Flora and Fauna
in Mughal Art, Verma, Som Prakash, 1999. Marg
Publications. 9. Puratatva.

Additional assistance : Jayant B. Joshi, Asawari
Sant, Juliana Clement, Dnyanesh Rajpathak,
Rajendra Kokate, Jitendra Khasgiwale, Sham
Deshmukh, Ali, Altaf Sayeed, Mohan Alampath,
Brent Ediss, Suresh Jani, Sharad Vyas, Umesh
Thanki, M. M. Momin, Sanjay S. Kharat, Vijay
Bodas, Surendra Karhade, Dilip Bhandare, Prashant
Deshpande, Maksud Shaikh, Deepa Tambe, Abhijit
Patil, Avinash Telkar, Chandrakant Shete, Suhas
Bartakke, Chandrashekhar Patwardhan, Uday
Hardikar, Shreedhar Metar, Sharad Gogate,
Anuradha Mascerenhas, Digambar Gadgil,
Mahendra Wagh, G. Y. & Kaustubh Limaye, Jayant
Dange, Vikas Gupta, Chandrashekhar Salunkhe,
Pradeep Patankar, C. M. Purandare (map), Anand
Abhyankar, Lalu Durve, Prakash Gole,
Dnyaneshwar Rayate, Mahavir Elure. Invaluable
help in the initial stages of the book was offered by
Sanjeev Nalavade, Sujeet Patwardhan , Asad
Rahmani, Late Shri. Jayantrao Tilak, Statira Wadia ,
Farookh Wadia, Smita Kumar , Kunda Pande &
Sulabha G. Sabnis. J. C. Daniel rechecked
correctness of the scientific data. We are thankful to
the staff of the Forest Department, the Irrigation
Department, the Bombay Natural History Society &
the Oxford University Press. Several tribal persons
residing in remote regions of the Western Ghats,
Kokan & Malabar, fishermen of the West coast &
various NGO's particularly Sahyadri Nisarg Mitra,
Chiplun, Prakruti Prayog Parivar, Kolhapur, Sahyadri
Mitra, Mahad, Nisargsewak, Pune, also deserve a
special mention for their hospitality & guidance.
Authors are most grateful & remain indebted to
several others who shared with them the wonders
and sublime joy of bird watching & whose names
are not mentioned here due to oversight.

BIRD MIGRATION :
Bird migration is studied by ringing the birds,
by implanting them with radio-transmitters, moon-watching, etc.
Several Grey Herons were rescued with the help of fire-brigade
and ringed recently by the authors in Pune district. One heron
was subsequently recovered in Kerala.
This is the evidence of local migration in the area
within the scope of the present book.

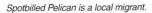

Spotbilled Pelican is a local migrant.

Pallas Gull is a long distance winter migrant.

FINANCIAL ASSISTANCE:

Govind S. Mudholkar; Arun and Vandana Joshi; ELA Foundation and in alphabetical order -
A.K. & Kunda Pande, Achyut & Aruna Umranikar, Arun Mudholkar, Aditi Pant, Ajay Brahmnalkar, Ajit Waknis,
Ali alias Naaz Photographers, Altaf Sayed, Ambegao Taluka Medical Association, Anil B. Pandit,
Anuradha Sowani, Ashwini Kulkarni, Aquil Khan, Arun Kinare, Arvind Inamdar, Arvind Mulgund,
Ashish Atre, Ashok Khandelwal, Bhimsen Joshi, B. N. Kulkarni, Bishwaroop Raha, C. S. Yajnik, Chandrakant Potnis,
Chandrakant Shete, Chandrasen Shirole, Chandrashekhar Patwardhan,
Charudatta Apte, Chetan Traders, D. V. Kirpekar, Dayanand Shetty, Deepa Nande, Eureka Industrial Equipment. Pvt.
Ltd., F. F Wadia, G. M. & Uday Shetty, Galore, Hari Parshuram Aushadhalaya,
Hemant Godse, Indumati & Manohar Apte, J. J. Oswal, Jagdeep Chhokar, K. S. Devadason & Sainik School Satara
staff, K. V. Pandit, Kunal Basu, Kurus Coyaji, Kiran Pande, Kishor Rathi Charity Trust,
Lala Telang, LTC Container Movers, Meenal Kamath, M. B. Bhide, Mukund V. Deshpande,
M. Y. Bapaye, Madhav Gadgil, Madhav Limaye, Mandakini Vartak, Milind Gadkari, Milind Godse,
MITCON Foundation, Mohan Panse, Mukund Rahalkar, N. S. Industries, Naseem Mulla, Nandkishor Gosavi, Nasik
Pakshimitra Mandal, Niranjan Sant, Nitin Sakhdeo, P. N. & Sudha Joshi, P. S. Chitale, Prabhakar Paralikar,
Pradeep Diwate, Prakash Joglekar, Prashant Deshpande, Pune Techtrol Ltd., R. D. Bhadbhade, R. N. Bhat,
Rajendra Deshpande, Rajesh Gadia, Ramesh Godbole, Ravi Kulkarni, Ravindra Joshi of United Printers,
Ravindra Khot, Richard Portwood, Rotary Club of Beeston (UK), S. M. Hardikar (in the memory of Smt. Malatibai
Hardikar), S. M. Koparkar, S. V. Kanitkar, Sachin Mohaniraj, Sadashiv Shivade, Saleel Tambe, Sanjay Gosavi,
Sanjay Gupte, Sanjay Shivade, Sanjeev Khurd, Sarojini Ranade, Satish Pande,
Satish Ranade, Saumitra Transport Services, Shailesh Puntambekar, Shital Ghodake, Shirish & Rajeev Yande,
Shirish Bhave, Shirish Gandhi, Shridhar Rajpathak, Shrikant B. Kelkar, Shrikant Ambardekar,
Shrinivas & Daya Joshi, Shrirang Shirole, Shruti Atre, Subhash . B. Hirve, Sunil Puranik, Suresh Bhalerao,
Suruchi & Nivedita Pande, Swami Vidyanand alias D. K. Pande, Tushar Dighe, Veena Padmanabhan,
Vidarbha Nature & Human Studies Center, Vidya Jagtap, Vijay D. Atre, Vijay Ranade, Vinay Joglekar,
Vindhya Nature Club, Indore, Vinod Chougule, Vinod Gadgil Pratishthan, Vishu & Smita Kumar,
Vivek Joshi, William Selover, Y. A. Ketkar, Y. B. Patwardhan, Y. V. Tawade, Yog Papers, Yogesh Khare,
Zafar Futehally (Newsletter for Birdwatchers)

Fairy Blue Birds.

Lesser Floricans.

SKETCH MAP OF WESTERN GHATS, KOKAN & MALABAR

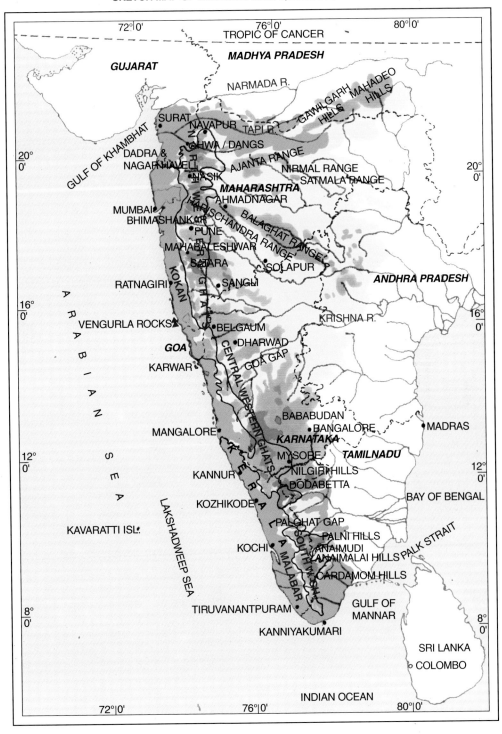